OUR REPUBLIC IN PERIL

September 1, 2004

Doug Townley – you've kept so many of us entertained monthly with your monthly tales that I thought you might enjoy this book as a gift from an old classmate. My God, where has the time gone?

appreciatively,

Roger McCarthy

OUR REPUBLIC IN PERIL

ROGER E. MCCARTHY

Red Anvil Press

Cover by Linda Japp
e-mail: ljapp@stny.rr.com

RED ANVIL PRESS
1393 Old Homestead Drive, Second Floor
Oakland, Oregon 97462-9506.
E-MAIL: editor@elderberrypress.com
TEL/FAX: 541.459.6043
www.elderberrypress.com
Red Anvil books are available from your favorite bookstore, amazon.com, or from our 24 hour order line: 1.800.431.1579

Library of Congress Control Number: 2003107559
Publisher's Catalog-in-Publication Data
Our Republic In Peril/Roger E. McCarthy
ISBN 1932762132
1. American Civilization.
2. American Culture.
3. Conservativism.
4. History.
5. Politics.

I. Title
This book was written, printed and bound in the United States of America.

Dedicated to my wonderful wife, Barbara,
my three sons, their families,
and the McCarthy Clan from coast to coast.

Acknowledgements

My heartfelt thanks to Judy Smith and Gilliam Eveland for their valued suggestions, to Frank Franzone for his efforts on the LBJ-McNamara chapter, and to Linda Shaver without whose help in putting this on discs I would likely still be with quill in hand. Again to Barbara McCarthy, my thanks for your extended patience, editing help, and insuring that the chapters flowed logically and maintained reader interest. You are a rare jewel indeed. My thanks to Linda Japp, %Linda Japp Graphic Design Studio, for your design of the cover for this book.

And to Anya Duke, I salute you. Your grit, determination, and perseverance in the face of ten year against heavy odds to achieve justice continues to be an epic story. With your former Green Beret husband serving as your eyes, and little legal assistance, you have indeed done the impossible. Yours has been a classic case demonstrating again how justice is far more often for the rich and politically connected than for the little guy with limited means. Your tragic case continues, but hopefully fairness and justice will finally be yours. Had the Media stayed on the story, you would likely have prevailed years ago.(See footnote in Epilogue for additional details.)

Contents

Prologue

So, who is Roger E. McCarthy, and why should anyone be interested in reading what he has to say? Reasonable questions, of course. He is a former 30 year career C.I.A. operations officer who spent 20 of those years overseas, primarily in Southeast Asia. He first served under Allen Dulles in 1952 and retired in July 1982 during Bill Casey's watch as DCI. He served as Chief of Station in three places, Deputy Chief of Station in two places, Base Chief once, and held three demanding Deputy or Chief of Branch positions in Langley, Va.* (Some of his associated experiences are cited in the chapters.) His concern? Five decades of abysmal failures by the Mainstream Media!

Because of an obviously failed and defective "Mainstream Media", our Republic is in peril. Media has long ignored its responsibilities, remains derelict in performing its duties, has abandoned the principles behind its origin, forgotten our heritage, our history, and cares little for or ignores our future. In case the Media folks don't understand:

• We badly need a Media that is fair, factual, balanced, and objective; not one that is tilted to the left, or that supports those who would damage or destroy our Republic or our Constitution. In sum, the time has come to heal yourself, Media. The Public has had enough of your sorry performances and expects you to help us to meet our goals.

• We expect our Bill of Rights and Constitution to protect us as our forefathers intended, not have either of them twisted or interpreted to fit the whims of those who do not share the ideals and beliefs upon which our country was founded.

• We expect our children to be properly educated in settings free of personal dogma or political agendas that others want or wish to impose on them.

• We ask only that opportunities be equal, fair, and available to all, with it left to the individual whether or not he or she can successfully meet the challenges and opportunities available. In other words, individual qualification is a paramount factor.

• We expect our elected officials to be honest and to represent us, not the special interests of others. We expect our government to be strong in will, moral in purpose, and able to protect us from all enemies, be they external or internal.

• We reject and denounce all persons in government who choose not to serve honorably and honestly, or who bring disgrace to their office(s) and our country.

• We also abhor and shame those in the private sector who would illegally enrich themselves at the expense of others, and we urge and seek their deserved sentencing.

• We cherish our freedom and do not want, nor will allow, any outside influences attempting to interfere with our chosen way of life in these United States.

• We encourage all people to have similar basic ideals, including embracing religious beliefs, but leave to them how best to achieve these, without interfering with others.

• We wish to live in peace and in health; and thank those who make this possible, just as we oppose and reject those who believe or would have it otherwise.

• We are patriots; proud of our flag, and of the many great accomplishments that our country has achieved since our forefathers came to these shores. We want and intend that our Republic endure for centuries more. We wish ill only to those who seek to destroy that which we as Americans have.

To achieve the above, we need the help of a forthright and fair Media, and the honest efforts of politicians. The latter is certainly not going to be a reality in the short term, but with the help of an honest Media, even the days of the wayward politician could be numbered. As it is, we have few reliable "watchdogs", a role that the Mainstream Media once had but has chosen to abandon.

Does my research clearly substantiate that Media has fallen by the wayside over these many years? Absolutely! Does Media share the values that our founding fathers intended to endure? Apparently not! Doesn't Media have a responsibility to and in this country? Certainly! Has it met this responsibility? No; far from it, as a matter of fact.

I write this as an 'outsider'. I have no connections to anything or anyone within the general term of Media, Mainstream or otherwise. Since my retirement I have watched the 'passing parade' from the sidelines with increasing dismay and disgust. My firm conclusion is that the single most guilty entity responsible for the regressions that have occurred in our country since the late 1950s is the Mainstream Media. At every critical juncture they have failed us. They try to hide their failings and shortcoming behind the 1st Amendment while spouting gobbledygook about being the 4th Estate and embracing "Freedom of the Press" whenever it best serves its purpose. Media has yet to admit how low it has collectively sunk. That a large majority of the public believes the Mainstream Media are doing a poor job doesn't seem to phase most in the profession, although ratings and circulations have dropped markedly in both video and print Media. Why? Simple - their record is truly sorry!

That I view the Mainstream Media in less than a favorable light is obvious; I make no pretensions otherwise. Do I damn all within Media? Certainly not, but few in the Mainstream Media have done anything commendable, while most have done much harm. Examples of when, where, and how Media has failed us follow in detail throughout the next eight chapters. That Mainstream Media has become an instrument of the Liberal Left is particularly irresponsible as well as disgusting, but the evidence is irrefutable. No other entity can be blamed for ignoring the multiple problems and attendant degeneracy that has occurred during these past five decades. For example, I seriously doubt that any of the following could or would have happened had the Mainstream Media met even its basic responsibilities:

• The ignorance of Indo-China's history in Washington, the extensive and involved supporting roles of China and the Soviet Union in Vietnam, and the numerous political and military mistakes that followed should have been exposed, and might well have prevented, or more fortuitously resolved, that war. (Chapters 1 and 2)

• American Prisoners of War/MIAs would not have been left behind in Vietnam. Nor would Sen. Kerry been able to "manage", then terminate, the 1990s Senate committee investigations of the POW/MIA issue in the shameful manner he did, to the dismay of other committee and staff members, and the POW/MIA families. (Chapter 3)

• Henry Kissinger would not have been so easily duped by the North Vietnamese nor able to sell-out South Vietnam in the Paris "Peace" negotiations, or convince anyone that he had negotiated anything resembling a "Peace with Honor". The Media should have exposed his duplicitous role with South Vietnam, Nixon, and Congress, plus his financial 'promises' to North Vietnam. His "Kissinger Associates" extensive commercial ties with China, including Gen. Haig's, and others, are listed. (Chapter 3)

• Lyndon Johnson's obsession with re-election politics while South Vietnamese and U.S. casualties mounted under Secy. of Defense McNamara's inept 'management' of the Vietnam War, along with that of other key political figures, should have been revealed by the Media for the shameful and costly farces they were. (Chapter 2)

• Jimmy Carter's multiple and serious foibles should have been exposed, especially some of which bordered on disloyalty, or worse. (Chapter 4) But the Media dozed.

• William Clinton and Hillary would not – and should not - have made it to the White House from Arkansas because of their many corrupt 'misadventures' in Arkansas. After they got to Washington, their continuing disregard for our Republic and serious misdeeds were not adequately pursued or exposed by the Mainstream Media. This country would not now be endangered had the Clintons' (and Gore's) "China Connections" been properly exposed. (Chapter 5) Yes, Mainstream Media's failure on this was near total.

• The extensive abuses by Congress of their elected positions, responsibilities, and unique perquisites, plus the extensive financial rewards provided by lobbyists and others to them, have been and remain in need of exposure by the Media. Congress's individual and collective foibles make for more indigestion than interesting reading. Specific examples are cited, as are some suggested reforms/changes. (Chapter 6)

• The questionable and highly suspect roles of such organizations as the ACLU, the NEA, our Labor Unions, and the United Nations remain in need of public exposure, but each seems to remain inviolate for reasons known only to the Media. (Chapter 7)

• The mystique of Mainstream Media is dissected in Chapter 8. Since the Media has chosen not to remedy itself, perhaps it is time for us to preach to that choir. The Media moguls, the badly tilted to the left voting records of those in Mainstream Media, and the particularly irritating and Liberal Left figures in Media are detailed. (Chapter 8)

Half a century ago, we could read a newspaper, or magazine, or watch a TV 'news' presentation and believe that what we were reading or watching was likely factual and largely unbiased. Sources cited in news articles or on TV were often identified by name and/or position. We seldom had to guess or demystify the identity of "a well placed source"; or "a high ranking official", or other such convenient smoke screens used to befuddle or deceive the reader or viewer...or to enable the "reporter" to hide behind as he/she gives their own view or opinion about a given topic. Those days are now gone.

Has Media not recognized the adverse impact on our Republic caused by those who constantly and belligerently demand (their) "Civil Rights", or see 'Political Correctness' as answers to most problems, or who immediately cry "Racism" as a first excuse, or insist on "Quotas" and perpetuation of "Affirmative Action" to gain unjust advantages, or those who opportunistically label many legal issues as "Racist" or "Unconstitutional", or hide behind "Freedom of Speech" in regards to most if not all issues? Has today's Media not noticed the obvious imbalance of these "angers", or the supportive rulings by many liberal activist courts? Has it focused on the youth of this country, including the so-called music and fourth grade level "*lyrics*" of the "*hip-hop rapsters*" who strut and jump around, grabbing their crotch, while urging violence against the "establishment", killing cops, and whine about the "advantages" that "whitey" has? Or does it come to mind only after a tawdry Super-Bowl show?

Is Mainstream Media not concerned with the failing education system across our country? Or not bothered by the blatant misdeeds in office by our politicians and by corporate officers? Is

Media not sensitive to the increasing abuses of power by most unions, or the basic agendas of those who would make this a country of agnostics? Or is there fear of those who see and whine and cry 'racism' for most all their problems? Do those in Media ever consider what it would be like to have a country without laws, or not have patriots serving in our military to protect us, or not have truly dedicated teachers to teach our children? And is there no thought that liberal judges are assuredly starting to end the freedoms our forefathers gave us, or that the "Liberal Left" is working diligently to eventually dismantle the Constitution the signers intended to sustain these United States? What concerns do the Mainstream Media have? Is it only profit? And equally important - what is your agenda? Are you just cloned sheep?

Simply put, 'Journalism' has deteriorated these past five decades to the point that it could easily give what was once known as "Yellow Journalism" a good name. Those now involved in slanting or "spinning" the 'news' usually do so in a manner that veils their bias. Half-truths, lies, deceptions, innuendos, etc., have replaced what once was a proud art form. Not all journalists/reporters in the broad category of "Media" are included in this sweeping condemnation, but the exceptions are, unfortunately, far too few. 'Entertainment Journalism' and sound bites are now the rule rather than the exception. This adds to the public's difficulty in trying to identify fact from that which is not.

Compounding this uneven landscape of print and visual media are the roles and performances of the major newspapers and networks. The New York Times, the Los Angeles Times, The Washington Post, The Chicago Tribune, The Miami Herald, ABC, CBS, PBS, CNN, NBC, among others, have largely tilted "Liberal Left" in their coverage, some far more than others.

Fortunately for we conservatives, 'Talk Radio' was re-born these last 15 years and has become increasingly popular, thanks largely to the commuters and the listeners who had become disgusted with the daily regurgitation of Liberal Left pap. Many who chose not to read or listen to what the major outlets offered began to search out the limited menu of balanced news broadcasts that differed from Liberal fare. 'Talk Radio' brought dialogue and the opportunity to exchange views, far different from the bias offered by the 'Ivory Towers' controlling the great majority of the main media networks. The gradual relaxing of previously stringent controls by the FCC facilitated the development processes and aided the expansion of 'Talk Radio' stations.

The leading benefactor of this relatively recent phenomenon was and is the listening – and, thanks to cable, - the viewing audiences. The individual who first rode this bucking bronco across the air waves to fame and fortune is Rush Limbaugh. His daily radio show reaches millions, as now does Michael Savage, Michael Reagan, Don Imus, and Sean Hannity. On TV we have Bill O'Reilly, Brett Hume, Hannity & Colmes, Joe Scarborough, Dennis Miller, Tony Snow, to name a few. Limbaugh's five days a week talk show still leads the 'Talk Radio' pack. Bill O'Reilly, of Fox News TV, with his 'No Spin Zone', also has a weekday 'Radio Talk Show'. What has made these shows so popular? Simple. Americans finally had the opportunity of hearing persons not pushing the Liberal Left agenda. They get a balanced view and can express their own views in return. Thus arrived the magic carpet for the listening audience.

Before and during this same period of years there have been other moderate and conservative personages such as: William F. Buckley Jr., Evans & Novak, Alan Keyes, Don Imus, etc., on 'Radio Talk Shows', or TV 'debates' across the country, but their voices and views were seldom heard on a regular basis, or during prime time. That deliberate imbalance has now changed.

Why have these very different personalities succeeded so well? A no-brainer answer: the hosts are well read, impressively prepared, strong in their straightforward views, deeply patriotic, well spoken, each a successful author, and willing and able to define the issues while exposing the shallow agendas, views, and spokespersons of the Left. Opposing views are welcome on their programs, but, more often than not, the Liberal Left caller or guest is effectively revealed for advocating or supporting views that have little or no validity beyond having a robotic-like liberal left pitch.

Recently, Mainstream Media have chorused the Liberals unhappiness because of not having "equal" draw, meaning they feel there are not sufficient Liberal outlets or spokespersons to counter the daily programs of Limbaugh, Savage, Reagan, Hannity, Scarborough, etc. Of course the 'Left' fails to acknowledge that for years William F. Buckley, Jr. effectively tied Liberals in knots single-handedly the few times he was given the opportunity to do so, or that for decades the major networks have spun the gospel of the Left nightly through those aging Liberal network spokesmen Walter Cronkite, Dan Rather, Peter Jennings, Tom Brokaw, Sam Donaldson, and Larry King. Who is the 'Left' searching for? A Liberal Left Pied Piper? In addition to those appearing as "newscasters" on ABC, CBS, and NBC, the "Left" has tried many spokespersons and different formats, ranging from PBS and PBR to tilted-left Ted Koppel, Michael Kinsley, Connie Chung, Barbara Walters, to James "Rube" Carville, and his fellow "Motor Mouth" Paul Begala, to Bill Press and Mario Cuomo, to name a few. The Left's desperation was such it even tried to resurrect the often less than coherent "Far Left" Phil Donahue. But this time Donahue was without his previous partner, Vladimir Posner, the dapper, smooth talking, fluent English speaking, Russian KGB co-opted mouthpiece, who used to appear regularly until the collapse of the Soviet Union. Even Dan Rather is on talk radio. Is the Liberal Left becoming desperate? You bet your Funk & Wagnall – or Webster's.

The transformation from a generally moderate Mainstream Media to the 'Liberal Left', and at times even further to the Left, was not an overnight happening. However, its decline and shame are now a reality. The fire and smoke belching Liberal Media locomotive that started slowly down the uneven tracks of the late '50s and early '60s became all but a runaway during the '70s and '80s and '90s, and basically continues at full throttle as we move further into the new millennium. For instance:

• In the early 1950s, North Korea and the People's Republic of China were our major concerns. In 1949 and 1950, Chairman Mao Tse-tung sent the People's Liberation Army into eastern and western Tibet without any justification other than to expand China's borders. Without warning, North Korea launched an all-out attack on South Korea in 1950. Three years later a troubled truce was signed keeping the border much as it had been before the initial attack. We still forget that two-thirds of all the American soldiers killed in Korea's war were killed after peace negotiations had started. So much for negotiations with the Communists! Little of enduring value was gained nor was a lesson learned the U.S. would remember. The Mainstream Media simply turned inward, if not isolationist.

• Next was a Middle-East problem: Egypt's President Nasser seized control of the Suez Canal, much to the consternation of England and France. The role of the U.S. in the ensuing imbroglio was largely that of peacemaker, but it was a controversial and uneasy pursuit. On the home front, there were State Department scandals surrounding Whittaker Chambers, an admitted long time Communist, and the accusation that Alger Hiss, among others, was also a communist. These were tempting morsels that the Liberal Media pundits delighted in pooh-poohing for the public. The accusations by the F.B.I. were accurate, and Sen. Joseph McCarthy's committee used them in its hearings regarding Hollywood, State Department, and other government offices having numerous active Communist members and Communist sympathizers within their respective ranks. Incredibly, the Mainstream Media largely sympathized with the accused. (Has the Media ever wondered about the loyalties of some of its own members? It should. There is no shortage of possibilities.)

• Vietnam's emerging problems stemmed from the uneven Geneva Accords of 1954, including the obvious implications for the United States that it would be expected to fulfill the role of supporting South Vietnam in order to stem the spread of communism in Asia. This too impacted the public's malaise as the Mainstream Media pundits attempted to put their respective but not always consistent - or accurate - spin on events in Vietnam. There was an increasing range of unbalanced and slanted reporting on Vietnam, with emphasis heavily on the negative. As practiced by the Mainstream Media, the negative sells - the positive does not. Wasn't there an old song "Accentuate the Positive, Eliminate the Negative". Long forgotten, obviously.

• On May 1, 1960, to the embarrassment of President Dwight Eisenhower, Russia downed a U-2 spy plane and captured pilot Gary Powers. A protesting Premier Khrushchev pounded one of his shoes on the lectern at the United Nations, declaring that "the Soviet Union will bury the United States". The Media was more outraged at our "violation of Soviet air space" than impressed by the intelligence gathering capability of the U-2, or concerned with the blatant threat by Khrushchev.

• John Kennedy became president in 1961. The Bay of Pigs fiasco against Cuba and Fidel Castro that followed in 1961 served primarily to stir various pots of public discontent, and increased Liberal Media's rhetoric. The Cuban Missile Crisis in 1962 and the ensuing U.S. Navy blockade of Cuba was an event that teetered on war with the Soviet Union, and was one of the more tense highlights during the "Cold War" decades. It was hard to remember that this extended period had begun with the Marshall Plan and its extended humanitarian Berlin airlift following WWII.

• The assassination of President Kennedy in Dallas, Texas in November, 1963, prompted wide ranging speculations and conflicting reporting as to who was behind the killing. The Media was interested primarily in 'selling' whatever stories they could, rather than pursuing responsible investigations. This added confusion from coast to coast and further decreased public confidence in the U.S. Government. The inconsistent reporting practices, including errors/corrections, by the Media simply compounded the situation. Conspiracy theories abounded (and still do) regarding Kennedy's assassination. Finger pointing and challenges became the rule, not the exception. The incomplete and seriously flawed "Warren Commission Investigation" of Kennedy's death raised more challenges concerning the accuracy of it, and further ate away at public confidence in our government and our politicians. Could the FBI, the C.I.A., or the Secret Service have been involved? Or the "mob"? Or L.B.J? Or Castro? The Oswald single-shooter theory failed the smell test then, and still does. And why was Robert Kennedy shot? Was Martin Luther killed for the reasons the Media attempted to explain? Who knows? The Mainstream Media began to pursue a more leftward lean during this period that has not been righted since.

• The Mainstream Media simply failed to do credible follow-up investigations, and remained more interested in profits and attracting listening audiences than in providing objective, factual reporting. Net result? More confusion, and more doubts raised than straight answers given. Surprising? Not at all! Sensational headlines gain attention as does the sensational "spin", regardless of the facts. The inept handling and "investigation" by the FBI of the shooting(s), and the rush by all, including LBJ, to designate Lee Harvey Oswald as the single assassin despite considerable conflicting evidence, deservedly spawned skepticism here and abroad. Our Media pretty much looked and shrugged at one another. And Vice

President Johnson became President, just as he had always aspired, and inherited Kennedy's War in Indo China.

• Johnson's less than strong leadership and his often inept but always politically influenced policies, both at home and abroad, plus his own less than clean record, more often than not simply drove more arrows of doubt into the body politic, irrespective of political persuasions. Johnson's poll based policies, McNamara's failures to gain any enduring military successes against the North Vietnamese, and no headway in peace negotiations, plus the growing domestic protests proved to be too much for Johnson. His poll ratings plummeted. Obviously, the negotiating lessons of Korea had been forgotten, even though the behind-the-scenes Chinese and Soviet 'hands' were well known to all the participants. And when would it dawn on Johnson's Secretary of Defense McNamara, the State Department, and our "all-knowing" Mainstream Media that it was not just Vietnam, but Cambodia and Laos that the Vietnamese, and Chinese and Russians coveted as well? Bottom line? Negotiations with North Vietnam were non-existent. The prospects of succeeding militarily <u>or</u> diplomatically under Johnson were negligible, and poll-watching Johnson finally chose not to seek re-election. In the midst of all this mixed-bag of developments, the Liberal Media enjoyed a field day, and increasingly covered the "war protestors", rather than examining the well financed and organized 'protest' movements more closely that were taking place world wide.

• Following Lyndon Johnson's decision not to seek reelection, Republican Richard Nixon was elected President over the "pacifist-light" Democrat George McGovern. In 1967, Henry Kissinger had attempted via Paris connections of the State Department to become involved in the negotiating arena regarding Vietnam on behalf of the Johnson administration, but the effort failed. Yet President Nixon appointed Kissinger as head of his National Security Council (and eventually Secretary of State as well) wherein Kissinger would serve as Nixon's chief advisor on Vietnam problems, including personally negotiating with the North Vietnamese. Vietnam remained a priority, but the Paris based negotiations remained dead-locked, with the North Vietnamese dictating both their scope and pace. Meanwhile, U.S. casualties in Vietnam mounted. President Nixon opted to resign in 1973 shortly after being elected to a second term, preferring resignation to facing impeachment charges stemming from the shameful revelations surrounding "Watergate". This demeaning and politically disastrous development added substantially to the already long list of government related calamities impacting a demoralized public. And it whet the appetite of a blood thirsty Liberal Left Media. Like the fabled Dracula, the Mainstream Media panted for more of the same political nectar that had become its favorite drink.

By this time, aspiring journalists in most media fields had noted that their more senior colleagues gained fame and recognition by pushing the negative and by abetting the liberal agendas, rather than pursuing the objective and balanced. The Liberal Media had achieved its desired power status, and the number of its practitioners steadily increased.

The subsequent caretaker period of Gerald Ford (1973-1977) in the White House was more a period of calm than one of enduring accomplishments, but it was largely scandal free. Under Ford's watch, with Kissinger as his Secretary of State, the painful and ignominious U.S. withdrawal from and total abandonment of South Vietnam was completed, with the Cambodian and Laotian dominoes also falling in 1975. Each of the three countries endured their respective experiences of harsh 'reeducation' camps for those that had not escaped. In Cambodia, well over a million persons were slaughtered, as were thousands in harsh 'reeducation' camps in Vietnam and Laos. Thousands of other Indo-Chinese people tried to reach freedom by escaping across borders or by boat from their communist victors. The seriously flawed concept of 'Neutral' or 'Coalition' Governments that had been pushed hard by the Liberals, the French, the Chinese (as a political-ploy carrot), much of our U.S. State Department, many in Congress, and the Liberal Media was unceremoniously exposed by the North Vietnamese as an intended fraud designed solely for the gullible. Because of multiple restrictive measures adopted by our shameless Congress in 1974 and 1975, the U.S. was forced to break its many pledges to millions in Indo-China and brought international disdain and embarrassment to our country. But this too was largely ignored by our Mainstream Media.

President Ford managed to get through his appointed term more or less unscathed, other than enduring the continued stains of Indochina that periodically surfaced regarding U.S. 'promises' (reparations) made during negotiations, which Kissinger blithely denied. At Kissinger's urging, Ford supported the give-away of the Panama Canal which Jimmy Carter subsequently completed, despite the loud and long protests by our military, some in Congress, – and the Panamanians - all of which were ignored. The role of the Media herein was again far less than what it should have been. Today we live with the long shadow cast by Carter's relinquishing control of the Canal, and the closing of the U.S. Military Southern Command in Panama that for decades had served as the eyes and ears of the U.S. on Latin America. In simple terms, the U.S. is out, having been replaced by China's controlled interests at both ends of the Panama Canal.

Jimmy Carter's rise from Governor of Georgia to having a four year stint in the White House was more the result of voter apathy and accumulated unhappiness with Washington politicians than it was an endorsement of him - or the Democrats. Carter's period in office was not one of enlightenment, but served to increase the country's malaise from coast to coast. While many cartoons and TV skits about the Carters during those four years were indeed exceptional, Carter is on the growing list of presidents that gave considerably less than impressive performances. While probably well intended, Carter's Middle East efforts were superficial – and doomed - as were many of his other foreign and domestic policy pursuits. Again, this was a period during which the Mainstream Media served the public poorly while trying to help the Liberal Democrats. With the majority of the public – and the free world - sighing in relief, in 1976 it was Ronald Reagan over Jimmy Carter by a wide margin.

The politically inspired attacks and grossly partisan activities by the Liberal Democrats during the two terms of Ronald Reagan stirred the political scene across the country. Again the Liberal and Liberal Left Media failed in the basic pursuits of objectivity and/or truth. The Democrats sought Reagan's scalp for most of his eight years in office, driven by impeachment dreams, but failed and fell on their collective political asses, so to speak, in their obvious efforts. The Reagan years were not easy ones, for he inherited many problems from the Carter gang, including: high double digit inflation; American hostages held in Iran; increased drug trafficking from South America through Panama and into the U.S.; Nicaragua, the Contras, and the Sandinistas; the Iran-Contra imbroglio; and the Cold War. A Democrat controlled Senate and House openly sympathetic to the Sandinistas – as were most of the Media - and against the Contras in the latter's quest for freedom – were less than helpful. The inexplicable and very restrictive 'Boland Amendments' that, as intended by the Democrats, adversely impacted the Administration's efforts to help unseat the Communist Sandinistas, created significant problems. Despite the expectations of most in the Mainstream Media and Congress, the Contras finally succeeded and the Sandinistas were soundly defeated at the election polls.

Reagan also provided U.S. support to Great Britain against Argentina's attacks on the Falkland Islands; he brought the Soviet Union to its knees by implementing and exercising multiple political and technical challenges that the Soviet Union was unable to meet technically or financially; he demonstratively challenged the Soviet Union to tear down the Berlin Wall; and for eight years he steadily rebuilt our military strength. He proposed the Self Defense Initiative (SDI) to the dismay of Moscow – which was derogatively labeled by Sen. Ted Kennedy and other liberals as "Star Wars". The Soviet Union faced financial and political collapse because of its inability to keep pace with the modernizing of U. S. military capabilities, but the Liberal Left Media was not supportive of Reagan's policies, nor was most of the Mainstream Media. Like most Democrats, Media chose to be derisive of Reagan's programs for reasons that remain largely a mystery. Simple 'sour grapes' is possibly a reason, but sulking is not an adult sport...except in Marxist ideology.

In 1988, Vice-President Bush handily won over Massachusetts Governor Dukakis, but was politically ambushed by a Democrat controlled Congress which pushed for and obtained a tax increase, despite Bush's campaign promise not to raise taxes. Iraq's invasion of Kuwait in 1990 resulted in U.S. forces responding with Desert Shield and Desert Storm, joined by a large coalition of supportive countries and the voice vote approval of the United Nation's Security Council. The 100 Hour War resulted in a stinging defeat of Iraq's military forces, a surge in popularity for President Bush, shame to those who had opposed Desert Storm, especially those nay-sayers who loudly claimed that the U.S. would be repeating the mistakes of Vietnam. Senator Ted Kennedy was one of the more outspoken opponents of Desert Storm, predicting that thousands of U.S. soldiers would come home in body bags. He was joined by Sen. Kerry and most other Democrats in the Senate vote against authorizing the U.S. to free Kuwait from the occupying Iraqi forces. The authorizing approval was by a single vote.

President Bush also used military forces to oust Panama's President Noriega who was flagrantly involved in drug trafficking. U.S. troops were also committed to Somalia both to prevent further slaughter there by communist supported forces, and to provide humanitarian assistance to the thousands that were starving. One of Bush's finest moments was on December 23, 1989 when the Brandenburg Gate in Berlin was opened, signaling the end of the physical separation between East and West Berlin, and leading to the first free national election in Germany since 1933. Equally significant was the end of the 74 year Communist dominance of Russia and the total collapse of the Soviet Union that followed on New Year's Eve, 1991.

The role of the Media throughout this period? Praise for the '100 Hour War' followed, but so did second guessing and questions as to: why so much force had been brought to bear on Iraq; why Saddam Hussein had not been removed; why the Patriot Missile system had not been 100% effective; why the Iraq military had been allowed to remain largely intact; what had the Coalition Forces really accomplished beyond liberating Kuwait; had the U.S. intended to stabilize Iraq; plus a litany of other negatives, including that U.S. interest had been driven only by 'OIL'. That the military action by Coalition Forces had been approved and restricted by the U.N. Security Council to driving Iraq forces from Kuwait seemed to be forgotten by the Media. Upon the withdrawal of U.S. and Coalition forces, the plight of the people and tribes in Iraq, especially the Kurds, continued to be muted, if not glossed over, by the Mainstream Media. Saddam Hussein was left alone to resume his brutish dictatorial practices. His attacks on the Kurds – and others - were basically ignored by the Media, as were the devastation and terrorist acts that Iraq had brought to Kuwait. Most of the Media basically shrugged and turned its collective attention to the ongoing struggle between Israel and the Palestinian Liberation Authority. Domestically, the Mainstream Media grew increasingly critical of the Bush Administration and very strongly supportive of the censorious liberal Democrats. Journalistic flag stomping had developed well beyond its puberty stage.

The poorly organized campaign waged by the Republicans to reelect George Herbert Walker Bush to a second term over the carnival-like candidates Ross Perot of Texas and William Jefferson Blythe Clinton, master snake oil salesman from Arkansas, all but closed the final chapter on any pretensions remaining in the Mainstream Media as to balanced and unbiased reporting. Even today it is rarely recalled or mentioned by the Democrats, or the Media Left, that Ross Perot's winning 19% of the popular vote likely provided Clinton (who garnered 43% of the popular vote to Bush's 38%) the election over Bush. Only 55% of those eligible voted in this Grade B, or less, soap-opera-like campaign. The ineptness by the Republicans continued unchecked with and by the selection of Robert Dole as the Republican standard bearer to run against Clinton in 1996, to the delight of Clinton-Gore (who received 49% of the popular vote), the Democrats, and the Liberal Media. (Chapter 5 on the Clintons is long, laconic and pithy.)

The disturbing role of the Media in the bench mark events summarized above are to note for the reader the increasing slide to the Left that has occurred over the years. The Indo-China War years first saw this development become a harsh and stark reality. For this reason I have chosen to revisit the Indo-China story in Chapter I, highlighting the nationalistic history of it for the reader. The less than responsible roles played by the Media and many of our high ranking politicians and appointed officials are also covered. The sorry performances of some, including the military, involved in the tortured steps along the way, are included, as are names of the more seriously flawed "journalists".

Chapter 6 highlights the Congressional "follies", including the diversified perquisites, appalling antics, questionable, and blatantly venal pursuits of many of its members. Some suggested changes are offered regarding congressional terms, cost savings to the tax payer for Congress, their staffs, etc. However, the 'Special Interests', Lobbyists, and PAC dollars clearly rule that roost!

Chapter 7 covers the organizations and front groups that pose serious threats to the well-being of our Republic. Why these "hate America" foxes are allowed in "our" chicken coop escapes me, but the dichotomy remains unchallenged by both our politicians and the Mainstream Media. Probably futilely, suggested changes are offered.

In Chapter 8, the Liberal Media are reviewed, or pilloried some might say, especially its leading lights. Since others have already succeeded in capturing many of the idiosyncrasies and frailties of the Media, and the personages involved, I quote and touch sparingly, primarily to provide stark examples of the questionable bent of the journalistically unclean. The list of the miscreants is not a short one.

The election fury of 2000 and the gore that Gore and his vast horde of lawyers created in Florida, the September 9/11 tragedy of 2001, and the invasion of Iraq in early 2003, including the capture of Saddam Hussein, are also briefly reviewed in Chapter 8. The "Mainstream Media" again performed at a less than commendable level, but that is more disappointing than surprising.

To underscore and blend with some of the points at hand, I have interspersed some of my personal experiences during my C.I.A. assignments in Taiwan, India, Peru, Vietnam, Laos, Burma, and Langley in most of the chapters. No "sources and methods" are divulged, and no previously unknown data are included, but I trust you will find them interesting.

To those who are not Liberal Left, I suggest you read on, not only to gauge for yourself how we got to this point, but to understand much you likely had little or no knowledge of before. I promise that you'll learn things you should not and won't forget.

To those who are "Liberal Left" – or worse – I suggest you <u>not</u> buy this book, but instead continue your whining ways, studying your Marxist texts, burning joss sticks to Lenin and Chairman Mao, maintain your closed minds, pursue your nefarious plotting, and practice such other bad habits you may have. You are likely already beyond rehabilitation or worth saving. Sorry.

Many of our allied countries have experienced problems similar to ours resulting from questionable political interests of their less than honest politicians. The lessons learned from the following chapters can generally be applied abroad as well. For example, France harbors a huge dissident population extending from its colonial days, which brings to mind that Ho Chih Minh and the Ayatollah Khomeni returned to their homelands after "hiding" in

France. Certainly there are similar dangerous personages/ radicals lurking there. Germany remains a politically disturbed and economically waning country, whose main interest appears to focus on being Europe's power base. Spain has faltered badly, but hopefully will recover from her leftist lean. Poland and Italy, despite active leftist opposition, remain strong beacons of hope, and examples to their neighbors, thanks to their people's support...but both countries need to keep a wary eye on those that would bring turmoil.

New Zealand, Australia, Canada, and England, all wounded repeatedly by their leftist bent politicians and media, have suffered politically and economically, yet have survived largely because of strong conservative leaders and a people steadfastly refusing to heed the rush to socialism. As even Tarzan and Robinson Crusoe learned, it's a jungle out there!

But now to our own problems, for they are many and serious...but correctable.

* some of Mr. McCarthy's other experiences are related in Frank Holober's "Raiders of the China Coast" (Naval Institute Press), Kenneth Conboy's "Feet to the Fire" (Naval Institute Press), and McCarthy's book on Tibet, "Tears of the Lotus" (McFarland & Co.)

CHAPTER 1

INDOCHINA

(The Deceived and the Deceivers)

Pres. Truman supported France's desire to resume its colonial status in Indo-China following WWII, but it was John F. Kennedy that led our nation into the Indochina conflict against the advice of many, including his predecessor, Dwight D. Eisenhower. Perhaps Kennedy was awed by De Gaulle and persuaded by the latter's insistence that the French retain some remnant of their halcyon days when Vietnam was once a part of the French Indo China colony. Certainly Kennedy had far different hopes, plans and intentions for Indochina than what Ho Chi Minh envisioned.

Reportedly, Kennedy believed U.S. forces should first be deployed to Laos, but given the political booby traps in the 1954 Geneva Accords, he was persuaded by his advisors that U.S. assistance be given to bolster the many needs of the South Vietnamese. Thus began the U.S. effort on its often painful journey into the multi-faceted land of political intrigues, traditional divisions, schisms, and religions of South Vietnam. Unfortunately, few in Kennedy's Administration, including the majority of the military, the State Department, Agency for International Development, the National Security Council, the C.I.A., etc., including the Media, had studied Vietnamese history, or the inglorious history of the French in Indo-China. But undeterred, the U.S. pressed ahead on all fronts in supporting South Vietnam, including the State Department that remained seized with the idea of stopping the spread of Communism in Southeast Asia. The upper levels of our military seemed to believe that the U.S. needed only to sound a bugle call, fly the flag, move in a few tanks, artillery, some aircraft, retrain some South Vietnamese troops,...and the North Vietnamese controlled Viet Cong and Viet Minh would assuredly not seriously violate the 17th parallel, the agreed line of demarcation established at Geneva in 1954. How could this conclusion be flawed? The reasons are many, as are the hard earned lessons. Countless books have been written on the subject and most make interesting reading. The more cogent points are particularly worth reviewing.

Only Hanoi, Moscow and Beijing referred to Ho Chi Minh's (born Nguyen Ai Quoc – translation: Nguyen, the Patriot) attacks on South Vietnam as a "War of Liberation". That in itself should have provided a large and obvious clue to all. Few Americans had learned that until mid-1884 China had for centuries been the protectorate of "An Nam" (as Vietnam was known in those early days) and that France, with assistance from the Spanish, had earlier (1868) militarily pressured and been ceded "protection of Cambodia" status by the "An Namese". The French military efforts against the weakened Chinese forces in An Nam succeeded primarily because the extended Taiping Rebellion in China had sapped much of China's military strength, precluding Chinese reinforcements being sent to An Nam. The poorly equipped An Namese simply were unable to fight effectively. Following their success in An Nam, the French attacked Laos and added her as a protectorate in 1893. It was thus that Vietnam, Laos, and Cambodia became French Indo China. French interests were primarily commercial, and of the three, Vietnam was by far the area that provided France most of the riches she cherished. The north of Vietnam was gradually developed as an industrial area, while the south became the rice bowl for much of Asia, as well as an important source of rubber from the abundant rubber plantations that were able to flourish there.

Unfortunately for the Vietnamese, the Cambodians and Laotians, France provided few schools, few roads, little nation building, and little training for laborers. French bureaucrats were in control at almost all levels of government, including the rural levels, and taxes were imposed and collected. Advanced education was available in France, but primarily for the privileged. However, over time some hundreds of Vietnamese, a fewer number of Cambodians -and Laotians- did receive higher education in France. Interestingly, there were nearly as many French civil servants in Indo China, primarily for tax collection purposes, as there were

British bureaucrats throughout India whose native population was 12 times greater than the Indo-China Colony. The meaningful and obvious differences between British and French colonial systems were that the British encouraged Indian representation in governing at all levels, as well as Indians serving in most branches of His Majesty's military. The British built schools, roads, and railroads in India. Training was provided Indians in management and commerce...and education was widely encouraged and provided. The French chose otherwise. Adding to the rather grim picture in Indochina was the practice of heavy taxation on all native commercial efforts, ranging from rubber plantations to the exporting of rice to fishing to light industry. "Gambling, opium dens, and prostitution were legally established in Viet Nam, and flourished particularly in the main cities of Saigon-Cholon, Hue and Hanoi. Other illegal trafficking was sanctioned by the French colonizers, and political and moral corruption was common. This was the picture from 1885 to 1954, and these myriad evils caused the Vietnamese to despise the French rule."(1) To understate it, the authoritarian and self-serving practices by the French in their colony had less than inspired the natives other than to cause them to become more determined to be independent.

Tiring of Chinese domination in and over An Nam, the An Amese (Vietnamese) had earlier often sought to shed Chinese control and to gain their freedom and independence. These efforts were seldom well planned, unified or coordinated, and so failed. Not until the French arrived with superior fire power were the weakened Chinese – and An Amese - finally forced to capitulate. However, the Vietnamese spirit of achieving freedom from foreign domination not only continued but increased during the years of French control. That France's occupying forces would be seriously challenged and defeated by the Vietnamese was only a question of "when", not if.

"The Republic of Vietnam existed as a quasi-independent state from 1947, and as a completely independent state from 1954."(2) The Geneva Agreements of July 1954, signed by a delegate of Ho Chi Minh, (but not by the United States or South Vietnam), officially recognized the political division of Vietnam at the 17th parallel, pending a national election under international supervision for a possible unification of Vietnam, North and South. The total defeat of the French forces at Dien Bien Phu by the Vietnamese under the command of Gen. Giap in June, 1954, signaled the demise of French control of Vietnam, Cambodia, and Laos...and a diminution in the exaggerated pretensions of France's President Charles De Gaulle. The Southeast Asia Treaty Organization in September, 1954, also provided for the protection of the free state of Viet Nam. The Joint Congressional Resolution of August, 1965, Public Law 88-408, authorized the President of the United States "...to take all necessary steps, including use of armed force, to assist any member or protocol state of the Southeast Asia Collective Defense Treaty requesting assistance in defense of its freedom."(3) This reflected the general concern of the free countries in Southeast Asia that the threat of Communism was increasing, not decreasing, and their concern that China and Russia intended that Communism would spread further.

That this concern would be so strongly expressed is not surprising. The signatories had seen what China had done to Tibet, what North Korea – and China - had attempted to do to South Korea, and how extensively both China and the Soviet Union had assisted Ho Chi Minh and North Korea with both materiel and manpower. "The (U.S.) dollar flow to back the French military campaigns in Vietnam grew rapidly from approximately $150 million per year in 1950 to over $1 billion in the fiscal year of 1954, when the U.S. was underwriting 80 per cent of the cost of the war...This equaled one-third of the entire American foreign-aid program and was by far its largest single component."(4) Nor had the political stance of the French in Indo-China during WWII, during which the Japanese used Indo-China forces to launch attacks throughout Southeast Asia, including the capitulation by the French to the Japanese in keeping with the pro-Vichy policy of the French, been forgotten by the neighboring countries.

Ho Chi Minh expected to be rewarded for his and the Viet Minh's efforts against the Japanese during WWII, including their having saved a number of downed U.S. pilots and having provided intelligence on movements of Japanese military forces to OSS agents. "The Viet Minh had cleared large sections of five of the northern provinces of Tonkin and engaged the full attention of the bulk of the Japanese 21st Division."(5) With the announcement that France would again return to Indo China, Ho's hopes were dashed. "Ho soon came to be acknowledged by most non-communists and communists as their outstanding leader in Vietnam's struggle for independence against the French and the principal symbol of the new Vietnamese nationalism."(6)

During the Korean War, it had been widely assumed that Sino-Soviet assistance was concentrated in furnishing assistance to Korea, thereby leaving the Viet Minh and Viet Cong forces to rely largely on their own resources. This was not the case. While Chinese volunteers were fighting in Korea, and Soviet munitions were being sent to Korea, both China and Soviet Russia sent men and arms to help their 'comrades' in Hanoi. A few interesting but little-known facts are cited here as examples of Sino-Soviet assistance given to Viet Minh forces:

On August 20, 1950, about 150,000 Chinese soldiers, traveling in civilian clothes, arrived in Kunming, Yunnan Province, to help the Viet Minh, as reported by the Chinese United News Agency in Hong Kong,...and confirmed by other neutral Chinese news services. On June 28, 1950, a news dispatch issued by the Kwong Wah News Agency in Hong Kong reported that two Chinese ships left Kwong Chau Wan for Haiphong, North Vietnam, loaded with munitions for Ho Chih Minh. (Author's note: Additional departures by Chinese ships carrying munitions to North Vietnam were reportedly carried in similar articles in Hong Kong papers in June and August.)

On September 21, 1951, the Chinese United News Agency, Hong Kong, reported that about 450 Soviet advisers had already arrived in Viet Minh headquarters. It was known to the Chinese Nationalist intelligence sources and to others that two-thirds of the Communists were Vietnamese, the other third being Chinese Reds."(7)

Editorials in several Chinese Communist newspapers said that China was powerful enough to defeat the imperialist countries on two fronts, namely the Korean and Vietnamese fronts.(8) These editorials were not intended as idle boasts.

"On April 26, 1954, with the Geneva Conference already in session, the bloody Viet Cong assaults on the French position (Dien Bien Phu) intensified, with large military forces and superior gun emplacements. The French had lost about 15,000 men, but the loss of French prestige was much greater than their military defeat. Above all, the French lost heart."(9) And "Three weeks after the close of the (Geneva) conference, Senator William Jenner told Congress: "The United States was outthought, out-traded, and outgeneraled."(10)

During WWII, Bao Dai was permitted by the French Vichy government to enjoy his An Nam Emperor privileges, and the astute Japanese used him in that position to pacify the Vietnamese. Bao Dai's abilities as a dilettante and play-boy far exceeded his bureaucratic ones, and he was quick to pass to Ngo Dinh Diem the title of Prime Minister of South Vietnam following the 1954 Geneva Accords. But to his credit, in August, 1945, Bao Dai had warned France: "Even though you may be able to re-establish French administration here, it would not be obeyed any more. Each village would become a nest of resistance; every former collaborator would become an enemy of France...The only way to safeguard French interests and the spiritual influences of France in Indo-China is to recognize the independence of Viet Nam and to renounce all thoughts of re-establishing French sovereignty or administration under any form whatsoever. We could easily be able to understand each other and become friends, if you would cease to pretend that you again want to become our masters."(11)

A National Referendum was held on October 23, 1955, to determine whether the South Vietnamese preferred a monarchy or a republic. More than 80% of those registered voted for the republic and some 66% of the candidates supportive of Diem were elected to the Assembly. The problems facing Diem and his Assembly were monumental. Not only was the country divided, but more than a million refugees had fled North Vietnam for the South. There was but a token (French trained) military capability, agriculture was in a sorry state, there were few consumer goods, and Viet Cong guerrilla activities were spreading. What Diem was able to achieve in a relatively brief period bordered on the amazing. "From 1955 to 1962, rice production rose from 2,800,000 metric tons to 5,000,000 metric tons. In 1955, rice exports were 70,000 tons. By 1960 they had risen to 340,000 tons. Without America's economic aid since 1954, Viet Nam probably would not have survived. But without Diem's determined and untiring efforts, South Vietnam would not have lasted beyond 1955 or 1956."(12) Fishing exports increased five times during this same period; textile industries increased, as did cement production, lumber, coal, silk; new roads and bridges were constructed and old ones repaired. There was no nationalization, nor threat of one, and land holdings were re-distributed to benefit farmers. "On October 26, 1963, hardly five days before his assassination, "President Diem reported to the National Assembly that the enrollment of students in government elementary schools had been increased to 1,174,020 (from 477,581 in 1954); the enrollment of students in government high schools was 198,749, and the enrollment at Saigon, Hue and Dalat universities was increased to 17, 345 – almost twice the number in 1960."(13)

In 1954 Senator Mike Mansfield spent two months in South Vietnam and on October, 15, 1954, submitted his "Report on a Study Mission to Viet Nam, Cambodia and Laos" to the Senate Foreign Relations Committee, concluding that President "Diem had a reputation throughout Viet Nam for intense nationalism and equally intense incorruptibility" and 'highly' recommended providing economic aid to Diem."(14) General Collins was selected as the 'special American Ambassador' to South Vietnam, and in turn he recommended aid for the resettlement of refugees, economic aid for South Vietnam, and military aid under a Military Advisory Group, headed by Maj. Gen. O'Daniel, in support of Diem. Air Force Colonel Edward Lansdale, on loan to the C.I.A. in Saigon (fresh from his successful stint in the Philippines where he had worked with President Magsaysay to neutralize the Communist Huk threat), Allen Dulles, Director of the C.I.A., and Secretary of State, John Foster Dulles, fully supported the plan."(15) Thus from this modest beginning did the United States take its first step onto what would prove to be an incredibly twisted path lasting nearly 20 years, largely shaped by our politicians - and the considerably less than objective Mainstream Media.

President Diem faced an almost impossible task. With the departure of most French field forces, it became only worse. He had less than an able army or air force. Religious and political organizations were many and varied. The French military was still present, but represented more a political presence than a military one. Some French were more interested in stirring trouble against the Americans than in helping to resolve problems. Regrettably, members of

Diem's own government and military were often engaged in coup plots against him, motivated by greed and personal ambitions. That Diem survived even those early years was a tribute to his unwavering determination to see Vietnam free and independent. This attribute impressed Col. Lansdale and led to the two becoming trusted associates. Lansdale was able to influence Diem to initiate a number of programs benefiting the Vietnamese people, much as he had done with Magsaysay, especially in implementing what was to become known as the Strategic Hamlet Program.

"The Communist aggression, infiltration, and subversion in South Vietnam had no front lines. The Viet Cong penetrated into hamlets and villages, into mountain areas, and along river banks. The American and Vietnamese politico-military strategists agreed on the adoption of the Strategic Hamlet idea which had worked so well in Malaya and the Philippines against the Communist Guerilla forces there. The project was approved by both President Diem and President Kennedy. They were encouraged by the "island" fortress of Father Nguyen Lac Hoa, who had often defeated the Viet Cong. This program, started in 1961, was designed to bring small isolated hamlets and solitary families together into larger (village) units. The Viet Cong had been kidnapping or killing the leaders of small hamlets and villages, terrorizing the inhabitants. By bringing a large group together, these tactics would not be so easy. The expense for the new 'homes', and transportation to them, was borne by the government. It was decided that 11,864 strategic hamlets would be established by 1963. Diem lived to see 8,600 hamlets established, providing protection for nearly ten million people. As Diem had pointed out, the Strategic Hamlets program had a very democratic base through the election of officers. Even the independent hill tribes, the Montagnards, became part of the program. The Communists had tried every possible means to get them to oppose the government, but by October 1, 1963, about 217,000 'mountaineers' from 436 different areas had become part of 210 Strategic Hamlets, and were determined to fight the Viet Cong. Furthermore, 10,570 former Communists had defected by September 1963 and joined the Vietnamese Government."(16) Unfortunately, of the 8,600 strategic hamlets that President Diem had established to protect the peasants, 5,000 were lost to the Viet Cong following Diem's demise on November 2, 1963.

The reference to Father Nguyen Lac Hoa and the 'Strategic Hamlet' program noted above, recalled my meetings with him in 1962 and 1963 in Binh Hung, An Xuyen Province, in IV Corps. In 1962, I traveled to Saigon from Taipei, Taiwan, to join a gathering of C.I.A. officers currently serving in Asian countries convened to brain storm what and how resources of our respective 'host' countries might be willing and able to assist South Vietnam's needs to meet the many Viet Cong and Viet Minh challenges. I was intrigued that one area in the Delta area of IV Corps in An Xuyen Province had survived countless attacks by the Viet Cong/Viet Minh. It had previously been visited briefly by Col. Lansdale, who had told Father Hoa he would try to get assistance to him. In the Saigon meeting I suggested that the Chinese Nationalists might well be able to fill Father Hoa's needs, especially given the "Chinese Connection", - the Father having come to Vietnam with some followers after fleeing Yunnan Province and the island of Hainan.

After two days of meetings, Cecil Cartwright, a C.I.A. case officer in the Saigon Station with whom I had previously served on Taiwan from 1952 – 1954, including on Quemoy (known as Chinmen, one of the off-shore islands between Taiwan and China), and I flew by C-123 from Saigon to Camau, the capitol of An Xuyen Province. Our pilot was a young, recently arrived, red-headed American who was a living caricature of 'Hot Shot Charlie' from the old cartoon strip, "Terry and the Pirates". His feet barely reached the rudder pedals - and brakes - but he was all spirit. We arrived in a cloud of dust on Camau's dirt field, thanked our grinning pilot, and boarded a less-than-new C-21 'Banana' helicopter crewed by Vietnamese, as the C-123 took off for the return trip to Saigon. Over Binh Hung, Father Hoa's village, our pilot circled down tightly (to avoid enemy fire) to the landing pad, where we were met by a well intended and enthusiastic village 'band' - consisting of two or three battered trumpets, an old tuba, two trombones and a couple of drums (all previously provided by Col. Lansdale) - most of the villagers, including many of the village 'militia', Father Hoa's personal security force known as the 'Sea Swallows', and by Father Hoa. After the arrival 'formalities', we joined Father Hoa in his modest if not meager quarters to discuss the purpose of our visit. In answer to my question as to what he needed most that I might be able to help arrange he answered: "Trusted and able working advisors...and supplies of all kinds." Cartwright and I quickly sorted out the details of that to Father Hoa's satisfaction.

General discussions continued over the next three days, including walks well beyond the village perimeter, and boat travel (this was an area of countless canals, all rich with fish) to neighboring villages that were under the general protection of Father Hoa. Members of his well armed 30 man bodyguard (with AR 14s and AR 15s that Col. Lansdale had provided the 'Sea Swallows') accompanied us without fail. (We Americans had to be content with M-1 Carbines provided by Father Hoa.) There were two 80 mm mortars and a single 155mm howitzer in well protected emplacements within the village perimeter that nightly fired a few rounds into nearby areas known to be Viet Cong/Viet Minh "active", primarily just to let the would-be evil doers know that not all in Binh Hung were asleep. The village was surrounded by two foot thick and, in most places, four foot high walls made of hardened mud brick, with sand/dirt bags, as were the 24 hour manned sentry positions. On the outside, well placed punji sticks and a variety of improvised booby traps completed the crude but adequate defenses. A single canal went through the middle of the village, with wire and booby trapped gates on each of the two openings. Many of the wood - with tin roofs - huts were on wood pilings, arranged along the banks of the canal and along the edge a large pool of water off the canal, from which the mud for the brick making came. All of Binh Hung ate well, thanks to abundant fish, rice, and some locally grown vegetables...and to the Father's bakery that produced excellent French bread. There was even a small parade ground which doubled as a drop zone for parachuted supplies and a soccer field that also served as a playground for the village's children attending the adjacent school house. There was also a strong holding facility for the 100-150 or so Viet Cong/Viet Minh prisoners – mostly young men - that had been captured, some of whom were awaiting transportation to a more functional facility for further debriefings/interrogations.

The areas immediately outside the protection of the Father were littered by multiple hundreds of pro-Ho and anti-Diem 'leaflets' affixed to bushes and trees as well as countless booby traps in the canals and in the dried out rice fields throughout the area. Most of the rice fields had earlier been flooded by the Viet Cong simply by breaking the dikes of the paddies, which allowed the flooding of them by brackish and salt water from the Gulf Of Siam The homemade trip mines in the fields consisted of glass or tin and contained small stones, broken glass, metal shards, etc., and dangerous because of resulting infections to inflicted wounds. It was obvious the 'enemy' preferred to save their grenades and/or more sophisticated mines for planned ambushes or attacks, but these improvised ones were capable of causing serious injuries.

During the talks and travel with Father Hoa, it was agreed that what Binh Hung needed most was an augmented Special Forces type team, with light and heavy weapons specialists; first aid technicians, an intelligence specialist, a psychological warfare specialist, boat handlers (outboard motors and repair), tactical communications specialists (including repair capabilities), demolitions specialist, ordnance specialist, survival specialist, interrogation specialist, teacher (to assist at the grade and middle school level), map and compass specialist, and a Team Leader tactician who could, if needed, serve as the Father's right arm. Other and obvious needs were more weapons/ammunition, medicines, whale boats, low to modest speed outboard motors, gas cans, field rations, tools, blankets, clothing for all ages, a portable printing press, tactical field radios, camera(s), binoculars, etc., etc. Father Hoa's desired time frame for the arrival of the training specialists was of course "next week", but after I outlined the process involved of first gaining U.S. approval, then approval of the Chinese Nationalists, then the screening and selection process of the volunteers, and the training and testing of them, we projected the arrival date of the Chinat team to be three or four months hence. In the interim, Cartwright assured the Father that he would see what supplies he could send him from Saigon...and did so.

To the above I should add that living in Binh Hung at the time as Father Hoa's 'advisor' was Bernie Yoh, who had also come to the notice of Col. Lansdale earlier in Saigon. Bernie Yoh was fluent in many languages, including Chinese, Vietnamese, English, and French. He was also an accomplished musician having played in the Shanghai Philharmonic Orchestra prior to Mao's succeeding Chiang Kai-shek. Bernie was a strong anti-communist and had met and become a friend of Father Hoa, who himself had become well known in Yunnan Province as the Kuomintang (Chinese Nationalist) "Fighting Priest". Forced by Mao's advancing Chinese Communist forces to seek haven in South Vietnam, Father Hoa had asked for and received land from President Diem in An Xuyen Province for his small band of Chinese followers and himself. There they had converted an originally desolate but potentially rich area, thanks to the rice crops and canals, to that which had become the village of Binh Hung, a miraculous beacon of freedom deep in Viet Minh controlled territory.

Despite the hand wringing by many in the State Department because of the concern about the possible reaction by the People's Republic of China to it, John Foster Dulles gave approval to the proposal as did the Chinese Nationalists, whose support and co-operation could not have been better. With the help of Maj. Chew Mon-Lee, a U.S. Army Major of Chinese descent who had served in Korea (where he had been awarded the Silver Star – and a purple heart – for exceptional bravery in the Yalu River action) and now fortuitously on detached duty with the C.I.A. in Taipei, we began the process of selecting volunteer candidates provided by the Intelligence Bureau of the Ministry of Defense. ©We first did a file selection from the more than two hundred files the Chinese provided, followed by a lengthy personal interview of those whose files had impressed us the most. We had no difficulty in selecting enough for two full teams, including two capable candidates for the position of the Team Leader who would take the first team to South Vietnam. Training took place largely in the coastal areas of southern Taiwan, which came closest to matching the flat lands of Binh Hung. Thanks to the previous training and experience that most of the candidates had, we were able to ready the teams, including rigorous and varied cross training to the extent practical, and to make final choices in just under three months. A couple of toes were stepped on, and a couple of rice bowls were slightly damaged in the process, but it all went far better than I originally thought possible...thanks largely to the strong backing of Pres. Chiang Kai-shek and our counterpart Chinese officials who wanted to attain a Chinese Nationalist participation in assisting South Vietnam.

With the 18 man team – and supplies - we had selected, Maj. Lee and I returned to Binh Hung by way of Soc Trang. As we waited at the airport for the helicopters that were to ferry us to Binh Hung, we received reports of an ambush that had just occurred near the airport, resulting in the death of the Soc Trang National Police Chief and some other Vietnamese government officials. This served to underscore to our Chinese team that they were now in an area that required all to be constantly alert to possible attacks, and that the enemy was not to be taken lightly.

Upon arriving at Binh Hung, we were again greeted ceremoniously, including the same village 'musicians' enthusiastically trying to play their version of the Chinese Nationalist Anthem, but it was not easy to identify it as such, nor, for that matter, was the South Vietnamese Anthem which followed. Father Hoa was delighted that we arrived ahead of schedule, and was equally, if not more pleased that we had brought items the villagers badly needed, including clothing, school supplies, some new soccer balls, ping pong balls and paddles, and plenty of Chinese tennis shoes. He had already made out a tentative schedule which called for two days of orientation in Binh Hung for the Team, followed by visits to other villages in the area over the next two weeks. He and the Team Leader hit it off immediately, helped by the fact that our choice for 'leader' had family ties in Yunnan Province, as had both Father Hoa and Bernie Yoh. After lunch, Cartwright, who had accompanied us to Binh Hung, boarded one of the helicopters to return to Saigon to continue with his other duties, but promised Father Hoa that he would be back as often as he could. The close bond that Cartwright had established with Father Hoa was apparent; Cartwright had regularly arranged critical support to the village. Their mutual respect and personal friendship served all of us well.

The Father and Bernie Yoh joined the Chinat team, Maj. Lee and me on the first trek to two of the nearby villages in the area, as did a good number of the 'Sea Swallows'. In one of the abandoned buildings we passed along the way we found a somewhat battered piano that was missing some keys...but that did not deter Bernie. He found a rickety box on which to sit and proceeded to entertain all of us for about twenty minutes with an enthusiastic and impressive assortment of tunes ranging from honky-tonk to semi-classics. He deserved and received our 'standing ovation'. (To his delight – and mine - I reminded Bernie of that day years later when he was working with Reed Irvine of "Accuracy in Media", and addressed a chapter meeting of the Association of Former Intelligence Officers of which I was president in Las Vegas. Unfortunately, Bernie died a few years ago.)

Chew Mon-Lee and I slept on old army cots in the roofed but otherwise open hut that also served as a covered platform located on the parade ground. Under the tin roof was a 'ceiling' made of old burlap bags salvaged from the air drops of rice to the village. Over our cots there was a large, sagging bulge in the burlap, indicating that either the burlap was simply giving way or that there was something quite sizeable there. Chew Mon chose to move his cot from underneath, although neither of us climbed up to see what might be the cause. The next morning at breakfast, we mentioned this to Father Hoa. After puzzling our description a bit, he laughed long and hard, and said he would tell us later. He showed us a sizeable bamboo cage near his quarters in which there was a very large python and a live but not very aware chicken. Father Hoa explained there had been two pythons, but one was missing, obviously the one that had chosen to make its 'home' in the burlap over our cots. He noted that the python in the cage was not hungry and therefore the chicken was not concerned; however, when the python became hungry the chicken would serve as a meal. He added that pythons also served to help keep the prisoners quiet, (a remark that intrigued me, but I did not ask how or why)...and said the chickens in the village raised a fuss if unwanted intruders tried to enter the village. Appreciatively, the second python was soon removed from our 'bedroom' back to its own.

A few days later the Team, Maj. Lee, six 'Sea Swallows' and I walked and went by boat to Hung Tau, then Nga Ba Dinh, and finally, to Cai Nuoc, some nine hours trek from Binh Hung. The trip to Cai Nuoc was hot and uneventful, but I remember moving my feet carefully whenever we waded waist high streams, fearful of stepping on spiked pungi sticks which could easily penetrate boots into one's feet. In Cai Nuoc we were told that a Viet Cong force was assembling across the wide canal and that an attack was likely during the night. Given our combined fire power I did not think such an attack was likely, but the guard posts were doubled 'in case'. The night remained quiet. On the return legs, we used different routes than we had on the way to Cai Nuoc, but were still ambushed twice along the way. Fortunately, neither one was effective. Two of the participants in the first ambush attempt were caught and told us that a larger Viet Cong force had expected us to return the way we had come, and had set up ambushes to "get the American". Since my U.S. Army friend looked Chinese, the conclusion was that I was the intended target. A second ambush attempt that day was more impressive, but also failed. While we were under fire from a hamlet to the northeast of us, a Viet Cong sniper, well concealed in a close-by pond to our rear, was captured after he missed his shot at me using a M-16. The round made a very sharp report as it went by my right ear. My first reaction was that Chew Mon, who was beside me, had fired his carbine, but then realized it had not been the sound of a carbine. We pulled the sniper from the water, where he had concealed himself in the reeds, and took him back to Binh Hung to join the other prisoners in Father Hoa's holding facility. Neither Chew Mon nor I was happy that the Viet Cong/Viet Minh were better armed than we were with our M1 and M2 carbines, but thankful that the better armed 'Sea Swallows' were with us.

Upon returning to Binh Hung, we learned that one of the two militia that had gone by boat earlier that day to Cai Doi Ngon, a near-by village, to get water for Binh Hung had been killed in an ambush, and the other one wounded. Father Hoa dispatched a small team of 'Sea Swallows' to find the guilty. They returned a day later, having killed two of the offending Viet Cong, and capturing one. The next day the head of the captured man was prominently displayed on a post along the path of the canal in the village. This less than subtle message underscored that it did not pay to serve the Viet Minh or Viet Cong, at least in the area of Binh Hung.

The Chinese Team adapted well and quickly settled into their respective roles of instructing and participating in all aspects of defending and strengthening Binh Hung and the neighboring villages. After two more weeks in the Binh Hung area, I returned to Saigon for another meeting while Maj. Lee returned to Taipei. In the meeting I was asked if I might arrange for boat crews for the Nasty and Swift gun boats that were being delivered to the South Vietnamese, and if there were additional Chinese that could be brought to South Vietnam to assist in other special projects. My response was positive and subsequently many more Chinese came from Taiwan to assist in a variety of South Vietnamese programs in many of the provinces. The first team that had gone to Binh Hung was eventually replaced by the second one. I later learned that a U.S. Special Forces Team had taken over the training and advisory role there.

The successes of Father Hoa and Binh Hung - and other similar Strategic Hamlet Programs - were seldom cited by or in the Media. That is not surprising, for few 'reporters' spent time in the outlying areas, or chose to report positive stories. Most preferred to concoct their own versions of more major events, with emphasis on the negative. There were exceptions to this rather blanket statement, but not many.

The Strategic Hamlet program was successful and well received in most areas it was implemented as intended, just as it had worked in Malaya and in the Philippines, although the circumstances were basically quite different. It was a winning effort, combining self-defense with applicable bits of nation building, with the military and civilians working together with a common goal. Unfortunately, it was disdained by many, including most of the U.S. reporters in Vietnam, for it was not seen by them as being militarily related or productive...or important. With Diem's death, the basic formula of the program was changed, but the Media failed to note this, or that the initiative had then swung to the Viet Cong/Viet Minh. From their vantage point in Saigon, where the Media stayed for the most part, that too is understandable.

Most of what Diem had accomplished in the few years he governed was simply not covered accurately by the American press. For example, during the time of increasing Buddhist demonstrations, respected Catholic priest, Father de Jaegher, urged Malcolm Browne of the Associated Press, and others:

"You are distorting the facts about Buddhist persecution. If you keep it up, you can get rid of Diem. You have the power to do it. But if you do, it will be American boys who will have to come and take his place, and they will die here." (17)

But who wanted to heed or listen to Father de Jaegher's remarks? Certainly not the media!

"As the Buddhists grew bolder in their demonstrations against Diem, the Saigon government tried to arrest several pro-Communist agitators among the Buddhist leaders. The main culprit was Thick Tri Quang, a Buddhist leader who had been trained by the

Viet Cong, and whose three brothers were reportedly still active Communists in the North. Quang was twice arrested by the French for Communist activity. When he was about to be arrested in Saigon, Quang fled into the United States Overseas Mission building. From there he was taken by car to the United States Embassy by American officials. The Saigon government asked for the extradition of this left-wing Buddhist leader on the grounds that the U.S.O.M. was not the American Embassy. Ambassador Lodge, however, rejected the Saigon Government's request."(18) But the Media did not report all the facts, just the portions they thought would 'sell'.

Lodge's decision remains unexplained, as did his motives at the time. "When Henry Cabot Lodge was appointed to replace Ambassador Nolting, it was known to the world that this was evidence of the displeasure of Washington towards the Diem Government. Within a few days after Lodge assumed his duties in Saigon (August 22), he reported to the State Department that Diem must be ousted. There was a lack of understanding on Lodge's part from the very beginning, and a conflict of personalities between him and Diem. Lodge had little background in Vietnamese culture and politics, although he speaks French fluently. According to some of the Vietnamese he acted as though he were the French Governor General in Saigon."(19)

The American press had a field day over the split between Madame Nhu and her father, Tran Van Thuong, a wealthy landowner, who was serving as the Vietnamese Ambassador in Washington, without bothering to determine the reasons for the split. "Thuong ...made a special trip to Saigon to ask Diem to resign, and to turn the government over to him by making him Prime Minister. The Ambassador told the (Vietnamese) cabinet that he had returned to Saigon with the full support of the U.S. State Department. ... Thoung's proposal was rejected. The majority said it was an act of treason and that he should be arrested. Ambassador Thuong hurriedly left the meeting and took a special American plane to Washington. There he declared that it would be useless to help the Diem Government. Thuong's actions against Diem were calculated to curry favor for himself with the State Department at Diem's expense."(20) Was the full story covered by the Media? The answer is 'no'.

"The majority of American newsmen found him (Diem) too blunt. They did not understand him, nor he them. He was bitterly attacked by reporters from leading newspapers. Important officials, including President Kennedy, were strongly influenced by these reports. Diem expelled several hostile reporters, but did not realize how powerful the influence of the press was in the United States. When asked by a reporter of the U.S, News and World Report on February 18, 1963, why he had expelled some foreign newsmen, he replied: "They have gratuitously insulted the institutions of my country. Why try to humiliate and defame us while we are fighting a terrible war for out survival and for the defense of a vital border of the free world?"(21) Did the Media ever admit to such? Of course not!

This hostile attitude of the U.S. press carried over into attacks on Diem by leading politicians, including Senator Fulbright who continually criticized Diem for violating the Geneva Agreement by not holding elections with the North (ignoring the fact that neither South Vietnam nor the U.S. had signed the agreement), or that the North had not held 'elections'. When President Kennedy publicly declared there needed to be a change of policies and personnel in the Government of South Vietnam, and with Ambassador Henry Cabot Lodge's all but open hostility towards Diem, Diem's days as President were all but over. Perhaps Kennedy was also persuaded that a change was in order by a message he received from John Kenneth Galbraith, the U.S. Ambassador to India, on November 20, 1961 which read, in part, "As I will argue, there is no solution that does not involve a change of government. To say there is no alternative is nonsense for there never has seemed to be where one man has dominated the scene. So we must look ahead very soon to a new government." President Kennedy had asked Galbraith to visit Saigon and submit his conclusions. This Galbraith did, prefacing the above quote with the notation "I have just completed three intense days in Saigon which, with CINCPAC (Commander-In-Chief, Pacific) talks, gives me a much better feeling for the tangled situation."* It is a bit mind boggling that such a brief visit by a person who barely grasped the problems in India, much less the intricacies of Vietnam, could write such shallow opinions. But Ambassador Galbraith had often used similar shallow – and worse - prose about India, Nepal, Tibet, and China from his lofty political perch in New Delhi. Unfortunately, often on such drivel are policies made.

Joseph Alsop, a regular and informed visitor to Viet Nam for more than twelve years, wrote in 1965 from Saigon: "It is almost comically silly to try to organize the kind of government in Viet Nam that will win the august approval of American editorial writers who know nothing of Asia and always seem to forget that the alternative is the bleak and ruthless tyranny of Asian Communism. There is no way out any longer, except to try and deal with the war crisis first, and to leave the political situation for later consideration. Dealing with the war crisis is the only way to create the essential conditions for comparative government stability. The government is unstable precisely because the war is going badly."(22) This was a rare piece of accurate, forthright reporting and comment.

Alsop was essentially using a rear view mirror at the time to make his case, but his words were all but prophetic. There were many problems for the young and struggling country, but impressive successes had been achieved. Unfortunately, political and religious and most Media opportunists could not see the forest for the trees.

The Buddhists were well organized and became increasingly bolder, inspired perhaps by the favorable press they were receiving in many parts of the world, including from the U.S. press in Saigon and at home, regarding their baseless claims that Buddhist interests were not being represented in the Saigon government. Their protests achieved even more attention following the self-immolation of a Buddhist monk in front of the Cambodian Embassy, followed by more immolations, all of which were given world wide press coverage. That these few suicides were miniscule in numbers compared to the annual number of suicides in most countries in the world - for example, some 20,000 plus in the U.S. - was not noted by the Media. The act of setting oneself on fire is certainly spectacular, especially when captured on film and shown repeatedly on prime time TV or featured on page 1, with accompanying headlines. Reportedly, some of the Media even fought the Saigon police to get shots of the "burnings", then protested to the U.S. Embassy that they were being interfered with in their efforts to gather "news". Macabre pursuits, but of course all in the process of bringing balanced news to the public! Like little boys at play with a new toy or matches.

"That Diem's Vice President was Buddhist and that his Foreign Minister was Buddhist, as were eight members of his 18 man cabinet, and that the military governor of Saigon-Cholon and the Army Commander-in-Chief were Buddhists did not get much mention, nor did the fact that of the 113 elected members of the National Assembly there were 75 Buddhists. (23) The Media apparently did not think these figures important, for they were not related.

"General Paul Harkins, of the U.S. Military Assistance Command in Vietnam, Frederick Nolting, Jr., U.S. Ambassador in Saigon, and John Richardson, CIA Chief in Saigon, lost most of their influence in Washington because they were allegedly "pro-Diem." Ambassador Nolting's report on South Vietnam at this time carried less weight than the reports of the young and inexperienced New York Times correspondent, David Halberstam."(24) Halberstam and other U.S. reporters unduly influenced the Administration and prejudiced it against Diem, and Kennedy finally insisted "There must be a change in politics and personnel in South Viet Nam."(25)

The 27 year old Halberstam arrived in Saigon from Africa in the fall of 1962 as Francois Sully of Newsweek was being expelled at the urging of Madame Nhu (for writing stories against the Diem regime). Jim Robinson of NBC was to follow shortly for incessantly writing unfavorable stories about President Diem and many members of his government. Halberstam joined many of the reporters in Saigon writing stories supportive of the Buddhist protests against Diem. He also joined with those reporters generally critical of Diem's policies and the military efforts by the South Vietnamese. As another reporter noted, Halberstam's 'expertise' was of his own making.

Although "...guns, rifles, bombs, TNT and other arms and explosives were found in some of the Buddhist temples, neither the Embassy nor most of the U.S. State Department would believe it. For months they had been saying that the Saigon government exaggerated the problem of Communist infiltration. Diem was accused of seeing a Communist "under every bed."(26)

"Some of the Buddhist leaders, such as "Thich Thien Hoa,

urged his fellow Buddhists to refrain from taking any action against the Government. He declared that the agitation of the Buddhists was politically instigated, and that it was not a religious issue. Such statements went unreported in the American press. There (were) about 3,500 Buddhist temples in South Vietnam, and only 14 of them became involved in maneuvers against the Government, but it was from these 14 that the *New York Times* and the Associated Press seem to have taken all their information."(27)

"On August 28, General Ton That Dinh, the military governor of Saigon-Cholon, who was a Buddhist himself and the son of a Bonze, reported to the press about what had been found in the temples. Along with the arms and explosives were documents that revealed plans for assassinating some Americans and Vietnamese Army officers. Again the American journalists who were anti-Diem were not interested in reporting the facts."(28)

The Times of Viet Nam, edited by an American, featured a story on September 2, 1963 that the "CIA Financing Planned Coup d'etat". The opening paragraph began: "U.S. Central Intelligence Agency was financing a planned coup d'etat scheduled for last Wednesday, reliable foreign source said yesterday...CIA agents in the Political Section of the U.S. Embassy, the Public Safety Division of USOM and the G-2 Section of MAAG, with the assistance of well paid military attaches from three other embassies, had prepared a detailed plan for the overthrow of the Vietnamese government. The CIA plan, it is said, had the blessing of high officials in the 'distressed' State Department'". The article went on to state that the CIA had helped build the Buddhist agitation and that funding ranging from $10 - $24 million was to be used to finance the coup."(29) Of course, the Embassy and the State Department denied the story, but the State Department continued to send instructions to Lodge that he interpreted as not being supportive of Diem. "Both Defense Secretary McNamara and CIA Director McCone felt that the "get Diem" group, led by Hilsman and Averill Harriman, had "pulled a fast one" on the other U.S. officials by sending the telegram.(30) ...The people in Viet Nam knew all about the telegram, because the American Embassy leaked it to the press."(31)

It is not easy to understand how or why these abrupt changes came about, for it appeared that things were going at least fairly well in South Vietnam, despite the wide variety of problems and serious challenges it faced as it grew as an independent and liberated country. But the power of a sulking Media set on destroying Diem was aware it had a strong voice in Washington. Whether it reported accurately did not matter, for who knew better than the Media? "Some of the American journalists had attacked the Diem government almost incessantly. David Halberstam of the New York Times kept repeating that Diem was most unpopular, and that there was great discontent among the people in Viet Nam. Halberstam maintained that the Buddhists "constitute 70 percent of the population" and "resented the "Catholic Regime" of Diem."(32) The Buddhists actually constituted about 30 percent of the population, and Diem's government was not "Catholic". Halberstam magnified the discontent of the people, and never mentioned the successes of Diem. (For example, his land reform, his strategic hamlet program, and the economic progress that had been made were seldom mentioned.) Malcolm Browne of the Associated Press wrote in a similar vein."(33)

Halberstam kept insisting that the Vietnamese soldiers would not fight, but he himself was young, without military training, and inexperienced. Certainly he was not qualified to pass judgment on military matters, be they U.S. or Vietnamese, or others. Experienced war correspondent Richard Tregaskis, pointed out that President Kennedy relied more on the opinion of Halberstam than on any of his official sources. Tregaskis, who is the author of "Guadalcanal Diary" and "Vietnam Diary", stated that the New York Times, through Halberstam, had more influence on U.S. foreign policy that at any period in the Times history with one exception: its reporting on Cuba by Hubert L. Matthews. It is too early to say which caused the more serious problem."(34)

General S. L A. Marshall, in commenting on Halberstam's reporting, explained young Halberstam's criticism of the White House, the Pentagon, the C.I.A., et cetera, by stating that "the boy has a love for shining targets...In commenting on Halberestam's praise of Lodge as a man "who really knew his job" Marshall pointed out that Lodge "must have been friendly to the young reporter."(35) And, unbelievably, "The following month (after Diem's death) Halberstam of the N.Y. Times and Browne of the Associated Press received the Pulitzer Prize for their reporting on the Buddhist crisis. General Thomas A. Lane (Ret.) wrote in his book on President Kennedy that the prize "should have been awarded in Moscow, at a review of the Kremlin Guard."(36) But these views never made it to the public at large thanks to the Media's 'protective' practices.

In his book, "The Making of a Quagmire", Halberstam uses the first 32 pages to tell readers of his free wheeling and dealing days in Africa, and then notes that the special access status he enjoyed in Africa was quite different than what he experienced in Saigon. Obviously, he concluded that his reports from Vietnam would have only minimal readership if he did not do something differing from the Embassy reporting, such as choosing to write highly critical articles. Thus the ambitious 27 year old embarked on his path to damn the South Vietnamese Army, denigrate President Diem and his policies, belittle most of the Embassy personnel, and to tilt if not increase the Buddhist turmoil to the extent possible. Halberstam describes his views as: "In general, a relationship of mutual respect exists between ambassadors and reporter in most underdeveloped countries; if anything, a reporter – and New York Times reporters in particular - may be treated too well. A reporter must always be aware that an ambassador in a small country which has no immediate crisis may view the press as the best way to break through State Department channels and get his country and its problems in the White House at breakfast...But in Vietnam these relationships simply did not exist. The split between the reporters and the mission was basic; it had foundered on the policy itself. It was not, as some were later to claim, the result of a bad press policy, or poor handling of the press, or inept news management." (37) Obviously, the problem was that Halberstam and his like- minded, ego-driven colleagues wanted "recognition", and to 'make' or 'shape' policy for Vietnam, not merely report developments.

The small African countries that Halberstam had earlier enjoyed opining about were indeed different from Vietnam, and obviously Halberstam did not want to be a small fish in a large and competitive pond. After all, he was a "N.Y. Times Reporter" on assignment, (emphasis added) not just another working stiff, or a simple government employee. He wanted a drum beat of his own creation...and had brought his own drum!

In "Making of a Quagmire", Halberstam noted "The split between the American press and the American Mission was very much the product of our country's traditional freedoms and attitudes. There was a conflict between these and the pressures of the Cold War, which often push the United States into complicated, difficult and essentially alien situations. We reporters were the heirs of a traditional American freedom: the right of a journalist to write what he sees, whether the news is good or bad for his country. We did not have to worry about the alternatives of policy making. The ambassadors and generals, on other hand, were the heirs of a new dilemma: the discord between this country's traditional instincts, and its duties and responsibilities in the Cold War."(38) How does one comment on such self-indulgent puffery and delusional ego?

It is interesting that Halberstam mentions nothing about responsibility or needing to provide balanced, objective, or factual reporting. Nor, understandably, does he admit using the slippery slope of deceitful reporting, or being hell bent on making a name for himself in order to impress his employer, and how desperately he wanted his articles read in Washington.

Halberstam also wrote: "They have us in caves," Father Hoa, the famous fighting priest of Camau, told me in discussing the new wave of such incidents (referring to the Vietcong (sic) ability to mine roads in the Delta), "and now they want to keep us there".(39) Had Halberstam visited Binh Hung, he would have known that Father Hoa was simply using a figure of speech for there were no caves in Binh Hung (it being entirely flat), nor were there caves in Camau, the An Xuyen Province capitol. In fact, there was no road in Binh Hung, and but a few dirt 'roads' in most of the other villages in An Xuyen Province in those years. I can only assume that this 'conversation' took place in Saigon during one of Father Hoa's visits there. Halberstam left Saigon in December, 1963, for the U.S. and a new assignment with the *New York Times*, but he left a distinctive trail of political destruction and turmoil behind that undermined South Vietnam and U.S. efforts there for years.*

"Max Frankel in the New York Times, November 2, 1963, succinctly summed up the Administrations' attitude towards Diem's overthrow: The Administration welcomes the Coup d'etat in South Vietnam, assumes that its policies helped to bring it about and is confident of greater progress now in the war against the Communist guerrillas It is conceded here, however, that the United States Government had created the atmosphere that made the coup possible."(40) So much for Ambassador Lodge's "process of political evolution" which he, and the Media, had pushed in those early years.

"The *Christian Science Monitor* said: The most urgent need of South Vietnam is to make a transition from military back to civil government at the earliest possible moment." —"The *Washington Star* pointed out that a "spokesman for the State Department denies 'categorically' that the United States government was in any way involved in this coup attempt." "The same paper also commented: "The people who will believe this can be comfortably housed in telephone booth."(41)

So much for fair and balanced reporting. Had our Media Left (and some State Department officers) had their way I have little doubt that Ho would have been a candidate for the Nobel Peace Prize and given a ticker-tape parade in New York City!

The period following Diem's death was one of peaks and valleys resulting from the sequence of military coups that followed, the emergence of contending political parties jockeying for position, and the various religious sects attempting to gain from both the military and political struggles. The backdrop for all these struggles was the Viet Cong and their unrelenting efforts to gain strength and expand their pockets of influence throughout the country. Replacing Diem, General Duong Van Minh became Chief of State...then General Nguyen Khanh assumed the position of Premier, but convinced Gen. Minh to remain as Chief of State. However, Gen. Khanh soon decided to assume Minh's powers as well. This in turn caused widespread demonstrations, prompting Gen. Khanh's acceptance of Gens. Minh and Khiem to serve with him in a triumvirate role. Gen. Khanh then appointed a largely civilian cabinet that was not acceptable to most of the military, so again there was a shuffling of authority. The "High National Council" attempted to form a civilian government under Phan Khac Suu, a Cao Dai leader, with Gen. Khanh remaining as Commander-in-Chief (with Gens. Khiem and Minh being assigned abroad), and Tran Van Huong, former mayor of Saigon, becoming Prime Minister. Gen. Khanh then influenced the Armed Forces Council to dismiss the Tran Van Huong government. Dissatisfied with this arrangement, Dr. Phan Huy Quat, the former foreign minister, formed another cabinet, but was overthrown in June when Buddhist, Cao Dai, and Hoa Hao elements agitated for more rights, prompting Catholic disagreements, which led to the formation of a new government under Gen. Nguyen Van Thieu as Chief of State, and Gen. (Field Marshal) Nguyen Cao Ky as Prime Minister. Clearly, someone should have thought of selling scorecards to help the attentive audience keep track of the constantly changing lineups, including identifying those warming up in the bull pen!

If the foregoing sounds like a poorly done Hollywood script, or something akin to the days of the slapstick comedy, with many players having a variety of roles and wearing various hats, I would have to agree. But sadly, to the South Vietnamese it was all too real, and millions of lives were at stake. If it is confusing to the reader, think of how it must have impacted the Vietnamese, be they in the cities, towns, villages, hamlets, or the remote rural areas in the provinces, or in any of the military forces fighting for their country. The North Vietnamese were elated with the news of Diem's assassination, and certainly hopeful their assets in the South would now become even more effective in causing additional discontent and turmoil. Meanwhile, unfortunately, the 'Black Hats', the Viet Cong/Viet Minh, made hay while these events played out. The U.S. – and foreign - Media had a field day in trying to guess who was doing what to whom, and what roles all the supporting figures were performing. That Media's role was less than helpful. It would not be inaccurate, especially in the continued stirring of political and religious pots, that Media's role was often suspect. The deadly struggle between good and evil in the field continued in all the provinces, despite the intricate political dance steps and maneuverings in Saigon, but ARVN confidence certainly had to be shaken

"Infiltration (of North Vietnamese army personnel) mounted to over 5,000 a month in mid-1966 but apparently declined somewhat by late 1966.Viet Cong initiated actions soared from 500 in September 1963, itself a record, to 3,500 in November and 500 a week five months later. Of the 8,600 strategic hamlets purportedly completed, about one in five was viable militarily. The U.S. urged completion of the hamlet project after proper consolidation, but Vietnamese leaders, wanting to reassure villagers, favored hamlets only in safe areas."(42)

"Taking advantage of these disruptive forces, Viet Cong activities mounted. Infiltration from North Vietnam of about 2,000 per month was matched by local recruitment. Increasing quantities of Communist-bloc weapons arrived by land and sea. In response, U.S. forces increased to over 22,000 by the end of 1964. General Maxwell Taylor's appointment as Ambassador in June 1964 indicated a stronger U.S. military stance, while the naming of career diplomat U. Alexis Johnson as Deputy Ambassador implied an equal concern for political and social reform. Following the attack by DRV PT boats on two U.S. destroyers in the Gulf of Tonkin in August, U.S. Aircraft bombed North Vietnamese naval bases. President Johnson took this occasion to assure the world that the U.S. would stay in Southeast Asia as long as the struggle required and would "meet aggression with firmness and unprovoked attack with measured reply."(43)

The so-called Tonkin Gulf engagement had far reaching effects, including Johnson being given unlimited powers to pursue the war. While the various stories regarding the two "attacks" on the destroyer Maddox were less than clear, Captain Herrick, her commander, concluded and reported up the chain of command that "It was simply a case of over eager young sailors caught up in the emotion of the moment." However, that is not the way it played out. "Lyndon Johnson gets the Admiral's message in the White House. He confides to an aide, "Hell, those sailors were just shooting at flying fish!" "He slips the message in his pocket and says nothing about it to the leaders of Congress. It all contributes to a smashing victory in the election."(44) The Media did not pursue or clarify the story, despite the many question marks, and simply abetted the reelection aspirations of Pres. Johnson. Few questions were asked, and even fewer answers given.

"To help counter this escalation (of Viet Cong/Viet Minh activities), U.S. forces in Vietnam 'officially' entered combat in June 1965. By early 1967, U.S. strength in Vietnam had risen to nearly 400,000 men, not including the Seventh Fleet operating offshore. In addition, 44,000 Koreans, 4,000 Australians, and small New Zealand, Philippine, and Thai contingents had joined the allied forces. The South Vietnamese Army (ARVN) had grown to 325,000 regular and 400,000 paramilitary troops, with most of both being volunteers. The enemy main forces numbered 50,000 North Vietnamese and 70,000 Viet Cong army regulars, backed by 150,000 Viet Cong irregulars, 40,000 political cadre, and 20,000 logistical support troops. In their "rear base" in North Vietnam, the "People's Army of Vietnam" (PAVN) had about 225,000 regulars, 200,000 paramilitary, and large trained reserves, which Ho Chih Minh officially called to arms in August 1966. Some 80,000 Chinese supplemented DRV manpower."(45)

It was something of a miracle that South Vietnam survived this uneasy period. Had the Viet Cong/Viet Minh forces been stronger, or the North Vietnamese Army been better positioned, it is possible that the struggle would have been all but over. Hand wringing had become a Washington sport in President Johnson's administration, reflecting that the reporting from the Saigon Embassy began to differ noticeably from that of the U.S. military commanders. Were we winning, or losing, or holding our own? It depended on whether you were reading political reporting, military situation reports, Media reporting covering what the religious activists were up to, or if Vietnamese coup plotters were at it again. It was a ready-made Media smorgasbord...and the fare was to their liking. Meanwhile, Washington struggled to juggle the Moscow, Beijing, Hanoi, and Saigon relationships.

"During 1965 and 1966, however, Viet Cong relations with the peasants deteriorated. When they launched their victory drive in 1965, the Viet Cong increased their taxation and conscription. Dropping their pose as persuasive peasant leaders, they took what they needed by force, and peasant hostility grew...The flood of over one million refugees to government-held areas cost the Viet Cong heavily, for abandoned villages produced no rice. They vented resentment by terrorist attacks on refugee centers and resorted to mass kidnappings to repopulate deserted areas. The flow of intelligence from peasants, an important clue to attitudes, markedly increased."(46) The Viet Cong had suffered heavy losses both in military engagements and from increased desertions. But again, the Media chose not to report successes in the South.

Another positive development was the performance of Nguyen Cao Ky. He soon pledged vigorous programs to reduce poverty, rectify illiteracy, eliminate corruption, reduce inefficiency, and reemphasize the rural reconstruction program. The U.S. readily supported these efforts. The Communists had assassinated multiple thousands of rural leaders, teachers, and government representatives throughout the country. Thousand more had been kidnapped, including doctors. "The problems of pacification are enormous, for a few organized guerrillas can terrorize large areas. Estimates of the population controlled by the Viet Cong in Vietnam at the end of 1966 varied, with about 20 percent accepted by many, while the government controlled about 55 percent, with the remainder contested. Government "control" is a relative term, meaning basically the existence of a functioning administration and reasonable security. However, a small group of determined people indistinguishable from the general populace can make any place unsafe at any time. Even in the United States a lone "mad sniper" can demoralize a city, and organized gangsters can baffle police for years.* Terrorist techniques, combined with a highly developed propaganda and indoctrination system, are very difficult to combat, especially when directed against isolated peasant communities. Critics of the pacification program expected the Vietnamese government to maintain greater security during an insurrection than is available at night in some American cities. To accelerate the pace of pacification, a shift of military emphasis was ordered. The burden of fighting Communist main forces was to be undertaken by the allied troops during 1967, while the Vietnamese regular army would support pacification in the lowlands by fighting Viet Cong guerrillas and working with rural development activities, thus simultaneously increasing local security and improving the people's welfare. While foreigners could provide technical and economic assistance, only Vietnamese could inspire, direct, and achieve an effective social revolution at the rice-roots level."(47)

With these agreed divisions of 'labor', the die was cast; i.e., the U.S. took on the military role it had originally not wanted. And not surprisingly, but unfortunately, their planning was still based on WWII and Korea. Put simply, one size does not fit all. The four main regions constituting South Vietnam were not at all the same politically, economically, militarily, religiously, ethnically, nor topographically. The programs that could work in IV corps, the Delta area of Vietnam, would not necessarily succeed in I Corps, the northernmost region of South Vietnam, or in II Corps either, with its vast highlands, forests, and famous resort areas in central and western Vietnam. It was physically located between I Corps and III Corps, with the latter being further to the south and close to Saigon. I Corps and II Corps had multiple settlements of Montagnards, the tribal or 'hill' people, disdained by the French and many Vietnamese, but who now wanted better representation in and support from the government. In addition, I Corps was a mix of all the religions in Vietnam, and Hue, the ancient capitol with its proud heritage and popular university, served as a caldron of political intrigue, as did Danang, the largest city in I Corps, second only to Saigon in population. While the Viet Cong was more active in the southern regions, North Vietnam's Main Force units posed the major threat to I Corps, extending from Quang Tri, which bordered North Vietnam along the 17th parallel to the north, to Quang Ngai, the southernmost province in the region. The Viet Cong/Viet Minh cadres did not miss these pronounced differences, just as they certainly added to Nguyen Cao Ky's headaches, and gradually befuddled the strategists. The South Vietnamese moving to handle the Viet Cong problem while the U.S. assumed to fight the North Vietnamese Army (and Viet Minh) looked neat on paper, but there were serious gaps and flaws that were not easily corrected. That the role of air power would be badly overlooked and misused would not become apparent until it was too late.

Most of the Media did not seem to grasp much, if any, of the above dynamics or the roles of the respective forces involved. Despite the carrot and stick style use of bombings of parts of North Vietnam, Hanoi continued to reject all offers of negotiations. "Throughout 1966, all peace efforts were rejected by Hanoi and Peking, with Moscow taking an increasingly hard public stand. In July 1966, the USSR rejected mediation efforts by Indian premier Indira Ghandi, U.N Secretary General U Thant, and (British) Prime Minister Wilson, saying the Soviets would not initiate negotiations or a new Geneva conference without Hanoi's assent."(48) While China remained less outspoken, it was evident to most that she was pressuring Hanoi to continue to delay, just as she had advised the North Koreans, which in turn led to costing 90,000 U.S. casualties during two years of (Korean) truce talks. "...Hanoi hoped for a break in American will to fight, believing the key factor in the war was not U.S. Military power, but the American rear...the political situation at home and abroad, the morale of its army and people, and its own ability to conduct the war."(49) This was another bench mark remark that went largely unheeded by both the U.S. State Department and the Media, and by Congress. Any thought that Moscow or Beijing would eventually prove helpful was but a fantasy. The shadow of 'protraction' as an art form was obviously at hand. But Johnson was reluctant to engage the enemy, and agonized over committing air power against the many military targets in North Vietnam. "Ambassador Taylor and General Westmoreland both beg Johnson to bomb North Vietnam again. He refuses. "The stakes are very high," he tells them. The view is growing in Washington that LBJ is a procrastinator."(50) The media equivocates...again.

However, the Media also continued to pursue their own version of developments on political and military matters. It was a rare event when a reporter visited Danang or Hue in I Corps; it was not as comfortable or convenient as gathering and 'collating' the "news" from sources in Saigon.

Dalat, a popular and cooler climate resort city in the highlands of II Corps was visited time to time by the Media, but Nhatrang, the Province Capitol in II Corps, seldom had visits by journalists, despite most restaurants there having the best 'lobsters' one could hope for, plus a variety of other sea food and marvelous French bread. Vung Tau, a large and popular beach resort area in IV Corps not far from Saigon, was popular, but probably difficult for the Media gang to explain on the expense account. So Saigon became the locus for the great majority of the U.S. Media. There they could get Embassy briefings, United States Information Agency briefings, regular U.S. military briefings, and briefings from South Vietnamese public relations and press persons, all on a daily basis. That most of the hotels in Saigon were air conditioned and well appointed, or that Cholon, an extension of Saigon, was more 'open' than Saigon, or that the bars in both locations were comfortable and increasingly well stocked (thanks to the post exchanges that had been established in Saigon), including many comely 'hostesses' in both areas, obviously had little or nothing to do with their limited coverage of the country beyond Saigon.. For instance, in 1966, there were more than 140 U.S. 'reporters' accredited to South Vietnam, with nearly 300 more to follow. I suggest this number might be described as an assembly of like-minded persons, as in 'choir' or perhaps just a 'gaggle'.

There were exceptions to this generalization, of course, and there were times when Media travel to Quang Ngai and Hue and Khe San and Nhatrang became popular, if only for brief periods. For example:

- Quang Ngai City became of interest when stories surfaced about members of Army Lt. Calley's patrol shooting some 100 civilians in the village of My Lai in Quang Ngai province. (Quang Ngai province is where communism first began in South Vietnam some two decades earlier. My Lai was also located in a "free fire" zone.) The attack on My Lai was in retaliation for some of Calley's search team being wounded from mines set near the village during the North Vietnamese 'Tet' offensive in 1968.
- Khe San, in Quang Tri Province, besieged in late 1967 by the North Vietnamese, but held, again became of interest to the Media when it was used as a launch base in February, 1971, for "Lam Son 719", a large ARVN search and destroy mission into Laos against North Vietnamese storage areas located there. The North Vietnamese learned details of this operation well in advance and moved in tanks, artillery, and thousands of troops to destroy the attacking South Vietnamese forces. The South Vietnamese suffered heavy losses of personnel, tanks and helicopters, but pressed on to establish fire support bases, but were unable to complete the plan. The operation was, at best, only marginally successful. Some 20 to 25,000 North Vietnamese were killed, primarily by B-52 strikes that struck massed NVA forces attempting to annihilate the attacking South Vietnamese forces. No U.S. reporters chose to ride any of the helicopters used throughout the operation in support of the operation, although Japanese and Korean reporters did so. Khe San had earlier been of interest when the North Vietnamese moved large numbers of troops into the area and shelled it heavily for an extended period prior to Tet in 1968, but very few in the U.S. Media chose to helicopter in to cover the gallant defense of Khe San by the dug-in U.S. Marines.
- Hue – the North Vietnamese seized part of the Citadel, the ancient palace and thickly walled grounds located in the city of Hue, at the outset of 'Tet' in 1968 and flew the North Vietnamese flag over the Citadel until driven out nearly a month later. Some of

the Media chose to denigrate the combined U.S. – Vietnamese efforts to dislodge the North Vietnamese, and to admire the North Vietnamese daring and tenacity in their suicidal effort to hold the Citadel.

• Nhatrang became of interest when the U.S. military broke the story of an investigation of Special Forces personnel having been involved in the demise of a Vietnamese double agent in II Corps. The thrust of the media coverage was very unfavorable to the Special Forces, yet supportive of the double agent and his duplicitous role. Aid and comfort to the enemy? Certainly not the Media!

Somewhere along the way our Media forgot what John F. Kennedy had said during his inauguration address in January, 1961 about the U.S. intention to "Bear any burden, meet any hardships, support any friend and oppose any foe to assure the survival and success of liberty."(51) Similarly, the Media did not seem to recognize that Lyndon Johnson was basically revising Kennedy's pledge to read "ask not what you can do for your country…but what your country can do for you!"(52)

Kennedy also addressed the 1962 graduating class at West Point in which he said: "Another kind of war, new in its intensity, ancient in it origins – war by guerrillas, subversives, assassins, war by ambush instead of by combat, by infiltration instead of aggression, seeking victory by eroding and exchanging the enemy instead of engaging him….When there is a visible enemy to fight in open combat, the answer is not so difficult. Many serve, and applaud, and the tide of patriotism runs high. But when there is a long, slow struggle, with no immediate visible foe, your choice will seem hard indeed."(53)

Johnson's decision to send more U.S. Military to Vietnam and to bomb North Vietnam was widely decried by the Liberal Media and protested on college campuses by both students and faculty members. The Rev. Martin Luther King described Vietnam "A White Man's War and urged Blacks not to participate."(54) That this view was refuted by other blacks both in and out of military service was not noted by the Media. And of course, some Hollywood personalities joined the calumny of praise for the communist side, and condemned the U.S. and South Vietnamese efforts. This was well covered by the Media.

Perhaps the sorriest reporting on Vietnam came following the 1968 Tet 'offensive' by the Viet Cong which began January 30. Walter Cronkite on CBS concluded that the Viet Cong 'uprising' in Saigon was successful and expressed his anger on the air. ABC and NBC followed suit. Not until three to four days later did the networks - and most newspapers - report that the U.S. Embassy had in fact not been penetrated, and that all 20 of the Viet Cong who attempted to do so had been killed. Their bodies were laid out inside the Embassy gates early the morning of January 30, and General Westmoreland showed them, and the hole blown in the wall by them to gain access to the grounds, to the Media corps, which was reluctant to believe him, or the Embassy Marine Guards that had fought off the attackers.

"Of the 65,000 Vietcong who revealed themselves in early hours of Tet...39,000 are now dead. Seven thousand more have been captured alive. Only 19,000 have replaced their black pajamas with civilian clothes and escaped. 19,000 communist VC soldiers remain in a nation of twenty million...and have no real hope of recruiting many replacements."(55) Regardless, Walter Cronkite continued to report the Viet Cong Tet Holiday attack to be a success - or an American failure. Like lemmings, the Liberal Media followed Cronkite's lead. As a result, Johnson concluded that U.S. confidence was beyond repair.

On February 27, 1968, Cronkite told his CBS-TV audience: "We have been too often disappointed by the optimism of American leaders, both in Vietnam and in Washington, to have any faith any longer in the silver linings they find in the darkest clouds. To say that we are closer to victory today is to believe, in the face of the evidence, the optimists who have been wrong in the past...To say that we are mired in a stalemate seems the only realistic, yet unsatisfactory conclusion." Cronkite went on to say: "Khe San could well fall, with a terrible loss in American lives, prestige and morale. And this is a tragedy of our stubbornness there."(56) This then led to his sermonizing about negotiations being the only way out for America. That he was 'wrong on all counts, but never in doubt' is an apt summary of such "analytical" reporting. Cronkite's seriously flawed "analysis" and comments contributed measurably to undermining the U.S. efforts in Vietnam and to fanning the flames of the well coordinated protests that followed across the country. Did he ever admit these grievous errors? No, but he did finally admit to being a life-long 'Liberal'...as though we had not known that for years.

Secretary of State Dean Rusk said: "We are taking the wrong turn at a fork in the road. The Tet Offensive is a very serious military setback for the North Vietnamese and the Vietcong forces. It is a great mystery to me as to how it is being translated into such a brilliant political success for the North Vietnamese – here at home in the United States. This is the first struggle fought on television in everyone's living room every day. What would have happened in World War II if Guadalcanal and the Anzio beach-head and the Battle of the Bulge or the Dieppe Raid were on television and the other side was not doing the same thing? War is an obscene blot on the face of the human race. But whether ordinary people can sustain a war effort under that kind of daily hammering is a very large question."(57) Add to this what Gen. Westmoreland had to say in Saigon: "We have as fine a military force as America has ever assembled – a force that could bring the war to a favorable end if political decisions do not prevent it. Doom and gloom reporting is giving the American public the impression that Americans are being defeated on the battlefield, swaying public opinion to the point that political decisions often take all the pressure off the enemy at a time when he is virtually on the ropes...Our worst enemy is the press...hundreds of reporters in Vietnam – all practicing, seeking and reporting news as they are accustomed to in the United States...all looking for sensational stories. The enemy controls the information on his side. We report only the off-beat, the unusual or bizarre. Our open society is vulnerable to manipulation."(58) And who are the manipulators? Well, certainly the Liberal/Liberal Left Media deserves recognition as being Charter Members!

In South Vietnam after Tet, volunteers for the Vietnamese Army were in such numbers that the size of the army was doubled; the numbers of desertions dropped to a new low; and draft dodging was down similarly. Pride in having kicked the crap out of the Viet Cong was high, as was outrage that the North would attempt such a sneak attack during Vietnam's nation-wide New Year's Holiday. The only downer countrywide was the successful seizing of part of the ancient palace fortress known as the Citadel in Hue by a North Vietnamese death force. Their occupation ended by late February.

U.S. and South Vietnamese forces chose not to bomb and/or destroy the Citadel because of its majestic role in the history of Vietnam. This decision necessarily resulted in prolonging the Viet Minh occupation of the Citadel...and cost more casualties to U.S. and South Vietnamese forces. But the Media preferred to ignore these points as well, while highlighting the 'ability' of the Viet Minh forces to persist.

The Media also failed to accurately report the true story behind the famed photograph taken by Eddie Adams, of the *Associated Press*, of South Vietnam's National Police Chief Gen. Nguyen Ngoc Loan's shooting of the Viet Cong terrorist that had just killed Loan's wife, three children, and another National Police officer during Tet. The terrorist had been seen escaping from Loan's home and been captured. The photo shows Loan in the act of shooting the terrorist in the head, and identifies Gen. Loan. The reason(s) the Viet Cong terrorist was shot, or that the Viet Cong had stacks of official death warrants signed in the north, but left blank (to be filled in by Viet Cong assassins), were omitted, but NBC repeatedly carried the camera shot of Loan shooting the VC terrorist on its evening news. The photograph remains one of the classic photos of all time, and Adams deserves praise for having captured it on film, but the failure of the networks to carry the full story was and remains irrational.

When refugee camps near Danang and Hue in I Corps were attacked and burned by Viet Minh/Viet Cong forces in 1969 and 1970 with heavy loss of life to the women, children and the elderly in the camps there was little or no interest evidenced by the Media, other than to note that perhaps the defenses of the refugee camps were inadequate. When 3,500 bodies of residents of Hue were found in the sand dunes outside of Hue after Tet, 1968, with their hands and legs bound by barbed wire, obviously summarily executed, little interest was shown by the U.S. press. Fortunately, media assets from other countries reported these massacres. More than 1,000 additional bodies were discovered shortly after. Only then did some, but not all, of the U.S. Media deem this noteworthy.

The *Wall Street Journal* wrote: "The American People should be getting ready to accept, if they haven't already, the prospect that the whole Vietnam effort may be doomed."(59) The Media and politics then joined in an outlandish plot. Pennsylvania Congressman William Moorhead hosted a party attended by Under-Secretary of the Air Force Townsend Hoopes and *N.Y. Time* reporter Edwin Dale, "Showing disloyalty to the President if not his country, Secretary Hoopes hints to the *Times* man that a Pentagon cabal is trying to resist a planned troop buildup in Vietnam. Dale passes the tip along to fellow "*Times*" reporters Hedrick Smith and Neil Sheehan – both former Vietnam correspondents. They contact their "sources" in Congress and the Pentagon. Within a few days they have a story. ...On Sunday, March 10, the "*Times*" publish its "expose" on Page One, under a three-column headline reading: WESTMORELAND REQUESTS 206,000 MORE MEN, STIRRING DEBATE IN ADMINISTRATION — But the story omits mention of the fact that 106,000 of the men are wanted for duty in West Europe. It also ignores the fact that Westmoreland wants the extra troops so he can immediately go on the offensive against the seriously hurting communists. ...The Westmoreland request is dead as soon as the article appears."(60) The formula of hatching a story, planting it, and then reaping the rewards of a "scoop" was successful then, just as it is today. The Liberal Left knows no bounds.

The offensive that Westmoreland unleashed in South Vietnam in late March, 1968, both ground and air, succeeded in demolishing some 10,000 North Vietnamese troops, with 1,200 enemy bodies in the Khe San area alone. However, the Media preferred to report on the growing protests by "anti-war activists" in the United States.

During my two years in Danang in I Corps (June'69-June'71), I cannot recall having met more than one Media representative. Nor was I visited by any high ranking U.S. civilian officials except for two or three mid-level ones visiting Vietnam from Washington. The Consulate and the CRDS (A.I.D.) offices in Danang were similarly fortunate. I did meet many senior military officers visiting I Corps, including Gen. Abrams and Lt. Gen Ray Peers.

The Army had given Peers the unenviable task of investigating the My Lai incident. I had first met then Col. Peers in Taipei, Taiwan in 1952, where he was on assignment to the C.I.A. as the head of "Western Enterprises, Inc." He was not happy with the My Lai investigative task he had been given, but like the outstanding soldier he was, he intended to leave no stone unturned – and didn't.

I regularly visited Quang Tri, Hue, Hoi An, Tam Ky and Quang Nam, the province capitals and key cities in I Corps, where C.I.A. personnel were assigned. Part of my responsibilities were to maintain liaison with the I Corps commander (Gen. Lam), the National Police and Special Police in I Corps, the Vietnamese Province Chiefs, Vietnamese Military Intelligence office, the commanders of the 1st and 2nd ARVN (in Hue and Quang Ngai), III MAF (under Lt. Gen. Herman Nickerson), the U.S. Consulate, (Terry McNamara), and the U.S. Army Military Intelligence Detachment (under then Col. James Williams and his deputy, Lt. Col. Al Buckelew). As the senior CAS representative in I Corps, I attended the daily early morning briefings prepared for Gen. Nickerson by his staff at his 24Corps Hqs. and met privately with him at least weekly. I also met weekly with Gen. Lam, with whom I occasionally played tennis and dined. The latter's main frustration was with Saigon, from whence President Thieu tried to manage all things, a common Saigon fault.

The C.I.A. Chief of Station in Saigon held monthly meetings of the ROICS (Regional Officer(s) in Charge), during which he would present a 'big picture' overview, following which the ROICS would give their respective regional reports. Seldom did my I Corps views match the optimistic ones given by most of my colleagues, nor seldom did I Corps 'intelligence' reports production match the numbers submitted by most of the other regions. We simply had few unilateral sources able to provide us substantiated intelligence, nor did the often fluid situation in I Corps lend itself to developing such sources, a view and problem shared by our police contacts, as well as by both the U.S. and Vietnamese military intelligence contacts. The Vietnamese police in Danang and Hue were targeted by the Viet Cong and more than one senior police officer was killed by an explosive device being placed under their jeep, as was one of the I Corps Province Chiefs. That there was a very capable centralized prisoner interrogation center managed by the U.S. Army located on the huge Danang Air Base was also a factor, for the few knowledgeable prisoners that our Province Reconnaissance Units (PRU) were able to capture were usually passed to the U.S. military in order that the information gathered from such sources could be quickly shared with field units.

The PRU were irregular but usually very effective forces...and Hanoi credited PRU with extensive successes against their Viet Cong assets. In I Corps we had five PRU groups, including two in Quang Ngai Province, one of which was outstanding, and one that was less so. It had become more or less habit that the PRU in each province would routinely "arrest" Vietnamese in rural areas as "suspect" Viet Cong, then after debriefing and investigation usually let the "suspects" go. The same would occur during the ensuing weeks, largely because the "reward' money was based more on numbers of "arrests" of suspects, than on quality captures. One of the two PRU groups in Quang Ngai Province was especially guilty of this, whereas the other one was a dedicated and professional group of "Bounty Hunters." The solution that got the results that I wanted was to increase the rewards for captured Viet Cong – or NVN Army personnel - plus no arrests of rural Vietnamese were to be made without an approved 'warrant' in hand in advance The number of "suspect" arrests went down markedly, but the numbers of authenticated numbers of Viet Cong captured increased...as did information reports from some of the captured. In two of I Corps provinces the PRU often nailed inattentive NVA personnel as well. These, along with Viet Cong leader cadres, went to the U.S. interrogation facility at the Danang Air Base. The priority concerns of the U.S. and South Vietnamese military were understandably in regards to 'order of battle' data, but captured documents addressing political matters were always highly prized, sought, and rewarded.

On occasion we would come by an operational gold nugget or two. One such case was from an unvetted informant who told one of our case officers that the Viet Cong had an underground command post facility just outside of Danang. With the support of Gen. Nickerson, who readily provided Marine CAP teams to assist us, the reported site was visited. One of the Marines kicked a can that was on a stick that had been stuck into the ground. He was yelled at because of the possibility of booby traps, but upon closer investigation – and digging – a sizeable underground complex was uncovered. Some 24,000 documents were recovered. No personnel were found, nor did the captured documents include any communiqués dated within the past 24 hours, but what was there proved to be a veritable gold mine of information. At our office facility in Danang we set up a 24 hour photographing capability that within a week produced, sorted, and sourced copies of each of the documents for Saigon Station, the Marines, the Army, for CRDS, the Consulate, the Vietnamese military intelligence service, the Vietnamese police, and for Gen. Lam. Not only were there details as to Viet Cong/Viet Minh 'nets' in I Corps and extending to some of the other regions, the documents revealed penetration nets into Saigon, including more than one into the palace as well as into the Saigon Special Police. There was both jubilation and consternation, with emphasis on the latter from some of the U.S. offices in Saigon. Earlier a similar net into the palace and military in Saigon had been uncovered, but Saigon assumed there could not be more. I have yet to collect on the bet that I made with one of our senior Saigon case officers that a very high ranking Saigon Special Police officer identified in the captured documents as a penetration agent would not pass a polygraph examination. I was correct...he didn't. To my knowledge, the media reported very little on the second 'net' story, or on the earlier one into key Saigon entities. Why? Hard to say.

The Danang Air Base, at one time the busiest airport in the world (takeoffs and landings) was often the site of 122mm rocket attacks by the Viet Cong/Viet Minh. Seldom accurate, the rockets did serve to frighten the entire city, for more rockets hit in the populated areas than on the airfield, including one that landed about 100 – 150 yards from the house in which I was living. My wife was visiting me at the time and was petrified, but quickly got under the bed as told. She never did distinguish between the sounds of incoming and outgoing, so usually got under the bed "just in case". Another time an errant rocket managed to strike a XXIV Corps munitions dump located across the river from our office complex, detonating some twenty four tons of ordnance in a spectacular series of explosions and air bursts. The resulting concussion blew the roofs off my living quarters and that of our nearby office complex. I awoke to a room full of dust, feathers, and starry skies above. There was extensive damage in the III MAF Hqs. Compound, plus one Marine officer killed and others wounded, that compound being located closer to the explosion area. (Had the Media been in Danang I have little doubt that it would have reported the bad guys simply "winning" another engagement.)

There was little change in Mainstream Media's role for the remainder of Johnson's term in office, with coverage shifting more to the protestors than the war in Vietnam. But when President Nixon ordered U.S. (and South Vietnamese) troops into Cambodia in response to Cambodia's President Lon Nol's request, the "*N.Y. Times*" said the "...mission is a virtual renunciation of the pledge to end the war" And the "*Wall Street Journal*" warned that it means deeper entrapment in Southeast Asia". Both were wrong. American and ARVN tanks drove communist soldiers away from their Cambodian base camps, and they found mountains of weapons and munitions hidden in underground bunkers, but none of this seemed to matter to the American newspapers and TV networks. Hour after hour they kept up their drum-fire of verbal assaults."(61)

Even coverage of the bombings in North Vietnam was distorted. Much of this was the fault of a French reporter in Hanoi who said "... American planes are laying bombs down like carpet right through the center of Hanoi...killing civilians intentionally and without discrimination." (62) The peace activists were claiming that 10,000 or more civilians had been killed by the bombings around Hanoi, but the Mayor of Hanoi put the number close to 1,500. Interestingly, Malcolm Browne (*N.Y. Times)* flew to Hanoi and confirmed the Mayor's report and said "The damage has been grossly overstated."(63) Not surprisingly, "U.S. television and newspapers continue to quote the same "carpet bombing report over and over again...despite the admission by the Mayor of Hanoi that civilian casualties are remarkably light – a tribute to the accuracy of U.S. laser bombs and the extreme care used by American pilots and crewmen aiming at military and warehouse targets."(64) In this he differed from the distasteful, disastrous (to U.S. policies), flawed performance of Harrison Salisbury.

In December 1966, Salisbury's articles appeared in the *New York Times* as his own personal observations made during a visit to North Vietnam; it later became known that they were based almost wholly on North Vietnamese propaganda pamphlets and on statistics provided by D.R.V officials. His description, for example, of "Nam Dinh as a cotton-and-silk town with no military objectives was wrong. In fact the city contained POL (petroleum, oil, and lubricants) storage, a power plant, and a railroad yard, and was surrounded by anti-aircraft and surface-to-air missiles. This ensured that Salisbury received no Pulitzer Prize. But his articles and their implication, drawn chiefly from North Vietnamese falsehoods, suggesting the United States was deliberately bombing civilian targets. carried worldwide convictions."(65) Perhaps from this came the media version of Hollywood cowboys wearing different colored hats - Malcolm Browne wearing the white one. But there is more.

"In the fall of 1966 or in the winter of 1966-1967, a delegation from the Czechoslovak government had arrived in Hanoi to advise the North Vietnamese on the modalities of carrying the propaganda war to the continental United States. The flow to Hanoi of grisly "Western" journalists, television "personalities". disaffected minority groups, "tribunals", "parliamentary" commissions, "concerned" writers, "independent" or "international" jurists, doctors, trade unionist, and other riff-raff, most of whose expenses were paid by the D.R.V. government, began and hardly stopped. The tide of hatred for America, pulled by falsehoods of Marxist, neo-Marxist, or Trotskyist professional manufacture, started thereafter to infiltrate the American Media and eventually the "intellectual" establishment."(66)

I would add to the above that while I was assigned to Langley in mid-1971 and 1972 tending to Vietnam, Cambodia and Laos matters, the C.I.A. learned from well placed North Vietnamese sources the anti-war efforts across the U.S. were not only being manipulated by a nation-wide activist coalition, including representation on and support from most college campuses, involving both faculties and students, but that it was also being very well funded by numerous national and international front groups, foreign entities and governments, including the Soviet Union and Cuba, using their diplomatic immunity protected pouches into the United States to fund the organizers. (Certainly no one thinks that 'spontaneous demonstrations' are really spontaneous - or cost free - do they? Perhaps the Liberal Left Media doesn't know that.) The demonstrations were well organized and coordinated...and, as many North Vietnamese leaders were to brag later, their successes in reducing popular – and political - backing in the United States for continued support to the South Vietnamese were instrumental in "winning the war" for North Vietnam. U.S. forces did not lose any battles against North Vietnamese forces in Vietnam, but the uneven political and Media contests are an entirely different story. The role of the Liberal Media certainly played a major role, and deserves the bow and accolades eventually proffered it by the People's Democratic Republic of Vietnam.

But the Mainstream Media does not deserve all the credit – or all the blame. Certainly there is enough to go around. There are many politicians who worked mightily to insure that South Vietnam would fall, and some of them still serve in Washington. The military's often doctrinaire handling of many of the challenges that Vietnam posed also deserves at least honorable mention. Near victory was turned into certain defeat on more than one occasion in each of the military regions, with some regions having suffered more than others. I previously used the term that "one size did not fit all", which was proven time and again. For example, tanks were useful in some regions, but in others not at all; the same for 'effective' artillery. The old school pursuits of massed forces against massed forces were of course long out of vogue, and modified efforts to reinvent that static tactic insured only unnecessary casualties. The over-use of Fire Support Bases made for inviting targets, as the French had learned previously. The U.S. often seemed to insist on walking in the footprints of the defeated French, and too often adopted an attitude that matched. Our State Department and other civilian self-taught "experts", occasionally including the Media, insisted on keeping their hands in military affairs with the result that seldom was the desired effectiveness achieved. Ambassadors or other Foreign Service Officers, or political appointees, seldom have the qualifications to determine or recommend "arc light" (B-52) boxes...yet too often they did. And the multiplicity of programs in Vietnam made only the semanticists happy. There were simply too many competing programs and not enough Vietnamese interpreters/translators to meet the never ending demands of various Interrogation Centers, Census Grievance programs, Hamlet Evaluation System, the Phoenix Program, Special Police, National Police, Pacification, Provincial Reconnaissance Units, Regional Forces, Popular Forces, etc. Similarly, military efficiency was judged and rated largely on the reported number of casualties inflicted on the enemy (KIAs, WIAs, and Bomb Damage Assessments, etc.) Woe to the officer who did not do better in these categories than his predecessor. The same held true for the Media; reporters were in competition for headline grabbing stories, with little or no regard to fact or fiction.

On June 30, 1970, President Nixon reported to the nation on the Cambodian incursions. "Ten major operations were launched against a dozen of the most significant base areas with 32,000 American troops and 48,000 South Vietnamese participating at various times. As of today, all Americans...have withdrawn, as have a majority of the South Vietnamese forces. We have scrupulously observed the 21-mile limit on penetration of our ground combat forces into Cambodian territory. ...My June 3 interim report pointed up the success of these operations and the massive amounts of supplies we were seizing and destroying. We have since added substantially to these totals. ...Here are some highlights. We have captured:

-22,892 individual weapons...and 2,509 big crew-served weapons;

-more than 15 million rounds of ammunition;

-14 million pounds of rice

-143,000 rockets, mortars, recoilless rifle rounds

-Over 199,552 anti-aircraft rounds, 5,482 mines, 62,022 grenades, and 83,000 pounds of explosives, including 1,002 satchel charges;

-Over 435 vehicles destroyed

-11,349 men (enemy) killed and about 2,328 captured and detainees."(67)

The Media failed to recognize these achievements, or the significance of the extensive routes and supply bases in Cambodia used by the North Vietnamese in supporting their forces in Vietnam. In regards to the latter point, the same could be said regarding the supply routes and bases through "Neutral" Laos, in obvious violation of the Geneva Accords. Apparently the Media did not judge these points noteworthy.

("On December 31, 1971, there were 156,800 U.S. military personnel assigned in Vietnam. Cumulative casualty figures read killed in action: 45,626; wounded, 302,375."(68) And the so called peace talks appeared stalled. "The figures released December 31,

1974 were: U.S. military personnel in Vietnam, "less than 250"; killed in action (for the year): 207; `cumulative: 46,370; wounded in action: 0 (for the year); cumulative WIA: 303,656. (69))

That Presidents Johnson and Nixon – and the 'master negotiator', Henry Kissinger – appeared unable to make up their minds became evident only because such was the case. Johnson was almost cowardly in his pursuit of negotiations, much preferring to devote attention to his political future, and favoring his 'Great Society' over the fate of Vietnam, Cambodia, and Laos. Nixon vacillated, and could not set a winning course largely because of political uncertainties and his misplaced belief that Kissinger would succeed in negotiations if given time. Neither Johnson nor Nixon was willing to listen to the advice of our military which, in simple terms - and as history had made clear - understood that strength and force, with cogent use of available air power, could win the day, diplomacy having already failed dismally. The respective "Rules of Engagement" established by Johnson and Nixon controlling military actions became a tangled and vacillating maize that in essence resulted in civilians mini-managing the war, while the Commander-In-Chief(s) looked over their shoulders at the political polls.

Unfortunately, the frustrations within the military did not result in one or more of the key flag-rank officers making their frustration(s) public, for to do so likely meant their being discharged…thus career concerns took precedence over truth and duty. This inexplicable failure on the part of many in key command positions over the years allowed the politicians and the self-appointed 'politically correct' to talk about lights "at the ends of tunnels" that did not exist. That the politically elected and/or politically appointed should leave the management and conduct of combat to those trained in such arts is not a new concept. Failed diplomacy does not mean that failed diplomats should tinker with military action, for that can easily result in military failures as well. "Yes" men in key positions in the military are not the desired products of our military academies. We have had far too many of them over the years, not only in the Vietnam years, but obviously during the Clinton Administration as well.

The Mainstream Media could not resist the temptations to conclude that their own role was key, and 'reporting' became an entertainment venue in which all could and did play. This was especially true of the Liberal Media as they recorded their perceptions of political and military shortcomings, including second guessing all those who they concluded as not having their own clear 'visions'.

The Cambodian Government collapsed on April 18, 1975. The picture of Ambassador John Gunther Dean's carrying the American flag under his arm as he left the Phnom Penh airport was featured on page 1 of most newspapers. It was the symbol of another fallen domino.

I had met Dean in Danang (Vietnam) where he was Deputy Chief of CRDS for I Corps. I liked him and enjoyed his openness. We often argued about the efficacy of 'coalition governments'; he thought the idea a good one...I did not, noting that never had it proved successful, and that eventually it had only caused more problems wherever and whenever tried. We resumed our friendship – and argument - in Vientiane, Laos, where he was Deputy Chief of Mission when I arrived in mid-1973. The French presence was surprisingly (at least to me) still large, especially their military advisory group. The French Ambassador was a strong advocate of a 'coalition government' that in his mind would consist of members of the Royal Lao Government and the Pathet Lao. (The Forces Armee Neutral (FAN) no longer existed, its members having fled, or joined the FAR, or followed their leader (Capt., then Col.) Kong Le to Sam Neua to join the Pathet Lao.) "Coalition" was the French solution for Laos; it was also the solution strongly pushed by the Pathet Lao and the North Vietnamese, China and most of the Media, following the disappearance of the so-called "Neutralists".

Prime Minister (Prince) Souvanna Phouma resided in Vientiane, while King Savang Vatthana and his family lived in Luang Prabang, the ancient capital of Laos located in MR 1. The Prime Minister was more tolerated than admired by the majority of the Royal Lao Government (RLG) officials, and held in even less esteem by the Royal Lao Forces (FAR). He was considered ambivalent, lazy, and inconsistent, more often than not being influenced by the last person to talk to him. That he insisted on being treated as 'Royalty' by his Lao minions was no small irritant. Not surprisingly, within Vientiane's Diplomatic Corps the French Ambassador had the most influence with Souvanna Phouma. U.S.

Ambassador Charles Whitehouse and his Deputy Chief of Mission, John Dean, had less influence, perhaps because of the less than successful U.S. pursuits in South Vietnam, but more likely because of a carry-over of Souvanna Phouma's strong feelings about U.S. Ambassador Parsons, about whom he had said: "He understood nothing about Asia and nothing about Laos. The Assistant Secretary of State is the most nefarious and reprehensible of men. He is the ignominious architect of disastrous American policy toward Laos. He and others like him are responsible for the recent shedding of Lao blood."(70) Harsh words indeed, but obviously deeply felt views.

"In the absence of hard facts to go by, newspapers normally known for their sense of responsibility fell back on the time-honored expedient of describing Laos as a never-never land where events followed no logic and were not worthy of serious attention in any case. A typical example described a "psychedelic political power struggle between fractured Centralists and splintered Neutralists, Rightists, and crypto-Communists, which might tell you a great deal about the significance of the military byplay if you could somehow wrap your mind around it, and having done that, take it seriously."(71) Now that is impressive 'reporting', is it not, or is it media analysis? Or simply a frustrated 'spin' of some sort?

U.S. relations at most government levels and with the military throughout the country were cordial, due in large part to the support that the U.S. provided to FAR (via the French military), and directly to the irregular forces (Groups Mobile) located in most of the Military Regions, as well as the extensive economic assistance provided Laos by the U.S. Agency for International Development. The period of internal turmoil began in the 1960s when a disaffected Captain Kong Le attempted a coup d'etat against the RLG. He subsequently defected to the Pathet Lao with remnants of his "neutral" force in the late 1960s. However, there had been a somewhat steady series of changes within the government and military forces in the intervening years, and the Pathet Lao, strongly supported by the North Vietnamese, continued to try to seize control of areas throughout Laos. The Geneva Accords had granted Laos "neutral" status in 1962, but this had in no way deterred the Pathet Lao, the North Vietnamese, Russians, or Chinese from pursuing their goal of making Laos into a communist state. The U.S. abided by the explicit terms of the treaty, but did manage to circumvent them where circumvention was possible. For example, the U.S. was precluded from directly supporting the FAR, but the agreement made no mention of the U.S. not supporting irregular forces.

The U.S. Government had a few civilian officers in Luang Prabang, Savannakhet, Pakse and in Long Tieng (with Gen.Vang Pao, the titular head of the Meo in northern Laos). The U.S. military presence in Laos was very limited, per the Geneva Accords Agreement, as was the role it could play vis-a-vis the FAR. The French could work directly with the FAR, but the role of the U.S. military was largely limited to meeting some of FAR's needs through the French. U.S. Training of FAR in Laos was prohibited, as was 'advising'. The U.S. Agency for International Assistance and Development had extensive representation in Laos, and its presence was all but overwhelming in numbers. Most of the AID personnel lived in a well appointed complex located at KM-6, a short distance (6 kilometers) north of Vientiane that resembled a small American town, complete with a K – 8th grade school, theater, tennis courts, softball field, soccer field, a swimming pool, gas station, and club. For some strange reason, this complex was not the subject of Media reports. Perhaps some of the Lao government restrictions placed on the travel of reporters kept them focused on Vietniane, or occasionslly, Luang Prabang. "Most of the press stories were written at the swimming pool of the new Lane Xang Hotel, which had become a hub of activity, where journalists, American pacifists on their way to or from Hanoi, and the wives of American prisoners of war who were intent on seeing North Vietnamese and NLHS representatives rubbed elbows and exchanged stories."(72)

Some three months after arriving in Vientiane, I received a phone call from one of FAR Gen. Kouprasith Abhay's staff officers at Camp Chinaimo alerting me that there was a coup d'etat attempt underway against the R.L.G. led by Gen. Thao Ma, the Air Force officer who in 1967 had unsuccessfully attempted a coup…and then fled to Thailand. My house in Vientiane was not far from Camp Chinaimo and shortly I heard the sound of explosions and the chatter of machine guns from T-28s. After making a couple of phone calls to other U.S. Embassy personnel to tell them of the attack, I headed for the Embassy to send a highest precedence message to Washington to alert those that needed to know. I then drove to Chinaimo to join Kouprasith and to report by radio to the Embassy what was transpiring. Thao Ma's and two other T-28s had succeeded in killing Kouprasith's G-3 in one of their bombing runs, along with two other staff personnel, They also inflicted considerable damage on three staff buildings, and Kouprasith's living quarters. I joined him under a huge tree adjacent to the parade ground that provided us excellent protection from the strafing planes. Kouprasith was on his radio with two of the platoons he had dispatched to move against Ma's 'forces' at the airfield and radio station. Some twenty minutes or so later the attacks on Chinaimo ceased, and Kouprasith's commanders reported that the radio station, which had been seized earlier and used to broadcast the impending overthrow of Souvanna Phouma, had been retaken,...and the capture of Thao Ma, who had made a controlled crash landing at the airport due to ground fire severing his plane's fuel line. Seeing Kouprasith's forces advancing, the other Lao Air Force T-28s that earlier had been commandeered by Thao Ma followers flew to Thailand, as did two helicopters and a light plane carrying others involved in the coup attempt. Some others escaped by boat across the Mekong River to Thailand. Of the original 60 to 70 that made the attempt with Thao Ma, about 40 or so made it safely back to Thailand.

Kouprasith's feisty and able deputy, Gen. Atsaphammavong, arrived with the captured Gen. Thao Ma on the back of a truck, guarded by two FAR soldiers. Thao Ma was unceremoniously shoved off the truck and put at the feet of Kouprasith, who told Thao Ma that again he had disgraced himself, again killed some FAR, including his G-3, and reminded him that he had warned Thao Ma the last time (Thao Ma's coup attempt in 1966) that should he return to Laos he, Kouprasith, would kill him. The lecture was emphasized by Kouprasith grinding the bottom of his boot on Thao Ma's cheeks, the ultimate of Lao insults. Thao Ma had little to say other than (Prime Minister) Souvanna Phouma was destroying Laos and that there had to be a change. Kouprasith said simply this was not up to Thao Ma, and turned away, motioning that the discussion was through. Thao Ma was then tossed back onto the truck bed and shot twice, first in the stomach, then in the chest, after which his body was dumped into a roadside ditch.

Those that had been captured at the radio station were also brought to face Kouprasith. He again delivered a strong lecture, challenging them to justify their actions to him. None did so. After they were taken away, I asked Kouprasith what was in store for the captured. His reply was straightforward: they would be questioned, and then would attempt to escape during the night, but would not succeed.

When I met with Kouprasith next day at Chinaimo I asked about the prisoners and he confirmed that indeed they had died while attempting to escape. He also said that he had received disturbing information that he wanted to discuss with me. He then showed me some of the leaflets that the coup makers had brought to Vientiane, their armbands, and some of the large amounts of U.S. dollars, Thai Baht, and Lao Kip that the individual had been carrying. He said that some of the coup makers had confessed that the C.I.A. was behind the coup attempt, had prepared the written material, and financed it. I told him that this was not true, for certainly "we" (C.I.A. Bangkok and Vientiane) would have known of any such plan and not supported it. I added that had the C.I.A. been behind the coup attempt, it would have been done professionally and succeeded, and not been the amateurish failure it was. I also reminded him that I would not have come to Chinaimo to join him had "we" been involved. He said nothing for nearly a minute, then with a big smile, slapped my leg and said "I believe you". Our bond became strong and remained that way until the day he and his family and staff had to flee to Thailand as the Pathet Lao conducted their own coup in early 1975.

The Bangkok papers made a big story out of the coup attempt, including some strong insinuations that the U.S. had been involved. Contributing to these suspicions was the unexplained move of Gen.Vang Pao and a large number of his troops southward from Long Tieng to not far from Vientiane two days before the coup attempt, then his abrupt return to Long Tieng immediately afterwards. Both Gen. Bounpone, FAR commander, and Gen. Kouprasith, deputy commander, suspected that Vang Pao was 'involved', but could not substantiate it. Unnamed wealthy Thais were suspected but no charges were ever made to my knowledge.

Despite the warnings to Prime Minister Souvanna Phouma by

FAR and the Group Mobile commanders in the regions, and the concerns of the King in Luang Prabang, regarding the consequences, the political march continued towards the formation of a Coalition Government. First, Pathet Lao forces were permitted into Vientiane in force, and were housed in a large renovated hotel located near the airport, while the accompanying political cadre took up houses and offices in Vientiane. (Flush toilets were a mystery to these troops, and soon all the hotel's toilets were plugged with bamboo sticks, or broken, requiring extensive repairs…and "potty" training.) Pathet Lao forces continued to gobble up more vital locations throughout the country, ignoring what few violations the ineffective International Control Commission (ICC) representatives noted and chose to protest.

In early 1975, the Pathet Lao shed its charade and expanded its seizures of territory in the regions, while Pathet Lao leaders in Vientiane demanded more government 'responsibilities' as due them in their coalition role, including insisting that U.S. A.I.D. personnel leave, that U.S. Military attaches leave, and that U.S. personnel be confined to Vientiane. As a consequence, many RLG officials fled to Thailand, as did many of the irregular forces, including Vang Pao who vacated Long Tieng on May 14 and flew to Thailand. Thousands of Hmong were evacuated from Long Tieng to Thailand, but some 10,000 were simply left behind bv Vang Pao.

The American presence in Lao diminished quickly…and the Laotians who had worked with U.S. offices and officials were fearful of not being able to leave before the Pathet Lao assumed total control. I remember driving to Chinaimo to say farewell to Gen Kouprasith, his family, and his staff. Many of them were in tears, having been defeated by political forces, and now facing refugee status. Most of them departed Laos by helicopters to Thailand, while others went to Thailand by boat across the muddy Mekong River separating the two countries.

Jim Sheldon, a C.I.A. colleague in Vientiane, organized a version of the old "underground railroad" to get RLG officials, RLG and Group Mobile military, Embassy personnel from South Vietnam, Thailand, Taiwan and others to Thailand by plane when possible, and by boat when aircraft could no longer be used. Some of the air shuttles were made with people in mail sacks or packing boxes to avoid detection by the Pathet Lao airport guards. With the cooperation of the Royal Thai navy, various launching sites along the Mekong River were used almost nightly, with Lao families carrying their few possessions in one or two suitcases. Hundreds of people were transported to safety, including Lao employees at the U.S. Embassy, heads of RLG departments and offices, and FAR personnel. There were some close calls and, because of some very heavy suitcases, some boats were nearly swamped, but fortunately the Pathet Lao failed to intercept this effort, nor was anyone lost or injured. Belatedly, but deservedly, Sheldon was presented an award by C.I.A. Director Bush for his outstanding services in these remarkable and sustained efforts.

Not all RLG officials left, some choosing to stay, despite the expected "re-education" camps - and harsh treatments in store. (Prince) Gen. Tiao Sinh, an outstanding and well educated army officer in MRI chose not to leave because it would not be in keeping with his heritage. He died in one of the re-education camps near Sam Neua, as did at least 15 other senior officers, either from the harsh living conditions, or from the exploding ordnance they were forced to clear from old bomb craters. Many others were badly injured, their fates not known, while others eventually died from exposure and sickness.

With the departure of dependents, and non-essential personnel (a State Department euphemism for "down to bare-bones staffing"), U.S. presence in Laos went from over 2,000 to about 17, including myself, three communicators, other Foreign Service officers, and four Marine Guards, in less than two months. Similar, but less severe, draw downs were conducted at the British and Australian Embassies. That all French military advisors were also required to leave, much to the chagrin of the French Ambassador, was very satisfying to me, for most of the French had relished seeing the hasty departure of the Americans, and had thought they would remain. A few U.N. personnel were allowed to remain, primarily so that the 'goodies' provided by that organization would continue to flow in. The only 'reporters' in Vientiane were Vietnamese, Pathet Lao, Chinese, a Japanese, and Russian...with the ineffective ICC representatives also present for reasons best known to the Pathet Lao. The Soviet mission increased in size, but otherwise, Vientiane had become a substantially reduced "diplomatic" community.

Not surprisingly, the 'peoples demonstrations' soon began, with the Lao residents remaining in Vientiane, led by civilian clothed Pathet Lao, throwing rocks at the U .S. Embassy as they slowly marched by. Fortunately there were no Nolan Ryans among them. There were similar 'spontaneous' marches or demonstrations requiring the people to attend pro-Pathet Lao rallies held on a weekly basis at the fairgrounds, or in the center of the city, with all cheering, on cue, the endless accomplishments and goals of the Pathet Lao. Attendance for the Lao was mandatory, and PL troops, or their willing supporters, checked each house to be sure all were at the 'rallies'. Many of the bigger houses in Vientiane, especially those belonging to the Americans or former Lao officials, became billets for the PL officials or troops, commandeered by the PL. My house near Chinaimo, for example, became 'home' for some 60 or more PL soldiers, utilizing the four bedrooms, the spacious living room, the dining room, the staff quarters (gardener and cook) and garage. It soon resembled a wasteland. After my family left in early 1975, I moved into a very modest one bedroom house.

Some of the hierarchy of the Pathet Lao moved into the houses at KM-6 which had previously been occupied by the AID people. The 'American' school was converted into meeting rooms and offices, and the school's auditorium used to discuss party matters. Some of the scores of vehicles left behind by the AID employees were made inoperable by their owners who poured sugar into the gas tanks rather than leave them for PL officialdom to use. Many vehicles were left untouched, ranging from high end models to "others". It became a spectator sport for we foreigners to watch the PL attempt to drive, for obviously few had mastered the basics. There was usually one requirement for the driver and that was speed - little else seemed to matter. Most managed to stay on the paved roads, but when confronted with the need to turn there were some impressive rollovers. Bumper cars could never equal this display and "Demolition Derby" had a new meaning. Eventually, the PL stopped the unqualified drivers because of the many injuries that resulted and the rapidly reduced inventory of usable vehicles. The Media would have enjoyed it, or perhaps criticized those of us remaining for not providing driving lessons.

Despite the large numbers of Pathet Lao troops and security

forces in Vientiane, we few remaining CIA officers managed to collect and send in daily reports to Langley on topics ranging from political developments, to occasional order-of-battle information, to PL policy decisions, to plans and intentions as best we could garner from the few sources left. Clandestine meetings continued to be held and information requirements were exchanged regularly under the noses of the PL. These accomplishments gave us a bit of an adrenalin rush, as well as considerable personal satisfaction, to put it mildly. It was seldom boring. Certainly it was interesting to experience and watch the script unfold. We all knew the ultimate ending of the 'play book', but it was interesting to watch the inevitable developments occur as the pages turned towards the inevitable end.

Thomas Corcoran, the U.S. Charge d'Affaires for most of this unsettled period following the Pathet Lao ascendancy, was a fine Boston Irishman with a wonderful dry sense of humor. He was a French speaking bachelor, a true gentleman, and could be gently persuasive or hard nosed, depending on the challenge. He did not back down from the Pathet Lao leaders or any of the other 'unfriendlies'. I still rate him as one of the better Foreign Service officers with whom I worked during my career. He died a few years ago, but he is remembered as a good man who excelled in a difficult situation.

An unexpected challenge that occurred in Vientiane remains in my mind as though it were yesterday. One early evening in June, 1975, I arrived at my new 'home' to find Gen. Soutchay sitting in my living room. We had met four or five time previously. He was an excellent Lao officer, having distinguished himself in southern Laos against both Pathet Lao and Vietnamese forces. He had been summoned to Vientiane by the Pathet Lao and was supposed to report for "duty" the next day. He was certain that he was going to be sent to a re-education camp, and wanted assistance to get out of Laos. His family was already in Thailand, having crossed the Mekong River earlier from southern Laos. After being flown into Vientiane he had been dropped off at a downtown hotel where he had left two suitcases before trying to find me. I told him to relax while I retrieved his suitcases.

I parked near the front of the modest hotel. Dressed in jacket and tie, I went past the uniformed PL guard standing nearby and entered the foyer. His suitcases were still where he said he had left them. I selected a paper from a nearby rack, and waited until the clerk behind the desk became busy with paperwork. Noting the two or three in the adjacent bar were watching a local TV program, I discarded the paper, took the suitcases out the door, down the steps to the sidewalk, and put them into the trunk - and drove away as the guard yawned.

I then contacted a trusted Lao friend who had worked with us on our "underground railroad" and told him what we needed to do. He simply said, "no problem", and we proceeded to work out the requisite details. I also cabled Bangkok, Thailand regarding Soutchay to insure he would be met on the Thai side of the Mekong River.

Soutchay and I left my house some 15 minutes ahead of rendezvous time, he in the car's trunk, and I behind the wheel. I waved at the "neighborhood watch" that the PL had put on my front gate from twilight to first light seven days a week since the PL takeover, and they smilingly waved back. (I had often given them cigarettes or other goodies, so had made 'friends' with my new found "guards".) The move of Soutchay and his luggage from my car to the rendezvous with my Lao friend, and thence across the Mekong River by small boat, went without a hitch. The only concern was how best to load Soutchay and his two pieces of luggage onto my friend's motorcycle ...but fortunately that proved not to be a problem, once balance, albeit precarious, was established.

Because of ever increasing numbers of Pathet Lao check-points in Vientiane, this was the last time this particular exfiltration route was used, but others were developed. A somewhat poignant but sad story involved my unexpected meeting with FAR commander Gen. Bounpone – and his wife – at one of the diplomatic functions in Vientiane shortly after Soutchay's successful escape. I asked Bounpone why he was still in Vientiane rather than having gone to Thailand. His response was that his wife wanted to sell their house and was having difficulty getting what she wanted for it. I expressed my exasperation in muted tones, but urged him to get out before it was too late, even if it meant leaving most, if not all behind. I learned ten days later that he had been sent to a re-education camp, but then been moved to China for medical treatment. Obviously his wife had not sold their house. Eventually

Bounpone was returned to Sam Neua to complete his "sentence", whatever the Pathet Lao had decided it to be.

The hill tribesmen (Meo/Hmong) continued their resistance against the PL for an extended period, although it was a losing cause despite the covert support provided through Thailand. Lao refugees continued to flow into Thailand, including the Hmong. From many refugees came similar reports of Hmong villages being sprayed with chemicals from planes flying low above their villages. The chemical was not identified, but many deaths and illnesses resulted from what became known as "Acid Rain" to those assisting in the refugee camps.

I reported these stories from our sources in Vientiane in late 1975 and into early 1976, but Washington was apparently reluctant to believe or publicize that the North Vietnamese were engaged in such attacks. As time passed, and more reports from refugees continued, it developed that the Russians had provided the Vietnamese the chemicals they used to spray on "the Hmong resistance" in Laos. Strangely, the U.S. Mainstream Media did not or chose not to relate these stories despite them becoming commonplace in the refugee camps in Thailand

When the take over of Laos by the Pathet Lao was fully completed in August 1975, it became the "People's Democratic Republic of Laos". The Indochina domino theory had played out as many had predicted. Ho Chi Minh's dream of having Vietnam, Cambodia, and Laos free of French and U.S. domination had been realized...but in reality, Vietnam had become the colonial power over its neighbors, while itself was totally dependent on Russia and China. The Media failed to note this ironic epitaph.

In 1961 President Kennedy said in a television address to the nation: "First, we strongly and unreservedly support the goal of a neutral and independent Laos, tied to no outside power or group of powers, threatening no one, and free from any domination. Our support for the present duly constituted government is aimed entirely and exclusively at that result. And if in the past there has been any possible ground for misunderstanding of our desire for a truly neutral Lao, there should be none now. Secondly, if there is to be a peaceful solution, there must be a cessation of the present armed attacks by externally supported Communists. If these attacks do not stop, those who support a truly neutral Laos will have to consider their responses. I want to make it clear to the American people and to all the world that all we want in Laos is peace, not war - a truly neutral government, not a cold war pawn; - a settlement concluded at the conference table and not on the battlefield."(73) But who remembers such lofty thoughts, much less other previous commitments and promises? Did the Media remind anyone? Hardly a word.

The Pathet Lao forced the King to abdicate in December, 1975 and to work as a common laborer on what had been his royal lands in Luang Prabang. But that was not enough for the Pathet Lao and their Vietnamese supporters. The former King and his son were eventually taken to Sam Neua to do hard labor, and, reportedly, there they died. He and his family deserved much better. But well before he was deposed, a life-size black statue of the King was officially presented to the Lao government by the Russians, probably in the belief that they would score points by doing so. The statue was placed on the grounds of a Buddhist monastery located not far from the Soviet Embassy. (I drove by the statue - and the Soviet Embassy - daily.) One of my Lao friends called my attention to what had happened to the statue a short while after the dedication, namely that two silver colored streaks began to appear on it running from the eyes down towards the feet, like tears. Eventually the 'stains' extended down the front of the King to his feet. There were no other visible faults on the statue. The Russians with whom we occasionally played tennis explained that this 'problem' was from a faulty casting, and that the statue was to be returned to Russia for repairs or a new casting. It was removed in early 1976, and not returned to Vientiane. The King's "tears" proved to be prophetic indeed. As he had predicted at his father's funeral in 1961, "I am doomed to be the last King of Laos."

Beginning in 1970, Nixon's opponents on Capitol Hill began to circle their political wagon train, using ever increasing restrictions which could only give assurances, aid, and comfort to North Vietnam, the Chinese, and the Soviet Union. For Instance: ®

• The Fulbright Amendment to the Armed Forces Appropriations Authorization for FY 1971 passed on October 7, 1970, specified that South Vietnam and other allied countries could not use their funds for military support or assistance to Cambodia. It also prohibited South Vietnam and other allied countries, such as Thailand, from transferring military supplies furnished under the act to Cambodia. In other words, Cambodia's neighbors were prohibited from improving their own security by assisting Cambodia with American equipment, the only equipment they had.

• The Cooper-Church Amendment to the Supplementary Foreign Assistance Act of 1970, passed on January 5, 1971, prohibited the use of American funds for financing "the introduction of US ground combat troops into Cambodia, or to provide US advisers." Not only were American combat troops prohibited, but American advisers were barred from training or instructing Cambodian units on the use of the American equipment we were supplying.

• The Symington-Case Amendment to the Substitute Foreign Assistance Act and Related Assistance Act, passed on February 7, 1972, placed severe restrictions on civilians serving in Cambodia. It limited the total number of "civilian officers and employees of executive agencies of the US government who are US citizens" to two hundred and the number of third-country nationals to eighty-five.

• The Second Supplemental Appropriation Act for FY 1973, signed reluctantly into law by President Nixon on July 1. 1973, prohibited the use of funds approved in the act to "support directly or indirectly combat activities in or over Cambodia, Laos, North Vietnam, South Vietnam or off the shores of Cambodia, Laos, North Vietnam, South Vietnam."

• The Continuing Appropriations Act for FY 1974, likewise signed into law on July 1 1973, prohibited the use of any funds to finance directly or indirectly combat activities by US forces "in or over or from off the shores of North Vietnam, South Vietnam, Laos, or Cambodia.

• The Foreign Assistance Act of 1973, which became law on December 17, 1973, provided that no funds authorized or appropriated under any of it provisions would be available to finance military or paramilitary combat operations by foreign forces in Laos, Cambodia, North Vietnam, South Vietnam, or Thailand."

Adding insult to injury, On May 1, 1975, the day after Saigon fell, the House of Representatives refused to approve a request by President Ford for $327 million for the care and transportation of the Indochinese refugees. Congressional leaders for weeks had argued against efforts to save Vietnamese, and now the House turned down refugee relief for those who had already been saved. What a magnificent display of statesmanship and moral strength by our patriotic, keen-eyed politicians in Congress! Thus did Uncle Ho – and the North Vietnamese Politburo - succeed in their long quest that Vietnam, Cambodia, and Laos be united under Vietnamese domination - not French, nor Chinese, or other unwanted foreign flags. In practice this simply meant that the three countries known for decades as Indochina were now part of a communist Vietnamese controlled triumvirate. However, all three would remain dependent on the largesse of the Russians, where Ho first studied and worked under Stalin, and the People's Republic of China, where Ho had joined the Chinese Communist Party after returning from France, and where he had helped form the French Communist Party. A 'triangle', so to speak. Thus the U.S. was forced to leave, having soiled itself in the eyes of the world. The War had been fought in the East, but was lost to inept, selfish and inept political interests in the West.

The U.S. dropped 7,975,000 tons of bombs on Indo China at a cost of $6 billion dollars. This was four times the total tonnage used in all theaters in WWII. Nearly one million Communist soldiers were killed, as were nearly 200,000 Vietnamese soldiers. And three freedom loving countries were lost, to say nothing of the thousands and thousands of civilians and refugees that were lost as well. Those figures should bother at least most, if not all, of those that deliberately chose to deceive.

Unfortunately, our Mainstream Media seldom addresses wrongs, or notes the lessons of history. They simply chose not to challenge or disagree with the guarantees of defeat that both Houses of Congress voted to impose by the shameful restrictions noted above. Politicians are seldom motivated by anything other than their own parochial interests, whatever those may be, and the Liberal Media was, and is, only too willing to go along with them, as long as by doing so it satisfies their own inclinations. Freedom of the Press? Certainly, but let's also consider the "Value of Constructive Dissent", as opposed to presenting biased and slanted views. This might lead to "Media" becoming truly "Fair and Balanced", as much of Fox News has applied so well. Is this asking too much? Probably – but not if the consumer demands it.

Socialists and Communists make such wonderful, reliable, and trustworthy friends and supporters of liberty and patriots. The same can be said of the recent vintage "ME" generation goals, with their eyes primarily on government dole. Never mind that this is the path to Socialism. Hasn't the Mainstream Media made all this clear? Strange – surely it is worth noting. Perhaps they have just been too busy with other matters.

All of us need to read history books, especially those narrating the many communist purges that have befallen the duped ...and note well the litany of abuses that accrued to the proletariat, the bourgeois, and often to those in leadership positions as well. We should not allow the "Liberals" and the "Liberal Left" and the "Liberal" Media or the "Liberal Left" in Academia to play the role of "teacher", or to be the sole source of knowledge. The Liberal Left at best is like Socialism 'light'. The extreme left is equally bad as the extreme right. In between there are many differences and problems that need to be identified and resolved. Ignorance is not bliss; it can only make us more vulnerable, and, more importantly, allow further injury to our country. It is not likely the Mainstream Media – or our politicians - will change soon; thus the future of this country is basically up for grabs. The words 'Responsibility' and "Truth" seem to have disappeared from our lexicon. We desperately need to bring them back and insist they be applied.

© Chew Mon-Lee and I formed a close working relationship during this project, and his many contributions went far beyond his Chinese language capabilities. Fearless, he took great pride in his work. We were to met again in Danang,Vietnam (I Corps) in 1970, he an Army Colonel, and I with the CIA. Sadly, less than a year later, I served as a pall bearer for his full military honors burial in Arlington Cemetery, Va.. A sad day.

* page 73, "Vietnam Voices" by John Pratt. Viking Penguin, 1984

* In Michael Lind's book, "Vietnam: The Necessary War", he notes that "In 1979, Halberstam, in an effort to excuse the many errors of fact committed by the group of journalists in Saigon to which he had belonged, claimed that "their political reporting was limited by the absence of serious skilled Asia experts in the American Embassy. The McCarthy era had taken care of

that"(pg.144)...and then "Halberstam's book (Ho) is perhaps the most sympathetic portrait of a Stalinist dictator ever penned by a reputable journalist identified with the liberal rather than the radical left."(pg.176) Halberstam also pays glowing tribute to Ho. "Ho had not only liberated his own country and changed the course of colonial rule in both Africa and Asia, he had done something even more remarkable; he had touched the culture and soul of his enemy." (pg. 178, 179 "Vietnam: The Necessary War.")

* The reader may recall the confusion and near panic caused by two "snipers" in Maryland and Virginia in 2002...and the successful ruses employed by pro-Hussein units against Coalition forces in Iraq in early 2003.

® These above and following amendments and financial restrictions are from Kissinger's *Years of Renewal,* pages 501 and 546.

CHAPTER 2

LYNDON BAINES JOHNSON

(Beyond the Peter Principle)

Following the assassination of President John Fitzgerald Kennedy in Dallas, Texas, on November 22, 1963, Vice President Lyndon Baines Johnson was sworn into office as our 38th president just prior to Air Force I leaving for the return trip to Washington, D.C. President Kennedy's body was in a casket in the cargo section of the same airplane. Jacqueline Kennedy rode as a passenger in her blood stained outfit, no longer the "First Lady". The "Camelot" sojourn had come to a sudden, brutal, and puzzling end in Dallas. The nation – as well as most of the world - was in shock and mourning, but "President" Johnson had finally realized his long awaited ambition.

A little more than a year earlier, on October 20, 1962, Communist China's troops unexpectedly invaded India along their common northern border. Inadequately armed Indian Army troops were no match for the massive attack, and India quickly sought U.S. and British military aid as the Chinese seized large tracts of Indian territory. The British demurred, largely for understandable reasons such as their Hong Kong colony. Perhaps to help his Harvard colleague, the U.S. Ambassador to India, John Kenneth Galbraith, President Kennedy ordered that covert assistance be provided the Government of India (GOI), which the Military promptly did. Following rapid coordination with the Indian Intelligence Service, the Indian Air Force Base at Agra and the Embassy, including providing details as to the equipment the Indian army needed, how and where deliveries were to be made, including route, and call signs involved, the multiple supply flights began.

As this was the first time that the U.S. was "On Stage", so to speak, in India for many years, it was obviously a test of how well the U.S. could deliver under pressure as well as an opportunity to demonstrate to the GOI that the U.S. was a capable friend and ally. (With the Chinese invasion, Nehru had dismissed his Defense Minister, Krishna Menon, an avowed leftist and outspoken anti-U.S. figure from the Indian state of Kerala.). The Indian Air Force was concerned that newspapers – or others – might possibly publicize the arrival of U.S. C-130 cargo aircraft ferrying the supplies, and so chose staggered arrival times for the flights. Two days later the first unmarked aircraft arrived precisely on time, unloaded its neatly marked and palletized load in less than 15 minutes, while two of its four engines kept running, and departed, much to the amazement of the Indians. The sustained air lift continued for some eight weeks. All went extremely well, with every crew being within 10 to 15 seconds of their pre-set schedules. Off-loading was all but immediate, and the planes departed as soon as it was completed. Indian ground crews then broke the palletized loads into their designated priorities, and the Indian Air force, using C-119s and C-47s, then ferried the supplies to the beleaguered Indian army units in the hills to the north. The Indian Air Force had no C-130s in their inventory at that time, and much admired – and coveted - the cargo capacity of the C-130s.

For reasons known only to them, the Chinese withdrew most of their troops from Indian territories on November 21. However, from this rapid response by the U.S. to India's request grew a joint effort with the Indian Intelligence Service that endured for years, thanks to President Kennedy's decision to quickly meet India's needs. It certainly was not a Johnson-like decision; there was no poll, no money, or lobbyists, or constituent votes involved.

Johnson had served 24 years in Congress, 12 as a Representative and 12 as a Senator, much of the latter as Majority Leader. Thwarted by Kennedy in his ambitions to become the standard bearer of the Democratic party, Johnson agreed to run as Kennedy's vice-presidential nominee, thereby providing the Democrats with the appearance of a united ticket. With the assistance of Illinois, especially the very suspect Chicago area vote, the John Kennedy-Lyndon Johnson ticket narrowly (2%) defeated the Republican ticket of Richard Nixon-Henry Cabot Lodge.

Following the burial of Kennedy, Johnson selected a committee to investigate the assassination headed by Earl Warren of the Supreme Court. The "Warren Commission's" report left unanswered a myriad of questions, and not all members shared its conclusion that Lee Harvey Oswald had acted alone, nor did the majority of the American or European public agree with that conclusion. The "single-shooter" theory simply did not pass the smell test. There were, and remain, many conspiracy theories, but the purpose of the "Commission" had not been to resolve the many questions but more to put the tragedy to rest in order that the country could return to normal by collectively putting the sorrowful event behind it. The Mainstream Media cooperated simply by not doing any investigations worthy of the name, despite the multiplicity of question marks surrounding the assassination.

Johnson was born August 27, 1908 near Stonewall, Texas in the southwest part of the state. Both of his grandparents as well as his father had been members of the Texas legislature, and his mother and father had been teachers. At the age of 17, he enrolled in the Southwest State Teachers College in San Marcos, Texas where he studied debate and public speaking, graduating in 1930.

"After teaching debate and public speaking for a year in a Houston high school, the opportunity came that drew Johnson into politics. U.S. Representative Richard Kleberg, one of the owners of the million-acre King ranch, invited Johnson to come to Washington with him as his secretary at $3,000 a year. From 1932 to 1935 – in the last months of the Hoover administration and in the exciting first years of the Roosevelt administration – Johnson learned how Congress worked. On a trip home to Texas in 1934 Johnson met twenty-year old Claudia Alta Taylor, called Lady Bird by her family and friends. The twenty-six year old Johnson conducted a whirlwind courtship, and they were married on November 17, 1934 – just two months after they had first been introduced."(1)

In Washington, Johnson sought and was given the job of administrator in the National Youth Administration (which had been established by President Roosevelt) program in Texas. He was the youngest of all the state directors. This was to serve him well, for in this capacity he organized projects that employed tens of thousands of youths. And in 1937, Johnson used these extensive contacts to help him run for the office of U.S. Representative in a special election required to fill the vacancy in Johnson's district resulting from the death of Representative James Buchanan. Running against nine other candidates, Johnson strongly endorsed President Roosevelt and won handily. That Sam Rayburn, the majority leader in the House, had formerly served with Johnson's father in the state legislature in Texas provided obvious advantages to the newcomer, as did his having campaigned extensively on the coattails of Roosevelt.

"In his early years in Congress, Johnson devoted much of his effort to pushing the building of federally-supported electrical services for farmers. He also was a supporter of public housing to clean up city slums and an active member of the Naval Affairs Committee. He did such a good job for his home congressional district that he was not opposed for re-election in 1938, 1940, and 1942."(2) In 1941, he campaigned hard for the senate seat to replace Sen. Morris Sheppard who had died. However, Johnson lost to Governor O'Daniel by some 1,300 votes.

Following the Japanese attack on Pearl Harbor on December 7, 1941, Johnson sought active duty as a member of the Naval Reserve in which he held the rank of lieutenant commander. "In 1942, President Roosevelt sent Johnson on a special (not further identified) fact-finding mission in the South Pacific. During this trip, a bomber in which he was flying was strafed by a Japanese fighter plane. Another time he was aboard a plane that crash-landed in Australia. General Douglas Mac Arthur, for reasons not publicized, decorated him with the Silver Star for "gallantry under enemy fire." "In July, President Roosevelt ordered Johnson and all other members of Congress to return to their legislative posts in Washington."(3) Interestingly, no other crew members on either plane received any decorations or commendations — only Johnson. His stories as to how he won the Silver Star grew and varied over the years; there was no doubt in his mind that he deserved it...a view not generally shared by others.

In 1948 Johnson again ran for the U.S. Senate. In a run-off against Governor Cole Stevenson, Johnson won by 87 votes, thereby earning the nick-name of "land-slide". A missing ballot box was rumored to have won Johnson the victory, but such things are part of Texas lore. He won over his Republican opponent in 1948 by a two-to-one margin. In 1953, Johnson was elected as Democratic leader in the Senate – the youngest ever floor-leader for either party.

The thrust of this chapter is not to rehash the political life of Lyndon Baines Johnson, but to surface what Johnson was in the eyes and minds of those that had known him best – doing in essence what the Media, especially the Liberal Media, failed to do for years. Particularly alarming is that Johnson was in essence the personification of what two succeeding Democratic presidents proved to be, especially William Jefferson Clinton. Johnson has been aptly described as devious, deceitful, dishonest, corrupt, irresponsible, a womanizer, a liar, a masterful and glib bull-shit artist, a user of persons, two-faced, not trustworthy, opportunistic, unreliable...and the consummate ambitious politician. How "dare" I make such a statement? Simple. Read the following quotes:

- "In a book about President Johnson published in 1982, his press secretary George Reedy wrote the following: "He may have been a son of a bitch but he was a colossal son of a bitch. He also possessed the finest quality of a politician."(3)
- Milton Kennedy (a fellow student at San Marcos) agreed with the "vivid description of Johnson sitting at his instructors' knees and drinking in avidly all they had to say...(then) says "Words won't come to describe how Lyndon acted toward the faculty – how kowtowing he was, how suck-assing he was, how brown-nosing he was...If he was obsequious to those above him, he was overbearing to those who were not.(4)"
- "Each edition of the San Marcos yearbook, the Pedagog, contained a section: "The Cat's Claw", which mocked student's foibles...Instead of a picture of Johnson, the editors used a picture of a jackass. The caption beside it read: "As he looks to us on the campus every day."..."The Pedagog was not the only publication in which the word "master" was applied to Johnson. In the College Star's humor column appears the following definition: "Bull: Greek philosophy in which Lyndon Johnson has an M.B. degree. "Master of Bullshit'- that's what M.B. means, says one of Lyndon Johnson's classmates, Henry Kyle. "He was known as the biggest liar on the campus. In private, when there were no girls around, we called him "Bullshit' Johnson".(5)
- "A master manipulator, Johnson turned the White House

into a locus of deception, using Pentagon money to fund improvement to his ranch, lying about the costs of running the White House, and stealing government furniture and other items from the White House. In doing so, Johnson enlisted the aid of Bill Gulley, one of those anonymous White House aides who continue from administration to administration. Among the many secrets Gulley learned were the day to day sexual activities of the presidents. "Johnson would pick women out of a crowd...and send an aide to be the pimp. Johnson would spot pretty secretaries in the White House, make a play for them, and if they went to bed with him, place them on his personal staff. Of the eight secretaries around him, only three were not having sex with the president, Gulley said."(6)

- "When Johnson left office, Gulley said he arranged for at least ten flights to fly government property to Johnson's ranch."(7)
- "Patrick O'Donnell, an Air Force One flight engineer, said he flew what he had been told were White House items back to the ranch. "We flew White House furniture back...I was on some of the missions. The flights were at 7:50 or 8:50 P.M. and early in the morning. I had three of the missions. I think he even took the electric bed out of Walter Reed Army Hospital. That was a disgrace."(8)
- "By his own account, Gulley helped Johnson swindle the government. Using a classified fund that was supposed to maintain classified bomb shelters, Gulley said he illegally spent the money on improvements to Johnson' ranch, including an extension to the landing strip, a theater, a roadway, a sprinkler system, and a water pump. It cost $125,000 to put the water pump in."(10)
- "He (Johnson) would screw anything that would crawl, basically, said William F. Cuff, executive assistant in the military office. He was a horny old man, but had a totally loyal White House Staff."(11)
- "Lyndon Johnson was furious. Johnson's wife, Lady Bird, had caught him having sex on a sofa in the Oval Office with one of the handful of gorgeous secretaries he had hired. Johnson blamed it all on the Secret Service. After the incident, which occurred just months after he took office, Johnson ordered the Secret Service to install a buzzer system so that agents stationed in the residence portion of the White House could warn him his wife was approach-

ing."(12)

• "Johnson had no more understanding of the Vietnamese, North or South, than he did of the man in the moon. In part this was the result of his provincialism, a trait he shared with most Americans. It was also the product of ignorance. Johnson prided himself on having read not more than six books since he graduated from college in San Marcos, Texas. A latent racism entered into the mix as well. The temptation to dismiss the South Vietnamese as insignificant "others" haunted the U.S. effort throughout the war; it inclined the president to discount their abilities and their powers, and it led to a great temptation to patronize."(13)

• "The hard fact was that Johnson had simply not had time to educate himself fully on the collapsing state of affairs in South Vietnam. Although he studied all the reports of his advisers and spent hours in briefing sessions with Secretary of Sate Rusk, Secretary of Defense McNamara, and National Security aide Bundy, he could not immerse himself in the war in Vietnam during the process of adjusting himself to the presidency and campaigning for re-election. His look was inward – to Congress, to the Great Society, to the election – not outward to a still small war in Southeast Asia."(14) •

• "Even before Laos was shunted off into a face saving cease fire and negotiations, Kennedy had approved short term plans to increase military and economic aid to Saigon. Vice President Lyndon Johnson returned from a survey of the area with a firm recommendation for a greater U.S. commitment, linked to reforms by the Diem government. The introduction of combat troops, Johnson said, was neither necessary nor desirable, at least for the time being. General Maxwell Taylor and Walt Rostow flew to Vietnam for their now historic survey. Reinforcing the course already set, their reports proposed reforms by the Diem government, a parallel U.S. sponsored counter-guerrilla campaign, a heavy infusion of retraining and equipment to put ARVN back into the war. Taylor and Rostow went even further. Taylor recommended preparations to send as many as 10,000 American troops to defend installations and for possible emergency relief of ARVN."(15)

• "The Secretary of Defense (McNamara) felt he had failed to seize his responsibility for Vietnam in the early months. The critical appraisal had been made by the White House through the

Johnson and the Taylor-Rostow missions. .For better or worse – and in the end to his own anguish – he became the dominant public figure on Vietnam policy beneath the President he served. In McNamara's case, as a close friend put it later, "he was trying to quantify the unquantifiable. He kept saying "If we can learn how to analyze this thing, we'll solve it."(16)

• "Where official American analysts differed chiefly over the record of this period was the point at which the decline was resumed. The government of the United States was trapped in a web of self-deception. Among those most deceived was the Secretary of Defense, a critical link in the policy making chain."(17)

• "History credits Johnson with being our most pro-civil rights president ever, but these transcripts will require at least some modest revisions. Just how great was Johnson's belief in racial equality? A whole lot less than we may have thought."(18)

• "The Great Society was more than Dick Goodwin's (with the Peace Corps prior to becoming a speech writer for Johnson) words said it was. Like the New Deal and the New Frontier, it was the badge of this particular President and this particular administration, of the uninhibited Johnsonian style, of the restless quest in which to use the presidential power, of the search for consensus, of government programs made by Johnson, as distinguished from those inherited from Kennedy. And of those programs, none was so much a symbol of the Great Society as the War on Poverty."(19)

• "The President told Congress in his State of the Union Address on January 8: This administration today, here and now, declares unconditional war on poverty in America. I urge this Congress and Americans to join me in that effort. It will not be a short or easy struggle, no single weapon or strategy will suffice, but we shall not rest until that war is won." "Such well meant but high flown rhetoric continued through much of the year, exciting vast expectations of help among the poor – particularly the impoverished Negro in the slums of the big cities. But Johnson's War on Poverty was more a slogan, a concept, a laudable desire than a carefully structured plan of action."(20)

• "The package had been put together too quickly to have much coherence. The excruciating problems of how the poor could best and most systematically be lifted from the rut of poverty were glossed over in the mad rush to put a program on paper and hurry it through Congress. But most important, Johnson had made the War on Poverty part of the national consensus. It was not only bad politics but quite immoral to question the propriety of the government's attack on poverty."(21)

• "Johnson's success as Majority Leader in cultivating a number of submissive correspondents led him to the erroneous conclusion that his press relations could be solved by the art of seduction. Hence, in the early months of the Johnson presidency, both before and after Salinger's departure, journalists had unprecedented access to the Oval Office, to the living quarters and the LBJ Ranch. This strict demarcation line between the Private and Public Person began to crumble during Johnson's freewheeling overseas trips as Vice President. But the problem was magnified beyond control in his early presidency when he tried to deal with fifty or more White House correspondents in the same style that he had dealt with a dozen Senate regulars. Stories inevitably were published that concentrated on the Private Person and displayed the new President as a whiskey-drinking Texas primitive, who told dirty jokes and mistreated his Secret Service bodyguard."(22)

From the sample quotations above it is abundantly clear that Johnson chose to place his priority on "The Great Society" and being re-elected in 1964, while leaving the "management" of the Vietnam War primarily to his Secretary of Defense. The "War" was of secondary importance to Johnson's quest for election, and he tried to wear the mantle of the moderate "peace President" to the extent possible. That the Joint Chiefs of Staff had few consultations or conversations with Johnson regarding Vietnam is shockingly clear. McNamara proved to be an effective filter. He was also dishonest in that he seldom accurately presented the views of the Joint Chiefs to Johnson, nor was he above badly distorting their views and recommendations when the subject of Vietnam was discussed privately with Johnson or Johnson's trusted civilian advisors. McNamara and Johnson adamantly refused to be pressured by the military into any effort beyond that best described as "gradually, or incrementally, increasing the pressure" on North Vietnam. Both apparently believed this would lead to North Vietnam agreeing to negotiations and ceasing their support of the Viet Cong. From these ineffective, if not feckless, efforts it is apparent that none of these key players knew much about Vietnam's history, or about Ho Chi Minh and his unwavering determination to follow Lenin's doctrine to total victory.* Why the Mainstream Media did not challenge or try to enlighten Johnson or McNamara – or Congress - remains a mystery.

"During this period Johnson privately shows himself determined, if possible, to defer irrevocable decisions on the war in Southeast Asia until the 1964 presidential election is over. Eager to win the presidency in a landslide, he wishes to appear neither soft on Communism nor frighteningly ready to take the nation into a war of unimaginable cost – even if this means leaving Americans confused about his inner inclinations. .But his approach kept Americans from fully knowing whom and what they were voting for in 1964. It also tragically foreclosed the possibility of a grand national debate that might have educated both Johnson and the American people as they faced one of the most important presidential decisions of the century - whether the United States should make a monumental commitment to war in Vietnam."(23)

For example, in a meeting with McGeorge Bundy on March 4, 1964, Johnson said "I just spent a lot of time with the Joint Chiefs. The net of it is they say get in or get out. And I told them, "we haven't got any Congress that will go with us, and we haven't got any mothers that will go with us in a war."… "I've got to win an election. Or Nixon (Johnson thought that Nixon or Goldwater would be the Republican nominee) or somebody else has. And then you can make a decision. But in the meantime, let's see if we can find enough things to do to keep the North Vietnamese off base, and to stop these shipments that are coming in from Laos, and take a few selected targets to upset them a little bit, without getting another Korean operation started. Taylor (Gen. Maxwell Taylor) doesn't think that there'll ever be a Korean operation. The Marine man does. The Navy man is between the Marines and Taylor. The Air Force thinks that it will be a Korean-type operation. The Army, well, didn't say much. But Taylor thinks that you've got to pick some specific targets. So I told him to make up a list of everything. He's got his ideas now, and let's see what he does after his trip (to Saigon with McNamara, Gen. Taylor, Secretary of State W. Bundy, John McCone (CIA) and David Bell, Director of AID.)"(24) To conclude that this was less than a tight, smooth running ship of state does not require a psychic reader. The White

House version of helter-skelter was obviously being played out bit by bit, with the disparate parts left largely to their own pursuits - while Johnson concentrated on the election. * Johnson did not have a high regard for the military; in fact, he fretted about how to keep them under control and constantly told McNamara that it was his job to hold the Joint Chiefs, and CINCPAC, to minimal incremental increases in Vietnam regardless of their recommendations and urgings to substantially increase military efforts across the board. * The Media? Largely M.I.A. - or only mumbo-jumbo coverage.

In a conversation with Sen. Richard Russell (D, Ga.) on May 27, Johnson said: "Now one of our big problems, Dick, the biggest, between us – and I don't want this repeated to anybody – is Lodge. He ain't worth a damn. He can't work with anybody. We get the best USIA man to put on all the radios and try to get 'em to be loyal to the government and to be fighting, and quit deserting and he calls in the USIA and says, "I handle the newspapers and magazines and radio myself, so the hell with you! So that knocks that guy out. Then we send out the best CIA man we got and he said, "I handle the intelligence. To hell with you!" Then he wants a new Deputy Chief of Mission. We get him to give us some names and we pick one of those, the best we got, send him out to run the damn war and he gets where he won't speak to the Deputy Chief of Mission! Then we get General Harkins out there and we thought he was a pretty good man and he gets where he can't work with him. So we send Westmoreland out. And it's just a hell of a mess. You can't do anything with Lodge and that's where McNamara's so frustrated. He goes out and they get agreements and he issues orders and he sends his stuff in there and Lodge just takes charge of it himself and he's not a take-charge man and it just gets stacked up there." Russell's response was: "He (Lodge) never has followed anything through to a conclusion since I've known him for twenty years...He's a bright fellow, intelligent fellow, but he thinks he's dealing with barbarian tribes out there, that he's an emperor and he's gonna tell 'em what to do. There ain't any doubt that he had old Diem killed out there so he could – Johnson interjects: "That was a tragic mistake. It was awful and we've lost everything."(25)

(Author's note: From the above, and in all other reference materials reviewed, it is clear that Johnson had argued against the idea of forcibly removing Pres. Diem from office and opposed Pres. Kennedy, the State Department, McNamara, Ambassador Lodge, and Ambassador Galbraith, – and others - including much of the Media - in their urgings and decision to overthrow Diem. Unfortunately, Johnson's views about Diem were ignored.)

A Draft Congressional Resolution, May 24, 1964, record of an NSC Executive Committee Meeting, same date, notes that "In recent weeks, while the Department of State and Defense worked on a plan for gradually escalating political and military action against North Vietnam, a group led by George Ball had drafted a resolution to be submitted to Congress that would endorse "all measures, including the commitment of force" to defend the South Vietnamese or Laotian governments upon their request. Johnson told his men that "Truman had made a mistake in waging the Korean War without congressional approval: only if Congress were in on the "takeoff" would it accept responsibility for a "crash landing." A May 24 NSC meeting debated the draft. McNamara insisted that "it be withheld from Congress for now, unless Johnson decided to use U.S. combat (not training) forces in Southeast Asia within the next two or three months."

Bundy wrote on June 10, "The immediate watershed decision is whether or not the Administration should seek a Congressional resolution giving general authority for action which the President may judge necessary to defend the peace and security of the area. The best available time for such a move is immediately after the civil rights bill clears the Senate floor. On balance, it appears that we need a Congressional resolution if, and only if, we decide that substantial increase of national attention and international tension is a necessary part of the defense of Southeast Asia in the coming summer."(26) But Johnson was not yet ready to undertake such a move; he first wanted to get parts of the "Great Society" plan passed, especially the "War on Poverty" portion. The "little war in Vietnam" would have to wait. Again, most of the Media missed or ignored the obvious.

An interesting aside to the concern over the passage of the Johnson's civil right's plan is that the southern led filibuster against the bill by the Democrats lasted eighty-three days. "Several minutes after the Johnson-Humphrey call (for a vote), Senator Robert Byrd of West Virginia began the final speech on the civil rights debate, insisting for fourteen hours that the bill "cannot be justified on any basis – legal, economic, moral, or religious." After Byrd sat down at 10:00 a.m. (June10) the next morning, Senators shut off debate by (a vote of) 71 – 29. An angry Richard Russell complained that Senators had been "confronted with the spirit" of a "lynch mob."(27) In that Byrd previously had been an active member of the Ku Klux Klan, Russell's comment was especially poignant, and hopefully embarrassing, to the ultraparadoxical Byrd.

In a June 18th discussion with McNamara regarding the replacement of Lodge as Ambassador to Saigon, Johnson said: "What we've got to bear in mind as the number one objective is holding there and not seeing South Vietnam and Thailand and Cambodia and all of 'em go under. Now in order to do that, we've got to get the man that will most unite this country. You're still gonna catch hell from the right-wingers. He's not going to alleviate any of that and in addition to that, you're gonna grab the left-wingers here. That troubles me. But I don't see anybody that will make it less, and he'd (Taylor) be more competent. I believe Khanh's gonna be pretty upset when this man goes out and he gets a second-rater he never heard of. Taylor can give us the cover that we need with the country and with the Republicans and with the Congress. We need somebody to give us cover with the opinion-molders (columnists). I don't know how to do that. That worries me. You see Bobby's (Kennedy) instinctive reaction – political job, not a military job. ...There's been a little feeling like that in the State Department. They've got the biggest bunch of leaks in the country. You know damn well they'll start giving backgrounders immediately about the military has won, the Army's taking over. It's all going to be a military operation. That's the type of stuff that the 'New Republic', and the 'Nation' and the 'Reporter', the 'New York Times', the 'Washington Post', 'Louisville Courier', 'St. Louis Dispatch' – that's what they're going to say. The greatest danger we face there now is losing (Prime Minister) Khanh and being asked to get out."(28)

It is ironic, if not strange, that Johnson, a Texan Liberal, would single out the Liberal newspapers that he did as likely being critical of his Vietnam policy and his decision to appoint Gen. Taylor to replace Lodge. Johnson wanted Taylor not only for his 'stature' as a General and his proven flexibility in the JCS, i.e., willingness to increase troop strength only 'incrementally', but because he believed the other possible nominees, including from the State Department, were "soft", or simply too politically focused. Johnson's previously expressed fondness for and efforts to cultivate the 'Press' contrasts sharply with the views he was expressing to McNamara. Also nagging Johnson was his concern that Vietnam and the slowly increasing number of American military serving there might become the main issue in the forthcoming election, rather than his "War on Poverty" and "Great Society" program. By publicly pushing for "negotiations" with the North Vietnamese, he hoped to impress the voters that he was the "Peace Candidate', rather than have any other candidate usurp that claim. A companion fear was that the disagreements between McNamara and the Joint Chiefs might surface, as well as the disagreements in Saigon between Westmoreland and Lodge over military matters. But Johnson's "luck" was to hold once again. The Media again was asleep at the switch.

The August 4, 1964, Tonkin Gulf "incident" served Johnson well, for now he could be a stand-up "Commander in Chief" as well as the man behind the "Great Society". The public and congressional support which had concerned him now flowed to him, reassuring his election by record-setting numbers. Few, if any, probed the merits of either the claims by Johnson regarding the incident or the extraordinary powers that Congress gave him following it. He was "home free", both with Congress and with the public, thanks to the gushing Media support given him. (It was at this point that I decided I could stomach LBJ no more. I had been a Democrat since my youth, as had my father, mother, and most of my relatives, but Johnson's deceits and perfidies had gnawed away to the point that the transparent Tonkin Gulf was the final straw. If this was what the Democratic Party had become then I wanted no more to do with it.) But to return to Johnson…

"In any event, Johnson called a meeting at the White House on September 9 to discuss the various proposals. In preparation for the meeting, William Bundy and Michael Forrestal drafted a working paper, "Courses of Action for South Vietnam", in which they summarized the emerging consensus of the president's top national security advisers. The paper called for the following military actions: 1. U.S. naval patrols in the Gulf of Tonkin; 2. covert operations by the South against the North (primarily by MACV, which had taken over most of the covert operations previously the responsibility of the CIA); 3. limited South Vietnamese ground and air operations in the corridor areas of Laos; and 4. a tit-for-tat response against the North in the event of any attack on U.S. units or any special communist attack against the South – but no continuing attacks on the North".(29)

But on November 1, "the NLF attacked the U.S. air base at Bien Hoa, killing four Americans and destroying several aircraft, again providing Johnson the opportunity to demonstrate his leadership and resolve against North Vietnam for such a "sneak attack", but Johnson and McNamara, after huffing and puffing a bit for the Media and Congress, again demurred, but agreed to send a few more troops to help guard U.S. installations. "Johnson and his advisers hoped that air strikes against North Vietnam would convince Hanoi to end its support for the South Vietnamese insurgency, so that the deployment of U.S. ground force would not be necessary. Operation Rolling Thunder (aptly named because there were few meaningful 'strikes'), a bombing campaign against the north that began on March 2, failed to achieve this result. After a pause on May 13 failed to inspire Hanoi to negotiate, the bombing (sporadic) resumed. U.S. ground forces began to engage in significant combat with Viet Cong units in the late summer and autumn of 1965. As U.S. troops poured into South Vietnam, the political turmoil in Saigon came to an end. A coup (had) brought to power military officers Nguyen Cao Ky and Nguyen Van Thieu, who would rule South Vietnam for the next decade, until its demise."(30)

However, Johnson and McNamara did approve limited air strikes against targets in North Vietnam above the 17th parallel (the DMZ) when attacks on U.S. installations prompted retaliation. "Rolling Thunder", the overly optimistic code name given to these strikes, was neither heavy nor constant, and often not worth the risk to the crews involved. Only on rare occasions were strikes approved above the 20th parallel. Targets, tightly controlled by Johnson and McNamara, were usually petrol facilities, warehouses, some railroads and bridges, and barracks. Even MIG airfields were off-limits. Photographs covering North Vietnam and all targets suggested by the Joint Chiefs were maintained in the White House for Johnson's review and reference. It had started out as a "94 Targets List" formulated by the joint Chiefs, but Johnson had designated many areas in the North as off-limits and not to be hit without his specific approval. Johnson and McNamara also controlled what targets could be hit, much to the frustration of the Joint Chiefs and CINCPAC. The target list provided to McNamara and thence to Johnson by the Joint Chiefs was to reach nearly 500 by the end of 1967, including airfields and port facilities, but this was more to keep the JCS satisfied than to actually authorize hits. Perhaps the Media had few "inquiring minds" to spare, for there was little attention paid to this strange dichotomy.

"Despite being under tight political control, however, the "Rolling Thunder" program was allowed to slowly grow in intensity, in geographic coverage, and in the assortment of approved targets. By mid-1965, the number of target packages had increased from one or two per week to ten or twelve per week, and the number of sorties rose to about 900 a week, four or five times what they had been at the outset. The Joint Chiefs were, from the very beginning, opposed to this gradual and limited bombing campaign. They argued repeatedly for a more forceful and continuous air campaign designed to force North Vietnam to its knees – to render it absolutely incapable of supporting the insurgency in the South. To accomplish this, the Navy and Air Force representatives on the Joint Chiefs sought from the president and secretary of defense complete freedom to pick targets in North Vietnam, the elimination of restrictive zones around Hanoi and Haiphong and the Chinese border (re-supply routes), and expanded authority to strike in Laos and Cambodia (supply routes). These requests for release from civilian control over targets were refused time and time again."(38)

"Between 1965 and 1968, China provided massive support for North Vietnam. In addition to supplying Hanoi with military equipment and civilian good, Mao's government dispatched antiaircraft artillery troops to the northern part of North Vietnam. Even more important, at Hanoi's request China sent engineering troops to repair and expand the North Vietnamese railway system so that it would not be disabled by American bombings. The Chinese troops also freed North Vietnamese regulars to journey into Laos and keep supplies moving down the Ho Chi Minh Trail(s). Between June 1965 and the end of 1969, the Chinese logistics troops built 20 new railway systems, 39 new bridges, 14 tunnels, and 117 kilometers of new rail lines and repaired 362 kilometers of existing rail lines. In all, China sent 327,000 troops to North Vietnam."(31)

And "...the Soviet Union already had surpassed China as the largest supplier of military aid to North Vietnam. Between 1964 and 1974, aid to North Vietnam amounted to 50 percent of the Soviet Union's aid to communist regimes (the rest was divided chiefly between Cuba and North Korea. "...As many as three thousand Soviet advisers took part in the Vietnam War between 1965 and 1968. They instructed North Vietnamese soldiers in the use of Soviet-supplied equipment, manned the ground-based antiaircraft system, and participated in combat actions against American and South Vietnamese...During the Vietnam War, as earlier during the Korean War, the U.S. Government kept the degree of Soviet military participation secret in order that public pressure would not make it difficult to keep this proxy war limited."(32) The above figures did not include the impressive amount of supplies provided the North and the Viet Cong by way of Haiphong, or the port of Sihanoukville in Cambodia. The number of Ho Chi Minh trails from the North to the South, primarily through Laos, continued to grow.

Unfortunately, the massive support provided by China and the Soviet Union included substantially improved aircraft defenses in North Vietnam. By the time that Johnson allowed increased air strikes above the 20th parallel, which had previously been only on rare occasions, North Vietnamese antiaircraft fire had improved in intensity and effectiveness. The Johnson and McNamara ploy of gradually increased pressure, punctuated by pin-prick bombings above the 17th parallel, and the 20th parallel, usually against relatively unimportant targets, had failed. As Sun Tzu stated in "The Art of War" - "If you know the enemy and know yourself; in a hundred battles, you will never be defeated. When you are ignorant of the enemy but know yourself, your chances of winning and losing are equal. If ignorant both of the enemy and yourself, you are sure to be defeated in every battle." The Johnson-McNamara version of the "Art of War" again proved Sun Tzu right. The Media remained somnolent.

"Johnson and his advisers thus gambled that without taking

exceptional measures they could hold public support long enough to achieve their goals in Vietnam. As a result, the United States went to war in July 1965 in a way that was uniquely quiet and underplayed. The president ordered his July decision implemented in a low key manner. He announced the major troop increase at a noon press conference instead of during prime time. It was even lumped in with a number of other items in a way that downplayed its importance. With the exception of several hastily arranged, typically Johnsonian public relations blitzes, the administration persisted in this low key approach until the summer of 1967.". "Throughout much of 1966 and into 1967, however, almost to the point of being conspicuous about it, Johnson followed the low key approach he had imposed on his administration, remaining, in the words of *Time* magazine, "strangely silent" on the war. He made few public appearances and fewer speeches. He spent three months at his ranch in Texas in 1966 and two months in 1967. In 1966 he held forty press conferences; the following year only twenty-one. Journalists later speculated that he was lying low as a way of dealing with his own sagging popularity. Increasingly, he was criticized for being the most aloof chief executive since Calvin Coolidge."(33) Never mind the sagging polls! Didn't a three month absence strike the Media as strange, especially given the mounting casualties and political problems in Indo-China? Hell'o, operator? I think we've been disconnected!

And this was the man that had previously openly sought the limelight, swayed members of congress, courted the Media, the ultimate back-slapper, but apparently the Mainstream Media did not seek or choose to learn what the underlying reasons were for these changes. Nor had the Media been unkind to Johnson…or McNamara. In fact it had been very supportive of the Johnson - Humphrey ticket. The Media chastised him for: speeding on Texas highways at eighty miles an hour while sipping beer; for lifting his dogs up by their ears for photographers; for shading the truth about the booming financial success of Lady Bird's Texas radio station; stories of his having manipulated the FCC for Lady Bird's radio station in Texas; and for his involvement with Bobby Baker, a former aide to Sen. LBJ. Baker, who was now a free-wheeling insurance salesman and entrepreneur who 'sold' Johnson a life-insurance policy, despite Johnson's heart condition, and then given gifts to

Lady Bird and Johnson. "In their book, "Lyndon", Washington Journalists Richard Harwood and Haynes Johnson wrote the following: "Lyndon Johnson floated in and out of our consciousness under many labels – New Dealer, Wheeler Dealer, Crude Texan, Friend of the Poor and Black, Warmonger, Tyrant, Fool, Imperialist…larger than life".(34) Yet there were few articles questioning the many changes in the man, or connecting them to the mounting protests against the Vietnam War in Washington and across the country, or addressing the lack of meaningful military progress in Vietnam, or in achieving any progress towards "negotiations" with the North Vietnamese. "Secretary of Defense McNamara might have filled the strategic void, but he was no more willing to take control than Johnson. When asked on one occasion why he did not tell the military what to do, as Churchill had done, McNamara responded that he was no Churchill."…Thus, as Stephen Peter Rosen has observed, Johnson and McNamara refused to do what the civilian leadership must do: "They did not define a clear mission for the military and did not establish a clear limit to the resources to be allocated for that mission. They saw their primary task as maintaining tight operational control over the military."(35) *

On 7 Feb., 1965, McGeorge Bundy drafts what will one week later become U.S. Policy. "We believe that the best available way of increasing our chances of success in Vietnam is the development and execution of a policy of *sustained reprisal* against North Vietnam – a policy in which air and naval action against the North is justified by and related to the whole Viet Cong campaign of violence and terror in the South. While we believe that the risks of such a policy are acceptable, we emphasize that its costs are real. It implies significant U.S. air losses even if no full air war is joined, and it seems likely that it would eventually require an extensive and costly effort against the whole air defense system of North Vietnam. U.S. casualties would be higher – and more visible to American feelings – than those sustained in the struggle in South Vietnam. Yet measured against the costs of defeat in Vietnam, this program seems cheap. And even if it fails to turn the tide – as it may – the value of the effort seems to us to exceed its cost."(36) * Only an uninformed and uncaring bean counter could have made such a remark as the last sentence.

"Toward what end was the expansion of the war to be directed? On April 7, 1965, President Johnson spoke at Johns Hopkins University in Baltimore and gave the answer that was to cause grave concern to some decision makers in the future. Lyndon Johnson wanted a limited war with limited resources in a limited geographic area for limited diplomatic objectives. The objective, as stated, was itself unobjectionable: an independent, noncommunist South Vietnam." It was the president's caveat that was to frustrate some, "We will do everything necessary to achieve that objective, and we will do only what is absolutely necessary." "It was the "only" that would cause trouble, because it would come to seem to some U.S. leaders to indicate a less than all-out approach to winning the war militarily."(37) The Media did not pursue this key point.

"Admiral Sharp, commander in chief of U.S. forces in the Pacific (CINPAC), a zealous proponent of airpower, held the view that a properly executed bombing program "will bring the enemy to the conference table or cause the insurgency to wither from lack of support." "A more cautious, but still optimistic, views was taken in a Special National Intelligence Estimate (SNIE) dated July 23, 1965. The SNIE estimated that a bombing program that included destruction of the petroleum facilities and military target(s) in the Hanoi-Haiphong area, combined with sustained interdiction of the lines of communication from China, could significantly reduce communist capabilities in the South."(38)

Despite his fear that such an air war would provoke the Chinese – or the Soviets -into a wider war, Johnson finally approved "Rolling Thunder" as an air campaign to attack the Ho Chi Minh trail system of supplies to the Viet Cong in the south. Originally confined to air strikes below the 20th parallel, and used previously only sporadically in retaliation for attacks by North Vietnamese forces against U.S. locations, military pressures, especially by the Joint Chiefs on McNamara and through the NSC on Johnson, led to approvals for more attacks north of the DMZ (17th parallel). Initially, the Joint Chiefs had provided the president with nearly one hundred targets for Rolling Thunder. "By late 1967 the number of targets numbered nearly 500, including airfields, ports and port facilities, lines of communications, power plants, and missile sites (in the North, Cambodia and Laos). To the Joint Chiefs, airfields, including MIG aircraft, and ports/port facilities, including

Haiphong and Sihanoukville in Cambodia, plus the large storage areas along the Trail, would have far more effect on the North than attacking POL storage areas and barracks below the 20th parallel."(39)

But to the chagrin of the Joint Chiefs, Adm. Sharp – and the combat crews flying the strikes - Johnson and McNamara still would not provide them the unrestricted approvals that the military sought. Approval authority was solely the president's and the Secretary of Defense. Within the White House, Johnson – and later Nixon – maintained a room dedicated to complete photographic coverage of Vietnam, Laos and Cambodia, including all suggested targets in the North. While there were relatively few "hard" targets, such as airfields (and aircraft), port facilities, military supply dumps, and the few industries that existed in the North, it remains unclear how Johnson and McNamara hoped to bring the North to the negotiating table without employing meaningful air power. While U.S. and South Vietnamese forces were following the Johnson-McNamara strategy of applying "incremental" pressure primarily below the 17th parallel, the North was able to maintain the flow of men and materiel to the South with little problem, thanks to the uninterrupted shipments from China and the Soviet Union into Haiphong and Sihanoukville. Our powerful Media chose not engage in this debate, but merely wrung its collective hands while providing the war protesters with increasing coverage.

Somewhat akin to the previous quote of Sun Tzu from his *Art of War,* our political leaders were well on the road to losing, not because of lack of military power or will, but more because they did not understand or recognize the resolve of the leaders of the North to win the political battle, in which they were aided and abetted by the extensive support from the Soviet Union and China. Nor did our politicians recognize this was part of the multi-front struggles, including psychological warfare, against South Vietnam, against Laos, against Cambodia…and the U.S., including influencing the Media, Congress and the U.S. public. The tunnels were without lights.

The Johnson administration chose to minimize what it actually knew about Chinese and Soviet assistance to Vietnam. Hindsight suggests that this was a public relations blunder because it permitted critics on the left as well as military and conservative critics to argue that the 'government' exaggerated the roles of the communist powers in Indochina. "You will ask why in relation to Hanoi the Chinese and Russians did better," John Kenneth Galbraith wrote in the *New York Times* on July 12, 1975 after the Soviet-equipped North Vietnamese Army invaded and conquered South Vietnam. "One answer is that they were wiser: No Chinese or Russian troops were sent; no great body of advisers debouched. They were not thrown out because they were not there." Again, Galbraith simply did not know what he was talking about, but apparently that seldom, if ever, bothered him.

The Chinese and Russian troops *were* there, and, in the case of the Chinese, in great numbers. According to the Chinese government's own calculations, between 1965 and 1973, 320,000 People's Liberation Army soldiers, not disguised but wearing Chinese uniforms were assigned to North Vietnam. The greatest number of Chinese soldiers present in North Vietnam at any one time was 170,000. The Media let such inane statements as Galbraith's go unchallenged.

"By the time the last Chinese troops left Vietnam in August, 1973, 1,100 are alleged to have been killed and 4,200 wounded (mostly antiaircraft teams)."(40) * In essence, Johnson and McNamara had reached the point that they had originally tried to avoid - another Korea in the sense that again China and the Soviets were supporting an enemy of a people that we were trying to defend, with Australia and South Korean troops assisting us.

The problem remained that Johnson and McNamara, and most of the other U.S, politicians in Washington and Saigon, did not recognize what the enemy was doing, or the totality of its resolve. Before 1968, the North primarily pursued guerrilla warfare tactics which the U.S. and South Vietnamese forces tried to combat using the outmoded French inspired use of enclaves while trying to bomb the elusive guerrillas of the North. After 1968, the North reverted more to conventional war tactics which were vulnerable to air attacks, but in both time frames the Ho Chi Minh trail systems went largely uninterrupted. Why? Intelligence, including order-of battle data, was less than impressive and usually old by the time the fighting units received it. There were not many good sources, and analysts often disagreed, or were more often wrong than correct in reading the tea leaves. South Vietnam resources often had better information and intelligence than the U.S., but were often mistrusted by U.S. entities. Precious time was routinely lost by U.S. analysts because of the length of time required to process the "take", including translation time. The emphasis was on providing quantities of data per the McNamara doctrine of getting field reports from all sources regarding the numbers of killed in action, the numbers of wounded in action, the amount of traffic on the trail system, the number of desertions of enemy and friendly Vietnamese, the number of prisoners captured and interrogated,, the number of hamlets pacified, the results of the Census Grievance efforts, the number of engagements/fire-fights, the statistics from the Provisional Reconnaissance Units (PRU) 'Phoenix' program, bomb damage assessment, disruption of lines of communications data, Viet Cong activity and attacks, main force (North Vietnamese Army) activity, political developments, etc, etc. The Washington appetite knew no bounds, but the "digestive" system often became clogged, a victim of itself. Quantative analysis may have worked at the Ford Motor Company, but war simply had too many disparate parts that could not be fit neatly into a production line chart. During this entire effort, the strategy of pacification, which initially had started so successfully, was lost in the shuffle until well into 1968. By then, because of political reasons both in Washington and Saigon, it was too late for resuscitation of that vital element.*

"The American dismissal of the French experience in Indochina and Nigeria, or of the British in Malaya, or of the warnings of Mao Tse-tung, or of Khrushchev, or of Ho Chi Minh and his lieutenants that Vietnam would be a different kind of war than the Americans contemplated on the 'North German plain' or experienced in Korea, is difficult to understand." "The collective memory of the army and air forces, the services that dominated the Pentagon, was shaped by the experience of using firepower to pulverize the conventional forces and infrastructure of Germans, Japanese, North Koreans, and Chinese."(41)

In late 1966, McNamara proposed and was authorized to install an electronic barrier along the entire DMZ and to place sensors in Laos along the Ho Chi Minh trail systems to detect and interdict North Vietnamese efforts to infiltrate men and equipment into Laos and South Vietnam. This effort became known as

"McNamara's Fence", "McNamara's Folly", and a number of other uncomplimentary names during the period of time it was tried. That the DMZ was to have an electronic barrier became public knowledge and was carried by the Media as an innovative idea well worth the effort and cost. Of course, the North Vietnamese also learned of the plan and made the fencing venture along the DMZ a hazardous one by simply employing artillery to interrupt the U.S. emplacement efforts. The installation of an effective "electronic barrier" proved to be impractical as well as excessively dangerous. The parachuting of sensors along the Trails that followed proved to be easier. However, dropping the sensors onto or near a large North Vietnamese supply or repair depot along the Trails also proved to have problems because of the difficulty of accurately pinpointing such areas, as well as the North Vietnamese often spotting the parachutes attached to the dropped sensors in the trees along the trails.

I recall attending an early morning briefing at III MAF in Danang (I Corps) in 1969 during which a recording of a sensor transmitting its own abduction was played. The early portion of the transmission clearly recorded the sound of trucks being repaired, including the chatter by Vietnamese as they worked on vehicles, including profane complaints about the conditions of the vehicles, and the large number of wheel changes required because of damage. Then one of the workers apparently spotted a parachute in the overhead canopy, and the sensor attached to it. He yells to others and the babble of voices increases as the sensor is 'captured'. After a brief discussion, the sensor is thrown into the back of a truck and driven away still transmitting, but more faintly as the truck moves further from the area. The tape of the 'capture' of a sensor (basically reporting its demise) was not particularly unusual, but it did reflect the inability of aircraft to quickly respond to the transmissions of a sensor and attack the "target". Availability of dedicated aircraft and poor weather were two factors working against the sensor efforts. In addition to the sensor effort, both MACV and the C.I.A. formed and dispatched indigenous "Road Watch" teams along the Trails to cover and report on movements of North Vietnamese personnel/materiel. In most instances, these efforts had very limited success, including that most teams preferred not to get close to the trails, or to simply pretend to send trail information. A few teams did very well, but not many.

"By August 10, 1966, The U.S. secretary of defense had concluded the following:

• The U.S. strategy of attrition was a losing strategy that played to enemy strengths and must be replaced.

• The political objective of not widening the war, and the military's procedure of asking for endless increases in troops and equipment, were fundamentally in conflict with each other.

• A military "victory" in the normally accepted sense of the term was quite unlikely through offensive military action – in fact, was an illusion that could be maintained only if turned a blind eye to the disconnects between predictions of the military commander and actual results on the battlefield.

• The nature of the conflict in Vietnam – bluntly, that it was a *war*, and an especially brutal and destructive one at that, without an end in sight – must be made clear to the ultimate arbiters of U.S. Security – the American people."

"After a series of conferences, the president approved on November 11 a new deployment program: Program 4, with an end strength of U.S. military personnel at 470,000, to be reached by June 1968 (as opposed to the original request for some 542,000 by the end of 1967). In explaining the reasoning behind the Program 4 decisions, McNamara posed the U.S. strategic dilemma as precisely as he, or anyone, ever had: "We now face a choice of two approaches to the threat of the regular VC/NVA (Vietcong/ North Vietnamese Army) forces. The first approach would be to continue in 1967 to increase friendly forces as rapidly as possible, and without limit, and employ them primarily in large scale "seek out and destroy" operations to destroy the main force VC/NVA units. "The second approach is to follow a similarly aggressive strategy of "seek out and destroy", but to build friendly forces only to that level required to neutralize the large enemy units and prevent them from interfering with the pacification program. It is essential to this approach that such a level be consistent with a stable economy in SVN, and consistent with a military posture the U.S. credibly would maintain indefinitely, thus making a Communist "wait us out" less attractive." "I believe it is time to adopt the second approach for three reasons: (1) if MACV estimates of enemy strength are correct, we have not been able to "attrit" the enemy forces fast enough to break their morale and more U.S. forces are unlikely to do so in the foreseeable future; (2) we cannot deploy more than 470,000 personnel ...without a high probability of generating a self-defeating runaway inflation in SVN, and (3) an endless escalation of U.S. deployments is not likely to be acceptable in the U.S., or to induce the enemy to believe that the U.S. is prepared to stay as long as it is required to produce a secure noncommunist SVN."(42)

"General Westmoreland indicated to the president that if the troops he requested were not provided the war would not necessarily be lost, but progress would certainly be slowed.,,, (he) concluded by estimating that with 565,000 men the war could well go on for three years; but with a total of 665,000, as he had requested, it could start to wind down in two years." "In essence, the president was told the United States needed vastly more troops in Vietnam than Secretary McNamara had deemed necessary and such a deployment was unthinkable without calling up the reservists. In his budget message to Congress on August 3, 1967, President Johnson at last disclosed plans to dispatch "at least 45,000" additional troops to South Vietnam in that fiscal year, bringing the total authorized strength in the war zone to a 525,000 American force. This was Program 5, officially published by the Defense Department on August 14. The press and the public, by now all too used to these sorts of announcements, seemed scarcely to notice."(43) The Joint Chiefs thought the number (525,000) insufficient...and McNamara was the only close advisor that expressed serious doubts about the war effort. "The illusion of an American "victory" on behalf of our South Vietnamese allies seems to have been abandoned by America's military leaders."(44) Didn't the Media in the U.S. and in Saigon think it odd that the heretofore optimistic Secretary of Defense was drastically changing his previous position, and that more thousands of men needed to be sent to Vietnam to salvage the effort there? Strange Media "coverage" indeed!

"Since March 1967, Hanoi and its NLF (Viet Cong) allies in the South had been planning the Tet Offensive – what the communist called the "general offensive/ general uprising. Hanoi sought a decisive victory in order to create the military, political and psychological conditions that would destroy the political foundations of the Saigon regime in the South and political support for the war in the United States. Dramatic, large-scale attacks across South Vietnam, they believed, might precipitate a general uprising, destroying the Saigon government and, for the United States, making it psychologically and politically impossible to continue to pursue the wear. Negotiations to end the war, they hoped, would follow."(45) *

The North's Tet offensive in late January 1968 was a disaster for Hanoi, but a psychological defeat for the U.S., thanks primarily to the distorted and inaccurate reporting by the U.S. Media.

The Viet Cong losses across the country were huge, and would require two years of recovery time to match their levels prior to Tet 1968. As noted earlier, only in the city of Hue in Thua Thien Province in I Corps did the Viet Cong and NVA forces enjoy an abbreviated success. After nearly a month of fighting, the North's forces were driven from the ancient Citadel in Hue where they had held-out against U.S. and South Vietnamese forces for some 28 days. However, public support in the U.S. for the war eroded quickly and measurably, primarily because of the widespread and hip-shot reporting by the Media on Tet, led by the liberal commentator, Walter Cronkite, who unilaterally decided Tet a success for the North. *

Despite the unique opportunities presented for further damaging the weakened forces of the North following Tet, Johnson chose not to do anything beyond a very modest augmentation of forces in South Vietnam. In his "...dramatic March 31, 1968 speech he announced a limit on the bombing (no bombing above the 20th parallel), a readiness to negotiate, and his withdrawal as a candidate for re-election, He seemed driven by a search for consensus rather than developing a coherent strategy to achieve a clearly defined objective."(47) Johnson's decision not to seek another term as president was largely prompted by his poll ratings, which had dipped into the low 30s, his frustration with the protesters, and fatigue that was affecting his health. He made a few changes among his advisers, including arranging for McNamara to go to the World Bank, with Clark Clifford replacing him as Secretary of Defense; Gen. Westmoreland was recalled to become Chief of Staff, with Gen. Abrams replacing him in Saigon. On April 4, Hanoi accepted the offer to 'begin negotiations', and representatives of the U.S.

and North Vietnam met in Paris on May 13, 1968; however, disagreements arose immediately as to the seating arrangements at and the size of the table – marking only the first in what would prove to be extensive, and effective, delaying tactics by the North. On October 31, President Johnson, obviously a 'beaten' man, announced a complete halt to all air, naval, and artillery bombardment of North Vietnam, handing North Vietnam yet another welcome respite from its failed Tet "offensive". On November 5, Richard M. Nixon won handily over Hubert Humphrey in the presidential election.

Thus the Johnson – McNamara era ended, and the Nixon – Kissinger era began. The ego and attitudes of Senator-then-President Johnson readily fit Kissinger, just as the Nixon determination to succeed would match McNamara's stubbornness in pursuing his (and Johnson's) sustained but highly questionable policy of quantified incrementalism. The Media remained uncertain of just what had happened in the past eight years under their favorite liberals, Kennedy and Johnson, and had no clue as to what was going to happen next. Only the North Vietnamese knew; understandably, for their own basic game plan had not changed. The U.S. resolve had been challenged and found wanting, especially politically, as well as militarily, under Johnson. Having delayed and rebuffed all previous efforts to negotiate, the North could now appear to be 'reasonable' in their agreement to engage in "peace" negotiations, even though their intention was to continue to delay and obfuscate, which served their immediate purpose given their heavy Tet losses. "One step forward, one step sideways, two steps forward, one step back, or sideways, etc.", - their Chinese 'advisors' had clearly outlined how effective the stalling and delaying tactics would succeed for them..

North Vietnamese losses had been high beginning in 1966, when "...the Communists lost a weekly average of over 1,000 killed and 150 prisoners lost, plus 385 defectors through the increasingly effective South Vietnamese "Open Arms" (Chieu Hoi) rally program. Uncounted others were killed in B-52 raids, and about two were wounded for each man killed in battle. Defectors, prisoners, and diaries of infiltrators told of the terrible toll from malaria and hardship. Communist losses, moreover, included highly trained cadre replaceable only by lengthy indoctrination, made difficult by allied attacks. Only by recruiting boys as young as 13 years and women in local forces did the Viet Cong maintain their manpower."(48) But Johnson and McNamara had steadfastly resisted the oft repeated urgings of their military advisors to bring the war to the North in meaningful ways. Precious time had been lost.

The North Vietnamese had resisted all efforts by international organizations, including the United Nations, world leaders, and private personages to join in negotiations. Without any significant pressure from within North Vietnam to do so this is not surprising. Their forces were progressing or holding their own throughout most of Indo-China, and Hanoi and Haiphong had suffered no serious damage. The Lao Dong Party was generally satisfied, the Chinese were satisfied, and Moscow was satisfied. DRV leaders felt no compulsion to change their formula for victory. There was no need for the North to listen to 'outsiders'.

"There was also the view of Hanoi, Beijing, and, to a lesser degree, Moscow that Indochina was a testing ground for Communist-led wars of liberation. This position was made evident at the Tricontinental Solidarity Congress in Havana in January 1966, when delegates of Communist and extreme left-wing parties of 82 countries set up machinery for coordinating liberation movements world-wide. A twelve-nation directorate, including Communist China and the USSR, created a 'Committee for Assistance and Aid to National Liberation and Fighting Movements', with a Secretariat, and a 'Tri-continental Committee to Aid Vietnam'. Havana became the center for these organs and for training guerrilla and political agents."(49) The Media failed to note this development - and its significance - as did the Johnson administration. But other foreign outlets, such as the New China News Agency (Beijing), the Nhan Dan (Hanoi), Radio Hanoi, and Japanese media carried the story, as well as the oft-repeated North Vietnamese assurances of the firmness of their intent to achieve victory regardless of the time or sacrifices required to do so, including the repeated statements that U.S. resolve could not match that of the North Vietnamese. Similarly, the North Vietnamese leaders never had any intention of "adopting" the Liberal's preference for a "Coalition" solution in South Vietnam beyond using the concept as a political ploy. Laos clearly remains an example of what the DRV politburo really had in mind regarding a "coalition", and reflects what they repeatedly said their intentions were. Our State Department eagerly grasped for the "Coalition" straw, but, not surprisingly, it proved to be the short one.

Few in our Media or in the Johnson or Nixon administration bothered to read or heed not only what the North Vietnamese said or did, but what the other communist countries said or printed. One could conclude that if it did not bear a U.S. Media source line or label, such as *AP* or *UPI*, or was not anti-U.S., our Media simply concluded it was not worth noting, reporting, or exploring. How intrinsic can most of our Media be? Have such attitudes or practices changed much over the years? More to the point: why had the Media not investigated the money sources or the organizations of the well coordinated demonstrators across the country? This was the least they could do! The pundits of the Left were clearly not interested in looking under those rocks.

In retrospect, had decisive military actions, including extensive air power, been exercised, Johnson and McNamara could have put the North on its knees in 1966, or in 1967, or in 1968, especially following the extensive losses the North suffered during their 1968 Tet attacks. A factoid worth repeating, but ignored by most.

Johnson's election blinders and political agenda prevented military pursuits before 1965 and 1966, and McNamara's snail-like "augmentation" plans, plus Johnson's focus on the "Great Society", ruled out meaningful military successes in 1967. The heavy Viet Cong Losses in 1968 and South Vietnamese gains in the field in 1968 offered an excellent opportunity, but Johnson and McNamara failed to recognize it, blinded by their parochial and political interests. Especially perplexing was that the Media did not address these points, perhaps because of the Liberal influences dominating the Media field. Even McNamara's point man in South Vietnam, Daniel Ellsberg*, originally a strong "hawk", had encouraged that air power be turned loose on the North, including making the North "into a parking lot". Gen. LeMay had made similar observations, but they were decried and rejected by the Liberals, including those in congress, as the rantings of a "mad man", as well as by Johnson and McNamara. Others also strongly endorsed the use of air power to deter the North from its obvious intention to subjugate all of Indo-China, but Johnson and

McNamara stubbornly refused to listen.

Johnson and McNamara deserve another blot on their escutcheon for their failure to respond to Navy Lt. Commander's Bucher's and his "Pueblo" crew's need for help in January, 1968 when the North Koreans attacked and seized their ship. Air support was but five minutes away, but none was authorized. Bucher and his crew were prisoners for eleven months and badly treated until released in December 1968. The Pentagon then tried to charge Bucher and his crew for 'having failed to defend their ship'. Unbelievable!

In general, the Liberal Media continued to purse their lips, wring their hands, and side with the "anti-war" demonstrators. Even the opportunity to withdraw during this same period with a modicum of respect and grace was lost. The less than testicular pursuit of the war would drag on, and thousands more would be killed and wounded, as had happened in Korea, but that too escaped the Media, Congress, Johnson, McNamara, Nixon and Kissinger.

The White House Follies merely continued its feckless run to an ignoble end, our escutcheon shamefully – and needlessly – disgraced. Under Nixon and Kissinger our Republic took even a more demeaning battering, from which it did not recover until the Gulf Wars in 1991 and 2003.

America deserved better, but doves have a way of despoiling the landscapes over which they fly, especially when they are in flocks. Some call them the Liberal Left, or the Hard Left. Others just call them Socialists. The doves simply do their "thing", regardless.

Unfortunately, in regards to Indo-China, later generations eventually inherit the inevitable and unenviable task of making the necessary repairs, while those on all sides that died, were wounded, served, or simply survived - will never know if it was truly worthwhile.

• In the category of Character and Integrity, Johnson ranked 37th in "Ranking the Presidents".

• Before the coup against Diem the Viet Cong controlled less than 30 percent of the territory in South Vietnam. By March 1964 they controlled between 40 and 45 percent. Page 15, "Vietnam: The Necessary War".

• "China, North Korea, and North Vietnam became Stalin's firm allies. Kim Il Sung and Ho Chi Minh were dictators with immense power who strictly toed the Kremlin's line and who in turn were heavily supported with Soviet arms, goods and services. Stalin gave them .both a great deal of thought and attention, constantly mulling over plans for unifying these two countries under the red banner and thus creating new opportunities for spreading Soviet Communism further into Asia. (pg. 8, *"Vietnam, the Necessary War")*

* On May 22 at the University of Michigan, Johnson gave his signature speech: "We have the opportunity to move not only toward the rich society and the powerful society, but upward to the Great Society. The Great Society rests on abundance and liberty for all. It is a place where every child can find knowledge to enrich his mind and enlarge his talent…where leisure is a welcome chance to build and reflect, not a feared cause of boredom and restlessness…where the city of man serves not only the needs of the body and the demands of commerce but the desire for beauty and the hunger for community…where men are more concerned with the quality of their goals than the quantity of their goods. But most of all, the Great Society is not a safe harbor. It a challenge constantly renewed, beckoning us toward a destiny where the meaning of our lives matches the marvelous products of our labor." pg 404, "Taking Charge" But no mention of Vietnam. Yet the Media failed to challenge this strange omission.

• According to the Pentagon, from January 1, 1961, through May 16, 1964, a total of 229 Americans had been killed in South Vietnam, 131 of these in combat against the Viet Cong. Pg. 365, "Taking Charge". "At the time of Kennedy's death, there were fifteen thousand American Advisers in the country, and fifty American soldiers had been killed..." pg 14, "Vietnam: The Necessary War".

• Robert Kennedy had sought the assignment as Ambassador to Saigon to replace Lodge, but Johnson – and others - quietly opposed him, selecting Gen. Taylor instead.

* 1964: U.S. military personnel assigned in Vietnam – 23,300. Net Change: +7,000. U.S. casualties: (KIA) Year – 147. To Date (accumulative): 267. WIA: Year – 1,039. To Date: 1,531. Pg. 183, "Vietnam Voices".

* "1965 31 December. U.S. military personnel assigned in Vietnam: Total -184,300. Net Change: +161,000. U.S. casualties: KIA – year- 1,369. To Date: 1,636. WIA: Year – 6,114. To Date: 7, 645. pg.223. "Vietnam Voices".

* "Soviet pilots were participating actively in flights of the Vietnamese Air Force in countering raids into North Vietnam. Soviet antiaircraft teams on the ground in Indochina downed dozens of U.S. planes during the Vietnam War." Pg. 89, "Vietnam: The Necessary War".

* "1966. 31 December. U.S. Military assigned in Vietnam: Total – 385,300 (Net change +201,000). U.S. casualties: Killed in action – (Year) 5,008. To date (cumulative) 6,644. Wounded in action: 30,093. To date: 37,738" Pg. 253, "Vietnam Voices".

* "Tet is the (annual) principal Vietnamese holiday celebrating the arrival of the Chinese lunar New Year. Saigon was, in fact, in its usual festive mood on the eve of Tet. For city dwellers, the war seemed remote. At least half the South Vietnamese Army had departed for home to celebrate Tet with families. Pres. Nguyen Van Thieu was celebrating in his wife's home town, My Tho, in the Mekong Delta" pg.364. "Argument Without End." Tet had been honored by both the North and South in previous years, and until shortly before the beginning of Tet when intelligence reports from various locations indicated that the north planned widespread attacks, it had been assumed that the traditional Tet cease-fire would again be honored.

* 1968. 31 December. U.S. military personnel assigned in Viet-

nam: Total 536,100 Net change: +50,500. U.S. casualties: Killed in action: Year: 14, 589.

To date (cumulative):30,610. Wounded in action: Year – 92,818. To date – 192,580.

* A Rand analyst (Vietnam Specialist), Ellsberg also worked as a civilian employee with Bundy and McNaughton at the State Department. He also served as a "fact-finder" analyst for McNamara in Vietnam. Originally a strong "hawk", he later reversed his earlier position and became a "dove". Interestingly, Ellsberg also worked secretly for Kissinger collating "option" papers into 1969. They had known one another and occasionally worked together since the late 1950s. It is ironic that the office of Ellsberg's psychiatrist – and Ellsberg's - later became "break-in" targets of Nixon's "plumbers".

CHAPTER 3

HENRY A. KISSINGER

("Indecent Interval")

Vietnam was not Henry Kissinger's finest years – nor ours. His record? Terrible! No wins, no hits, no runs, multiple errors, and many U.S. POWS left *behind*, not simply left on base. (This is addressed later in the chapter.)

The Paris setting for the lengthy negotiations that continued for well over three years between Kissinger and the North Vietnamese was certainly appropriate, for his was the world of Europe, but he was no match for his North Vietnamese counterparts. Had he forgotten or not known that Ho Chi Minhwas a hard corps Stalinist? And that Ho had studied and worked under Stalin in Russia many years before and also been a member of the Communist party of China under Mao? Or that Le Duc Tho – and the other top leaders of the DRV – had very strong nationalistic feelings and were as dedicated, and determined, and ruthless communists as Ho? Apparently not, for the East Wind came to Paris, wore down and simply blew away the West Wind in purpose, pace, resolve, determination, and patriotic morality. It was no contest.

How had this German (Bavaria) born, Harvard educated (and faculty member) cum-statesman, come to be euchred? Perhaps the adjectives used by many of his associates to describe him provide a clue or two. Some have referred to him as insolent, devious, obsequious (to those ranked above him), disdainful and domineering (to his subordinates), arrogant, abrasive, elusive, dishonest, etc., while a few others defend him as brilliant, stoic, unflappable, etc., a rather complex combination of personalities, to understate it.

For example, "One heard an enormous amount about him, what an extraordinary arrogant and vain bastard he was, said (Kissinger's colleague) Professor Stanley Hoffmann". John Conway, a fellow graduate student, said "People reacted badly to him because he was so pompous", and that "He was, in fact, almost as brilliant as he thought he was."(1)

"Kissinger's conservative realpolitik, as reflected in his dissertation, was based on the principle, taught by realists from Karl von Clausewitz to Hans Morgenthau, that diplomacy cannot be divorced from the realities of force and power. But diplomacy should be divorced, Kissinger argued, from a moralistic and meddlesome concern with the internal policies of other nations. Stability is the prime goal of diplomacy. It is served when nations accept the legitimacy of the existing world order and when they act based on their national interests; it is threatened when nations embark on ideological or moral crusades."(2)

When asked how he would choose between a legitimate state that pursued unjust ends and a revolutionary one that had justice on its side, Kissinger replied "If I had to choose between justice and disorder, on the one hand, and injustice and order on the other, I would always choose the latter."(3) Kissinger also wrote "A "revolutionary" situation occurs when a leader such as Napoleon does not accept the legitimacy of the international order. In such cases, he argued, negotiations are futile."(4) These views seemed to have been discarded during his extended negotiating efforts with the North Vietnamese. Equally telling is Kissinger's description of Bismarck: "The root fact of Bismarck's personality...was his "incapacity to comprehend" any amoral standard outside of his own will and ambition."(5) Many have said the same thing about Kissinger.

Then how did Kissinger come to be National Security Advisor to President Nixon, especially after having demeaned Nixon during the months before the election? He had been a long time favorite of Nelson Rockefeller, but it was Henry Cabot Lodge, Nixon's running mate in 1960, and then Ambassador to Vietnam in the early '60s, and a Harvard alumnus who had been impressed by Kissinger, that strongly recommended Kissinger to Nixon for the key post. Nixon intended to run foreign policy from the White House and apparently was confident that in Kissinger he had the accomplice he needed to circumvent the State Department in developing policy. Never mind that he had no negotiating experience.

The New York Times editorials, including those by James Reston, lavished praise on him, as did Evans and Novak. The Washington Post also endorsed Nixon's choice, as did William, F. Buckley, Jr. Would they do this again? Perhaps, at least by the Washington Post.

"One of the most perceptive comparisons of Kissinger and Nixon came in a 1973 speech by Thomas Hughes, a State Department veteran who had departed to become president of the Carnegie Endowment: "Both were incurably covert, but Kissinger was charming about it. Both abhorred bureaucracy, but Nixon was reclusive about it. Both engaged in double-talk, but Kissinger was often convincing. Both were fiercely anti-ideological, but Nixon had recurrent relapses. Both jealously guarded against any diffusion of power, but Kissinger dispensed balm. Both were inveterate manipulators, but Nixon was more transparent. Both insisted on extremes of loyalty, but Kissinger endeared himself to his critics. Both had a penchant for secrecy, but neither uniformly practiced what he preached. Both were deeply suspicious, but Kissinger was irrepressibly gregarious. Neither was widely admired for truthfulness, but Kissinger excelled at articulation. Neither worshiped the First Amendment, but Kissinger mesmerized the press."(6)

Sadly, the Agreements that were finally signed in Paris in January, 1973, following more than three years of "negotiations", were essentially those that had been rejected years earlier. By the end of December, 1972, the U.S. killed in action numbered 45,926, meaning that 14,000 more Americans had been killed while Kissinger 'negotiated' in Paris. The U.S. first began to seek "negotiations" in 1965, but the North Vietnamese were disinterested. That the U.S. actively, if not anxiously, sought "negotiations" had been quickly and correctly translated by the North Vietnamese as meaning ultimate victory for them, and "defeat" for the U.S. and the South Vietnamese. They were in no hurry, for time was on their side, as their Chinese 'advisers' (Mao Zedong and Zhou Enlai) repeatedly told them. The advice was from the PRC playbook: "fight, fight, talk, talk," or…"two steps forward, one step backward, or one step sideways, another step (or two) forward, one step sideways, in order to steadily advance." The oriental equivalent of the ancient water torture, or just effective negotiating?

That the anti-war demonstrations that swept the country in both 1968 and 1969 had a major impact on Johnson, Nixon, Kissinger, and Congress, played a major role in the U.S. seeking "negotiations" is simple fact. Thus it came to pass that after months of bickering, on January 16, 1969, the U.S. and the DRV announced agreement had been reached on a 'round conference table' for the Paris Peace talks. The Media failed to define this historical accomplishment as a step sideways, backward, or forward, being content to note that encouraging progress was finally being made. By what or whose measure was not explained.

"When Nixon and Kissinger took office, the U.S. had 536,000 troops in Vietnam. Americans were being killed at a rate of about two hundred a week. (About 31,000 had been killed to date.) The cost of the war to U.S. taxpayers was running about $30 billion a year (in 1969 dollars). Nor was the proverbial light visible at the end of the tunnel."(7) And there had been no real or meaningful progress in negotiations with the North Vietnamese.

On March 1, 1969, two battalions of the North Vietnamese 174th Regiment and three battalions of the 148th Regiment launched an offensive in northern Laos. Were these part of the two steps forward, and one backward, concept? Certainly it underscored that the North Vietnamese were intent on keeping the multiple Ho Chi Minh supply trails open to South Vietnam through Laos and Cambodia. On 29 March, 1969, U.S. military personnel in Vietnam reached peak strength of 543,340. From that high point began Nixon's gradual draw down of American forces.

The July 15, 1969 letter from President Nixon to Ho Chi Minh said in part: "My speech on May 14 laid out a proposal which I believe is fair to all parties"..."But we stand ready to discuss other programs as well, specifically the 10 point program of the N.L.F." ... "The time has come to move forward at the conference table toward an early resolution of this tragic war. You will find us forthcoming and open-minded in a common effort to bring the blessings of peace to the brave people of Vietnam."(8) On August 30 President Ho Chi Minhs' reply to Nixon's letter is received in Paris. "The war of aggression of the United States against our people, violating our fundamental national rights, still continues in South

Vietnam. The longer the war goes on, the more it accumulates the mourning and burdens of the American people. I am also deeply touched at the rising toll of death of young Americans who have fallen in Vietnam by reason of the policy of American governing circles. Our Vietnamese people are deeply devoted to peace, a real peace with independence and real freedom. They are determined to fight to the end, without fearing the sacrifices and difficulties in order to defend their country and their sacred national rights. The over-all solution in 10 points of the National Liberation Front of South Vietnam is a logical and reasonable basis for the settlement of the Vietnamese problem. In your letter you have expressed the desire to act for a just peace. For this the United States must cease the war of aggression and withdraw their troops."(9)

By December 31, 1968, 30,610 Americans had been killed in action in Vietnam. On September 3, 1969, Ho Chi Minh died. On September 16, Nixon announced a further withdrawal of 35,000, in addition to the earlier announcement of 25,000. By mid October, 1969, 38,887 had been killed in action in Vietnam. On 14 October, Premier Pham Van Dong published an open letter to American anti-war protesters which is read into the Congressional Record. It openly invited additional "peace and justice-loving American personages" to launch a powerful 'offensive' throughout the United States in the fall to demand the Nixon Administration to put an end to the Vietnam aggressive war and to immediately bring all American troops home. It ends with the wish that "May your fall offensive succeed splendidly" and is signed, "Affectionately Yours, Pham Van Dong, Premier of the DRV Government."(10) The Liberal Media across the country supported this letter like vultures feasting on carrion, as did the well organized "activists".

On 15 December, Nixon announced a further withdrawal of 50,000 U.S. troops by April 1970. On 31 December it was announced 9,414 had been killed in action for the year, bringing the total to 40, 024.*

Kissinger presented some of his views on Vietnam in an article for Foreign Affairs before his appointment in which he stated: "We lost sight of one of the cardinal maxims of guerrilla war: the guerrilla wins if he does not lose. The conventional army loses it if does not win. After the Tet offensive, the U.S. could no longer achieve its objectives within a period or with force levels politically acceptable to the American people."(11) However, America could not simply cut its losses and withdraw, Kissinger said, because the U.S. needed to maintain its "credibility". But Kissinger, according to both his friends and foes, really felt otherwise. If preserving credibility was the main goal, then the U.S. did not have to save South Vietnam indefinitely. In Kissinger's mind, the U.S. merely had to achieve a "decent interval" between America's withdrawal and South Vietnam's collapse. During 1968, Kissinger frequently said in private talks and seminars that the appropriate goal of U.S. policy was a "decent interval" of two or three years between the withdrawal of U.S. troops and a communist takeover in Vietnam."(12) In putting this diabolical formula together, Kissinger's Machiavellianism surfaced, but was not noted or questioned or damned by the Media.

But clearly South Vietnam's obituary notice was being formulated, and Kissinger's subsequent 'goals' and game plan had been made clear. In a way, this approach tied-in with-his reported habit of repeatedly challenging his colleagues during discussions or debates by asking what was to be gained, or what was the goal intended in three, or five years, time if their point of view or recommended course of action was to be followed. Obviously, he did not ask the South Vietnamese, or President Nixon, that question, nor did the Media or Congress ask him.

"Kissinger, however, quickly adopted a variation of Nixon's line that the path to peace in Vietnam went through Moscow. This led to an approach that Kissinger called linkage: American policies toward the Soviet Union on various issues – trade, arms control, Vietnam, etc. – should be linked."(13) Both Kissinger and Nixon missed the point that China was more of a political and ideological linkage than Moscow, the latter being the main logistics source for North Vietnam. At least some of the Media raised the points that the issues were more East and West in political nature than arms issues, and that arms control was apart from political issues.*

But Kissinger was intent on linking Moscow to Washington (to Vietnam) and sent Cyrus Vance, one of Pres. Johnson's Paris negotiators, "on a secret mission to Moscow to propose a package deal on Vietnam and arms control. The State Department would not be informed."(14) Kissinger's proposal included a cease fire, mutual withdrawal of U.S. and North Vietnamese troops, and a role for the NLF in South Vietnam, which far exceeded any previous proposal – and of course was made without the knowledge of the South Vietnamese. The implied threat was of tough military measures if Hanoi did not agree. There was no response from Moscow, nor was Kissinger able to gain Moscow's support for his many additional requests for assistance in influencing North Vietnam in 1969. The grand linkage scheme was buried without honors, with Kissinger's ego failing to recognize that Ho Chi Minh was linked to Beijing politically, not Moscow.

Nixon's and Kissinger's trip to Western Europe in February, 1969 proved to be of benefit in two ways, thanks to France's President de Gaulle. It was he who recommended to Nixon that he establish relations with China, and that negotiations with the North Vietnamese be pursued in Paris, perhaps secretly…an approach that appealed to both Nixon and Kissinger. De Gaulle also noted that "There was a great opportunity to be exploited; the Kremlin's growing paranoia about China. They are thinking in terms of a possible clash with China, and they know they can't fight the West at the same time. Thus I believe they may end up opting for a policy of rapprochement. To work toward détente is a matter of good sense: if you are not ready to make war, make peace."(15) Nixon and Kissinger were quick to seize and use 'détente' as their own 'logo', but without any attribution to de Gaulle.

It is interesting to compare Kissinger's version of the 'China' idea and that of the not-so-secret Paris negotiations with the above reference. He deprecated the State Department and Secretary of State William Rogers, who he characterized as: "Rogers had spent his career in the legal profession; he had not dealt with foreign policy in any systematic way. His general view of foreign policy never strayed far from that of the New York Times editorial page, which was also close to the dominant opinion in the State Department" "Marginalizing the Secretary of State was the worst way of dealing with Nixon's dual purpose of establishing White House preeminence and obtaining State Department support for a new approach to foreign policy. The Foreign Service, sensing the discomfort of the Secretary and explaining its exclusion from decision-making to itself with the myth that its views were somehow being kept from the President, resorted to massive leaks or procrastination – the idea being to give Nixon a second chance to see the light."(16)

That Kissinger fully endorsed Nixon's ploy regarding the NSC - State Department embroilment from the first day is not stated, although he pats Sec. of State Rogers on the head with platitudes and wrote - "In retrospect, I am not proud of the way I participated in Nixon's deliberate effort to marginalize the man who at the time was considered by most observers to be the President's closest friend."(17)

As to China, in 1954 Beijing and Washington had agreed to maintain contact via their respective ambassadors in Warsaw, but with the onset of the Cultural Revolution in China there had been no contact for a number of years. Nixon reactivated this channel, according to Kissinger, which subsequently led to China's indicating she would welcome an American emissary. However, China shut down this channel in protest to American 'incursions' into Cambodia in 1970, but then chose Pakistan as a cooperative alternate, which led to Kissinger's "secret" visit to China in 1971. There is no mention by Kissinger of de Gaulle having suggested using "détente" with the Soviets, or of de Gaulle's advising the U.S. to seek contact with China. Obviously, copyright or plagiarism does not apply to all "diplomats".

As noted in the previous chapter, infrequent talks with North Vietnam's Le Duc Tho had begun in Paris under the Johnson Administration with Averill Harriman and Cyrus Vance the designated negotiators. Little, if any, progress had been made. Kissinger decided that he was the one person who could achieve success, and persuaded Nixon to assign him this task. Gen. Vernon Walters, a talented army linguist, was in Paris as the U.S. Military Attache, and was instructed to arrange Kissinger's meeting(s) with the North Vietnamese. On August 4, 1969, Kissinger met with Xuan Thuy, who first gave a 45 minute lecture outlining Vietnam's history and struggle against 'foreigners', then said that the U.S. would have to withdraw its troops unconditionally."(18) He said that the only 'outside' force in South Vietnam belonged to the U.S., that North Vietnam was not interested in discussing a mutual withdrawal, and that it would not change this fixed position. These same points had been stressed previously by the North Vietnamese, both publicly and privately, which seemed "baffling and devious", according to Kissinger, nor did he seem to grasp that they actually meant what they said. This latter apparently was a strange concept to him. Although Kissinger issued a warning that if there was not progress in these talks by November the U.S. would "consider steps of grave consequence", "Kissinger also noted to Xuan Thuy the unilateral withdrawal of 25,000 Americans from South Vietnam and that the U.S. "was willing to withdraw all of its troops without exception from Vietnam, and was willing to accept the outcome of a free political process.(19) As part of that process, the United States realized that neither side could be expected to give up at the conference table what had not been conceded on the battlefield; that a fair political process must register the existing relationship of political forces; that the United Stated States would differ from the North Vietnamese on how to achieve this, 'but that neither side should be asked to accept the proposition that it can be defeated without noticing it."(20) "The second lecture we received was the impossibility of separating military from political objectives and ...that we settle the military issues by way of a cease-fire, POWs and so forth, and leave the political evolution to Vietnam, as long as it was done by reasonably democratic processes." Xuan Thuy added that "Your policy is to withdraw troops by droplets and to prolong your military occupation in South Vietnam. We doubt your goodwill...and insisted on a "Ten Point Overall Solution", introduced by the NLF on May 8 that called for a complete U.S. withdrawal, the release of all prisoners of war, and the creation of a coalition government."(21) Xuan Thuy added that Hanoi might consider additional secret discussions with Kissinger. So ended this initial meeting.

"There was little justification for the secrecy fetish about the Paris talks – other than to cut out the American public and the State Department bureaucracy. In fact, it served Hanoi's purposes to have the true American bargaining proposals remain secret for months while it drummed up public propaganda in favor of its own positions. In addition, a visible and well-explained effort by Kissinger to negotiate a Vietnam settlement might have calmed some of the anger of the anti-war movement."(22) But Kissinger's many concessions that followed, his duplicity, and his deceits, could not have stood public inspection either at home or in South Vietnam - or elsewhere in the world. But perhaps these factors prompted his insistence on total secrecy. No sunshine, no bothersome shadows, no risk of unfavorable publicity: quite simple, or so it seemed to appear to Kissinger.

Nixon also sounded the implied (November) warning signals, stating that "It will be the policy of this Administration to warn only once". He had Kissinger contact the Soviet Ambassador to the U.S. (Dobrynin) to repeat the same warning, linking the lack of progress on Vietnam with the arms control talks with Moscow. In mid-October, Kissinger advised Nixon by memo that the massive air raids planned by the Joint Chiefs should not take place, having "...concluded that no quick and decisive military action seemed attainable." Nixon agreed, but both Nixon and Kissinger later admitted that this decision had been a bad mistake. "When asked by Bill Safire (NY Times) at the time of the January 1973 peace accords what he would have done differently, (Kissinger) responded: "We should have bombed the hell out of them the minute we took office...."We could have ended the war much sooner, if we had been willing to do in 1969 what we ended up doing in 1972". And South Vietnamese President Thieu later said "You Americans missed many opportunities to settle the war in our favor, and that was one of them."(23) Kissinger's "hindsight" is more than suspect.

While the exchange of "views" in Paris netted no real progress, "...this first meeting opened the door for the secret parallel negotiating meetings that would follow between Kissinger and Le Duc Tho. The story of Paris would henceforth be a dual tale of the official talks (the plenary sessions), with South Vietnam present, and the real talks (private meetings) between the United States and the North Vietnamese. Nixon and Kissinger's obsession with secrecy would have disastrous consequences. Aside from a handful of NSC aides, Kissinger, the president, and Ambassador Ellsworth Bunker would be the only "main players" aware of the parallel diplomatic track. Until January 25, 1972, the State Department, the Joint Chiefs of Staff, the CIA, and the Defense Department were kept out of this loop. Even worse, although President Thieu had been assured by Kissinger and Nixon that he would be consulted and kept abreast of any private discussions, especially on issues that pertained to the internal political processes of South Vietnam, he was (generally) "informed" but never "consulted".

Nixon had set up a process by which he could cut a deal directly with the enemy and keep much of his own government and even his ally in the dark about it. In the end, he would tell a different story to each side – to Thieu, to his advisers, and to the public. The use of threats would have no effect. In fact, the greater the threat, the more rigid Le Duc Tho would become. Nixon and Kissinger firmly believed that when the threat became an act, as it did in the punishing B-52 bombings of May and December 1972, it would bring the communists into compliance. But by then the balance of forces had changed in Hanoi's favor (including) several crucial U.S. concessions. All along, their (Hanoi's) plan to stall was aimed at dividing the U.S. from Saigon and world opinion against the U.S. It succeeded brilliantly."(24)

In effect, the die had been firmly cast by Le Duc Tho, but Kissinger would not, or failed to recognize this fact. The charade of the secret meetings continued, despite the same litany of demands by the North Vietnamese…and their successful nibbling away at the U.S. proposals. Not only were the South Vietnamese kept in the dark, but Kissinger far too often deceived Nixon by reporting that Le Duc Tho was becoming less strident in his position and conciliatory in their Paris discussions. "As was his style, Kissinger relied heavily on ambiguity to camouflage the concession (unilateral withdrawal), and make it palatable to South Vietnam's supporters. Kissinger made no clear public explanation at the time, even on background, that the demand for North Vietnam's withdrawal was being dropped."(25)

Nixon had earlier made public his plan to "de-Americanize" the war; he termed it "Vietnamization". "Perhaps most important, Nixon changed the political objective for U.S. intervention from guaranteeing a free and independent South Vietnam to creating the opportunity for South Vietnam to determine its own political future. Vietnamization and negotiation were Nixon's twin pillars for achieving an honorable peace."(26) Or was "Vietnamization" simply a ploy designed by Kissinger - and Nixon - to cover the decision to gradually abandon South Vietnam? Obviously, Vietnamization was a process that should have begun by the mid 1960s, not part of a ballyhooed political effort in the early 1970s.

In response to a ten point NLF proposal by the NLF on May 8, 1969, on May 14 President Nixon unveiled a major eight-point peace initiative for ending the war in Vietnam. "The cornerstones of Nixon's new proposals were simultaneous or mutual withdrawal of both U.S. and North Vietnamese troops from South Vietnam and maintaining the current government in South Vietnam. The speech also contained important language for the North Vietnamese: "A great nation cannot renege on its pledges. A great nation must be worthy of trust." Nixon told his national audience that "the record is clear as to which side has gone the extra mile for peace. We have gone as far as we can or should go in opening the door to peace; now is the time for the other side to respond." Hanoi replied the next day. "The plan of the Nixon administration is not to end the war but to replace the war of aggression fought by US troops with a war of aggression fought by the puppet army of the United States." (27) Blunt and to the point, Hanoi, was telling Nixon and Kissinger that there would be no concessions and that North Vietnam was more than willing to continue to wait.

"Yet America's deepest challenge was not the technical aspect of withdrawal; it was a moral issue even when presented in realistic guise. Nixon would not abandon the tens of millions who, in reliance on the words of his two Democratic predecessors, had cast their lot with ours. We argued on the basis of the national interest that, as the leader of the Western Alliance, our credibility toward friend and foe was at stake. But beyond "realistic" considerations, Nixon also had laid down a moral marker that defined the limit of his strong desire to extricate the United States from Indochina. He was prepared to display considerable flexibility in negotiations with Hanoi – going beyond what had ever been suggested, even by the "doves" in Johnson's presidency. There was only one concession we would never make on geopolitical and, above all, on moral grounds – to impose a Communist government on peoples who had joined the anti-Communist cause in reliance on America's word. That, however was the demand from which Hanoi would not budge and which the peace movement came increasingly to insist."(27) Thus does Kissinger in hindsight attempt to explain the problems experienced in negotiating with the North Vietnamese. He credits Nixon and himself with taking the high ground, but is more concerned that the reader understand the credibility concerns he and Nixon faced. It doesn't sell, Henry.

A more frank explanation was that Kissinger was meeting secretly with the North Vietnamese in Paris to negotiate "…an American troop disengagement while informing Thieu only after the fact. The U.S. entered the talks in 1969 seeking mutual withdrawal and no coalition government, and ended the talks in 1973 accepting unilateral withdrawal of its own troops and no coalition, but a guarantee by Thieu to step down before the election. Throughout, the North refused to recognize the government in the South as legitimate and swore to destroy it. In the end, South Vietnam would be served a death warrant, and the North would be rewarded for stalling."(28)

"According to Haig (then Kissinger's deputy on the NSC),"I am absolutely, categorically convinced that, had we done in 1969 what we did in the end of 1972 that conflict would have been over and we would have gotten our prisoners out and our objectives would have been met. We should have moved much more decisively in applying our power to the source of the problem. I think in 1969 when the new President came in, had he immediately started to bomb Hanoi, mobilize our forces here at home, put the Soviet Union on notice, and done all the things that should have been done in the Johnson administration, that there never would have been those additional three years of conflict and that there would have been a negotiated settlement and withdrawal of North Vietnamese forces from South Vietnam, where they were the invaders."(29) But both Secretary of Defense Laird and Secretary of State Rogers were inexplicably opposed to any bombing of the north or of the Ho Chi Minh trail systems, which without doubt gave Nixon considerable pause. And, not surprisingly, neither of the dual negotiating efforts in Paris had made any progress by the end of 1969. "The United States would call the talks two party; the communists would call them four party (the Democratic Republic of Vietnam, the National Liberation Front of South Vietnam, the United States and Saigon). The United States called them the Paris Peace Talks; Hanoi the Paris talks. For months, nobody spoke the same language."(30) But Americans and Vietnamese continued to die daily, none on either side aware of what the negotiators were quibbling about, and there were very few insightful comments or questions by the Media.

Amazingly, Kissinger chose to characterize his first meeting with Le Duc Tho and Xuan Thuy as "a significant meeting. We had a frank exchange of views. They basically accepted our proposed procedure for future private meetings, dropped their preconditions for substantive negotiation, and gave the impression of being much more ready for business than before."(31) Quite likely he was the only one present that would have summarized the seven hour meeting this way...but seldom did Kissinger describe accurately the tone or contents of the discussions to Nixon, or to Bunker, or to anyone, especially the South Vietnamese.

"President Thieu was never adequately briefed on the talks by Ambassador Bunker; Bunker received his information from Kissinger and did not always know the full score. The secretaries of state and defense, joint chiefs of staff, and other bureaucratic entities were also kept out of the loop, viewed by Kissinger and Nixon as untrustworthy and potential sources of leaks."(32)

Kissinger made no headway with Le Duc Tho. He obviously thought he could persuade him with Teutonic logic wherein he was trying to separate the military struggle from the political, as he had earlier done at Harvard in discussing Bismark. Le Duc Tho was not impressed. "And so it went, on and on, throughout 1970 and into 1971. Reading the transcripts of the meetings is like being transported into an Alice-in-Wonderland world. The United States had already accepted the principle of total withdrawal of American and allied forces, along with a precise schedule for this withdrawal. It had guaranteed that it was not committed to the maintenance of any political force in power once a settlement was achieved. Kissinger even tabled proposals and methods for consulting the will of the people. Kissinger told Le Duc Tho that he was prepared to discuss the relationship of elections to the elements of political power, the relationship of military to political authorities, the relationship of executive and legislative power, and how minorities are to be protected. It all sounded so noble."(33)

But Le Duc Tho was not buying; he believed that the U.S. was trying to use military pressure to negotiate from a position of strength. "...through what has happened in Vietnam, Laos, and Cambodia, particularly your actions in Laos and Cambodia, I think that you are still actively carrying out the Vietnamization plan, prolonging the war and expanding it to the whole of Indochina. In Cambodia, it is wondered who has organized the coup to wipe out the independence and neutrality of that country and to rig up a reactionary group to power. No one other that the U.S. You said that it was not a threat to Vietnam, that you were not involved in Cambodia. This is completely contrary to the facts. As a Vietnamese saying goes, 'You cannot hide an elephant with a basket.'...You will reap what you have sowed. You have tried to incite national hatred between the Vietnamese and the Cambodians peoples, but the people of Vietnam, Cambodia, and Laos have got traditional cohesion in opposing the French during nine years and long been associated together, you will not be able to break this cohesion. Today, the people of Vietnam, Cambodia, and Laos are still more determined to fight and will surely win victory. That is why the peaceful solution of the problem of Vietnam, Laos, and Cambodia depends on you and not on us."(34)

On February 8, 1971, more than 16,000 ARVN (South Vietnamese) troops crossed into Laos to interdict the Ho Chi Minh trail systems and to destroy North Vietnamese troops and storage areas in the "panhandle" area. It was an ambitious effort involving more than a hundred tanks, air assaults by helicopters carrying ARVN Marines, paratroopers, and infantry, as well as mobile light artillery units. The plan had been put together in Saigon and was (theoretically) TOP SECRET, known only to a select few. Earlier, when I asked the Saigon Chief of Station (COS) if there were any special reporting requirements expected of me during Lam Son 719 from I Corps, his only response was to ask how I had learned of the pending operation. I replied that the I Corps Commander, Gen. Lam, had told me of it about two weeks earlier during one of our weekly meetings, including that he had described it as a Saigon conceived and drawn plan to test the 'Vietnamization' process, while disrupting North Vietnamese supply lines into South Vietnam through Laos. The COS said that he did not envisage needing an input from my Danang office.

U.S. Army Gen. Mal Zais, who had earlier succeeded the III MAF Marine commander Lt. Gen. Keith McCutcheon upon the latter's departure as the senior U.S forces I Corps commander, also raised the topic of Lam Son 719 with me at our weekly meeting upon my return from Saigon, probably anticipating that I had information I could share with him. I told him this was not the case at all and that I had been kept in the dark by my boss. Zais said he fully shared Gen. Lam's concern regarding the security of the operation, including the likelihood of leaks out of Saigon, because of penetrations of many of the key offices in Saigon, including the ARVN. I agreed, noting the extensive cache (24,000 documents) of information we had acquired during a joint operation near Danang with a Marine CAP team conducted a year ago. The documents confirmed the existence of a second extensive network of penetrations extending from Hanoi through the SVN provinces into ARVN, the South Vietnamese National and Special police, and GVN offices in Saigon. Zais said that Lam's information indicating that the infiltration route for Lam Son 719 via Khe Sanh into Laos was already being reinforced by NVA units was particularly worrisome. I told him that I would check with one of the communications intercept installation to see if anything similar had been detected along that line.

I contacted an intercept site the next day and was told that there had been a substantial increase in coded 'traffic' in the general Khe Sanh area in the past week following a gradual build up over the past month or so, but, unfortunately, because the site did not have the capability to decode the take, the intercepts had been sent on for translation, and no return reading had been received as yet. I passed this information to Gens. Zais and Lam, to the ARVN Military Intelligence Service in MR I, and to the Saigon COS. They thanked me, and the latter said he would check further. The ARVN MIS Chief in MR I told me that he had received increasing but scattered reports of North Vietnamese reinforcements moving into the general area near Khe Sanh in Laos, but had not as yet been able to put together a definitive report for Gen. Lam. He said he had also relayed this information to his ARVN counterparts in Saigon, but had heard nothing in return. It was evident that Gen. Thieu and the joint South Vietnamese and U.S. planners were coordinating and controlling the Lam Son 719 operation from Saigon, including their selecting the ARVN strike forces to be involved. With exception of I Corps ARVN units to be employed, other I Corps personnel were being largely kept in the dark regarding details until the very last moment.

Preceding the attack, ARVN and U.S. air strikes and artillery pounded the Khe Sanh area, following which ARVN tanks, personnel carriers, and ARVN helicopters carrying the ARVN assault troops moved into Laos and established the first Fire Support Base

(FSB). ARVN air and artillery units continued to provide coverage for the advancing troops. Unfortunately, the North Vietnamese had known most if not all of the details of the attack plans well in advance and had established multiple ambush sites, including extensive and well positioned artillery and anti-aircraft positions. ARVN was able to establish the first FSB only under very heavy pressure, including withstanding a night-long attack by North Vietnamese sappers and soldiers. At dawn, some 300 North Vietnamese bodies were counted in the wire protecting it. The second FSB, established further into Laos, experienced similar challenges, as did the few other FSBs that followed.

"The North Vietnamese used the first two plenary meetings on February 11 and 18 to protest the (Lam Son) raid and then boycotted the Paris talks for two months...and on February 17, Nixon declared: "Tonight I can report that Vietnamization has succeeded."(35) It would have been far more accurate had he said that Lam Son 719, and the concept of Vietnamization, while still a still a work in progress, had proven the mettle and excellent potential of all ARVN units. Indeed, ARVN forces fought well despite the many disadvantages and obstacles they faced from the outset.

The plan, and its objectives, while laudable; was not successful because of the North Vietnamese having known the details of it well in advance. This was a bitter reality for all involved, especially within ARVN. Documents found on captured and dead North Vietnamese officers revealed that Lam Son 719, including ARVN unit designations and their tasks and targets, was known in detail by NVN forces, including maps, routes, and projected timetables. Close to 40,000 North Vietnamese troops, including heavy artillery, were waiting for the ARVN forces. The losses were heavy on both sides: "...the North Vietnamese, determined to protect its crucial Southern supply lines, sent 36,000 regular troops into battle. The North Vietnamese lost over 20,000 troops, but ARVN lost half of its force. Furthermore, 168 helicopters were lost, with another 618 damaged. Fifty-five air crewmen were killed in action, 178 wounded, and 34 were missing in action. The high number of aerial losses stemmed from the effective North Vietnamese use of antiaircraft artillery. Bad weather and fierce North Vietnamese counterattacks combined to force ARVN to withdraw in 60 days.

"Lam Son 719 destroyed the cream of the South Vietnam army and was far more serious and detrimental than was believed at the time. Our handling of that was very bad, observed Alexander Haig, who had been sent to Vietnam by Nixon in the middle of the operation."(36)

Not mentioned in the above statistics was the loss of more than 100 tanks by ARVN, most of which simply ran out of fuel, or abandoned due to intense artillery fire, or could not be repaired during the heavy fighting. There was no close air-ground support by U.S. fighter aircraft, but the SVN Air Force was deservedly lauded for its effective and sustained efforts. U.S. helicopter and crews flew countless missions in support of the operation, ferrying men and equipment into areas near where FSBs were established, and helping to ferry out the wounded. Massed North Vietnamese troops were hit by B-52s during the latter stages of the operation with devastating effect. According to SVN soldiers returning to South Vietnam from Laos after being separated from their units described the stench from thousands of North Vietnamese bodies in the Tchepone area as all but overwhelming, and credited the B-52 strikes as having been extremely helpful in preventing SVN forces from being overwhelmed by the superior firepower and numbers of NVA that had lain in wait for the attack.

About the beginning of the third week in February, I received a cable from the COS, Saigon, asking if I could provide him with bi-weekly "situation reports" on Lam Son 719. I said I would provide what Gen. Lam was willing to share with me, and suggested he try to get Gen. Abrams to task the Air Force to provide whatever it had in the way of reporting from U.S. helicopter support efforts, other overhead resources, as well as intercept 'chatter'. His not surprising response was that no U.S. helicopter crews were involved in Lam Son 719, but any information I could get would be welcome.

I contacted Gen.Lam and he invited me to go with him to Khe Sanh the next day. The relatively small area of Khe Sanh was a scene of organized confusion, alive with military personnel, vehicles, and the air all but filled with helicopters. There was also an area for the Media, with representatives from most of the neighboring countries, and from Japan, Korea, Thailand, Singapore, Indonesia, etc., if memory serves me correctly. Most of the helicopters returning from Laos not only had wounded on board, but a few had soldiers sitting on the skids. The wounded were quickly off-loaded, and those not wounded were taken to a separate area. After an up-date briefing by his officers, Gen. Lam talked with the Media. He answered their many questions, most of which concerned stories that ARVN units had been overwhelmed or overrun, as evidenced by the soldiers on the skids of the returning helicopters, as had been reported in the most recent "*Time*" magazine article complete with a photograph of soldiers hanging on the skids of the helicopters (ostensibly) returning to Khe Sanh. Gen. Lam then pulled out a copy of a "*Time*" magazine from the Korean War showing an identical photograph, with U.S. soldiers sitting and hanging on the skids of a helicopter, and asked if anyone had seen this repeated on all the helicopters ferrying troops in and out of Khe Sanh. He said that to his knowledge only one foreign "reporter" had accepted the standing invitation he had offered earlier to all reporters to ride the mission helicopters. He then singled out a Japanese reporter for his bravery for having flown repeatedly with the ARVN crews...and repeated the offer to any in the Media to go on the helicopters. None present accepted his offer, but many seemed upset that Gen. Lam had "embarrassed" them.

I have recounted the above primarily for three reasons: 1.) to underscore that for two months ARVN troops, including their Marines and Airborne, fought very well despite the overwhelming odds of men and armament against them; 2.) to register that while Vietnamization had become a popular buzz word in Washington, insufficient training and time had been given ARVN to prepare for such an ambitious, complicated, and extended operation as Lam Son 719; 3.) to underscore that governmental security in Saigon had proved to be a fantasy. The North Vietnamese had penetrated nearly all levels of the government, the police, the military, and the politicians to what should have been an alarming and suspected degree to most, but obviously was not. President Thieu and his inner circle trusted very few, and for good reason. The same held true in most if not all of the military regions. Simply stated, there were very few secrets safe in Saigon, the locus of the heart of the country. That the U.S. Media had a field day deprecating Lam Son 719 and "Vietnamization" goes without saying.

In July of 1971 the Le Duc Tho - Kissinger meetings took a

back seat while Kissinger traveled to China to confer with the Zhou Enlai about a possible summit meeting between Nixon and Chairman Mao in Beijing, which was readily arranged. Interestingly, when Pham Van Dong was in Beijing on September 19 for a meeting with Zhou Enlai, it was agreed that there was a need "...to step up the diplomatic struggle...We have to influence the anti-war public opinion in the U.S. that includes not only the people at large, but also the political, business, academic, and clerical circles to ensure a stronger support by them. The world public opinion has been mobilized."(37) But Kissinger and Nixon were too flushed with their China success to think that their self-adulation might be a bit premature and possibly a little hollow. The Vietnam War seemed to drop in priority in Washington.

Pham Van Dong would also meet with Chairman Mao, who would tell Dong, "Negotiations have been going on for two years. At first we were a little worried that you were trapped. We are no longer worried...You are fighting well on the battlefield. Your policy for diplomatic struggle is correct....The Americans still want to come to Beijing for talks. It is what they propose...Kissinger is a stinking scholar. I have read the report about the meeting between comrade Xuan Thuy and Kissinger. The last part of it is very funny. Kissinger is a university professor who does not know anything about diplomacy. I think he is not someone who can compete with Xuan Thuy, even though I have not met Xuan Thuy."(38)

There was one benefit from Kissinger's visit to China that impacted his discussions with the North Vietnamese. Shortly after Kissinger left, Zhou Enlai went to Hanoi to reassure North Vietnam's leaders that China remained fully supportive. However, he urged that Hanoi accept a compromise that permitted the survival of the Thieu government, noting that the 'Saigon puppet regime' was more of a long term issue than a current one, and not difficult to resolve. When Le Duc Tho finally accepted this in subsequent meetings with Kissinger, the latter quickly concluded that the North had just made a bell-ringing "concession"...and crowed it as such to Nixon. His potty training as an insightful, perceptive, and effective negotiator with the East was another work very much in progress.

His personal achievement count-down was underway and Kissinger saw himself as one of history's leading figures for having established historic relations with Communist China; for having improved relations with Moscow; and now as having the once-in-a-lifetime opportunity to complete the cycle by adding North Vietnam to his resume of "negotiating" accomplishments. To Kissinger's delight, both Time and Newsweek featured him on their covers, and billed him as "Nixon's Secret Agent". Indeed, "...the U.S. was steadily disengaging from the most misguided war in its history. Of the 540,000 American troops in Vietnam when Nixon took office, 410,000 had already been brought home, and the withdrawal of another 70,000 had just been announced. None of the remaining forces would be combat units. The number of U.S. battlefield deaths had dropped to 10 per week, down from an average of 280 per week in 1968. American society, it now seemed, would survive not only the war, but the most threatening antigovernment uprisings in this century."(39)

"By January 1972 the U.S. had conceded on almost every major point, including, at least implicitly, that any cease-fire would be a "cease-fire in place", that is, North Vietnamese troops would stay in the South if they were already there. What came next was predictable: The North could not get the U.S. to get rid of Thieu for them. They did not intend to stop fighting until they regained the South. Thus, they had an obvious strategy: stall the peace, pour forces into the South, and strike a deal only after a cease-fire in place was in essence a "victory in place". For months, Le Duc Tho kept Kissinger's hopes alive that more serious talks were about to begin, but it was all a ruse. On Easter Sunday, March 30, 1972, North Vietnam launched the largest attack of the entire Vietnam War. It was a conventional military attack designed to inflict a crippling blow against ARVN forces and would last six months."(40)

Nixon had seized "Vietnamization" as his escape vehicle. Kissinger had handed the North the keys to the kingdom (South Vietnam) by offering a cease-fire-in-place. President Thieu continued to be kept in the dark as to what Kissinger was conceding to the North, including the cease-fire-in-place concessions that the North had so readily agreed to. Thieu and the South were starting to realize that they were little more than mere shelf items in the Kissinger pawn shop. When "negotiations" faltered, or to at least make good on a threat to convince the North he was not just all talk, Nixon would call for limited bombing runs against targets in the North – but these were more of a nuisance and irritation to the North than being militarily or politically effective.

President Thieu was not alone in his concerns about Kissinger. "A month before Kissinger's Moscow trip (April 20th), Al Haig had been given his second star. More importantly, he had won the favor of Nixon and Haldeman with his advice on how to handle the temperamental Kissinger. Haig's conduct during his weekend at Camp David reaffirmed what both Nixon and Haldeman had been feeling: Kissinger could be emotional and uncontrollable, but Haig seemed stable and reliable; Kissinger put too much faith in negotiations, but Haig was made of sterner stuff; Kissinger was willing to make concessions that would appeal to the left, but Haig knew the importance of hanging tough. In discussing Vietnam, Haig privately disagreed with Kissinger's advice that the bombing of North Vietnam remain sporadic; instead, he recommended more concentrated assaults. He also told Nixon how important it was to make sure that he (Nixon) survived politically." (41)

"On April 1 Nixon responded (to the attack by North Vietnamese forces) by ordering bombings within 25 miles of the DMZ. By April 14 he ordered air strikes up to the 20th parallel. The Paris talks were again suspended."(42) On April 16 a new operation, "Freedom Train", sent five B-52s to attack North Vietnam in the Hanoi-Haiphong area. Four Soviet ships were hit by mistake, but the Soviets chose to do nothing – sending a clear signal that they did not intend to cancel the summit. On April 20, Kissinger secretly departed for Moscow, where the reception he received was exceptionally cordial. The Soviets told him they wished to assist in a Vietnam settlement and eliminate any obstacle to the summit. Kissinger promised Brezhnev that the U.S. would accept a cease-fire in place, as well as the presence of North Vietnamese troops in South Vietnam – even those troops that had entered after the offensive. Kissinger asked Moscow to pass this message on to Hanoi. Now it was official."(43)

In his May 2 meeting in Paris with Le Duc Tho, "Kissinger asked about what he had conveyed through the Soviet Union. "Through your allies, we have put forward issues to be discussed, and your allies have conveyed them to you. I am surprised that you make no comments on those questions." Le Duc Tho answered "We have on many occasions said that if you have any questions you should directly talk to us, and we shall directly talk to you. We don't speak through a third person. If you have anything to say now, we are prepared to discuss it with you." Kissinger "later wrote that Le Duc Tho was not even stalling. Our views had become irrelevant; he was laying down terms."(44) This frank assessment of Le Duc Tho's remarks was one of the few totally accurate ones made by Kissinger.

According to a study of airpower during the Vietnam War by Mark Clodfelter, "from April through October 1972, 155,548 tons of bombs fell on North Vietnam."(45) "Linebacker seriously disrupted and damaged North Vietnam's lines of communication. Most significant, both sides claimed victory from (Hanoi's) Easter offensive. In the short term, the offensive was clearly a military defeat for the North Vietnamese and would cost General Giap his job. By spreading its troops across three disparate fronts and attacking in massive waves, the North Vietnamese were spread too thin and proved unable to pour enough firepower on any one point to win major territory. Many North Vietnamese units were annihilated and some battalions were reduced to 50 men. The North Vietnamese forces were vulnerable to air attack in logistical resupply areas for gasoline, tanks, trucks, and rail. Hanoi's casualties from its conventional offensive were staggering, losing over half of the 200,000 troops committed to the fighting. The Linebacker campaign also had made headway in reducing the supplies funneled along the Ho Chi Minh Trail, although the trail was not crucial because of the nature of the conventional attack. Although Hanoi never retained control over a provincial capital, the North Vietnamese did gain ground along the Cambodia and Laotian borders and the area south of the DMZ. Hanoi remained in control of this territory for the rest of the war, and in 1975 it would use this territory to launch a successful attack on Saigon. As with the war as a whole, then, the story for the North was one of massive sacrifices yet eventual success."(46)

The 'secret' talks in Paris resumed on July 19 and again on August 1, with the latter described by Kissinger as "the most interesting session we have ever had", but there seems to be no basis for his remark other than his own presentation of a "new twelve-point plan". After another meeting on August 14, Kissinger wrote Nixon

"They have an agonizing choice. They can make a deal with an Administration that will give them a fair chance to jockey for power in the South, but refuses to guarantee their victory. Or they can hold out, knowing that this course almost certainly means they will face the same Administration with a fresh four-year mandate that reflects the American people's refusal to cap ten years of sacrifice with ignominy. During the process we have gotten closer to a negotiated settlement than ever before; our negotiating record is becoming impeccable; and we still have a chance to make an honorable peace."(47) Kissinger's mastery of fantasy over fact was never better defined. The September 15 meeting in Paris was similarly distorted by Kissinger. "They were defensive; they professed eagerness to set the earliest possible deadline for an over-all settlement; and they have never been so eager to have early and frequent meetings. They repeatedly and almost plaintively asked how quickly we wished to settle and there was none of their usual bravado about how the U.S. and world opinion were stacked against them. For the first time in the history of these talks I sensed they were groping for their next move and the tack was devoid of an apparent, clear-cut strategy."(48) Could not it have been play-acting to see just how gullible Henry really was? Or that another trap was soon to follow?

Kissinger's sugar coating of his reports to Nixon regarding his negotiations in Paris with Le Duc Tho had not escaped the President. "...Nixon did not share Kissinger's faith in negotiations. Throughout 1972, when Kissinger would send back reports about the promising nuances of a new Hanoi offer, or, even worse, give an optimistic analysis of the atmosphere at a meeting or the food served, Nixon would fill the margin with question marks and sarcastic comments. "Nixon never really agreed with Kissinger that a diplomatic solution was possible" Kissinger's aide Peter Rodman recalled."(49)

Hanoi requested a two day meeting on September 26 and 27, "...the location was switched to the suburbs of Paris in a house that belonged to the painter Fernand Leger, a member of the French Communist party. Le Duc Tho first asked if the United States wanted to prolong the negotiation until after the (November) election or if it wanted to sign a comprehensive agreement on October 15. "Yet after eleven hours, there was no significant progress, and no agreements of any kind were reached."(50)

The next Paris meeting was scheduled for October 8, and Kissinger thought it would produce a breakthrough. "He flew there with Haig (now a four star general, having been promoted from a two star general only six months earlier) in tow because the general had a firsthand sense of what the traffic would bear in Saigon…and because I don't trust him behind my back anymore."(51) Also, while not using the word 'reparations', "Kissinger pledged the U.S. to provide aid to both Vietnams in order "to heal the wounds of war". (52)

"This was the accord Kissinger had been working for, and he accepted it subject only to the resolution of a few tiny details. Either out of arrogance or optimism, he did not make it clear to Le Duc Tho (and probably not to himself) that the whole package was contingent on President Thieu's approval. Amazingly, Kissinger did not send a copy of the peace accord he had just drafted – a document that would determine the fate of South Vietnam and its leadership – to Saigon. He did not even inform President Thieu about the basic provision of the deal, or even that a deal had been struck. Worse yet, he sent Bunker a cable from Paris on October 12 containing information for Thieu that was deliberately misleading: "My judgment at this juncture would be that they appear ready to accept a cease-fire in place in the near future."(53)

The North Vietnamese delegation sent a summary of the agreement to the Politburo in Hanoi" "We have achieved four objectives (end of U.S. Military involvement, cease-fire in SVN and NVN, withdrawal of U.S. force, recognition of the existence of two administrations, two armies, two zones of control, recognition of the democratic liberties in SVN – moreover U.S. acceptance of the responsibility for post war reconstruction."(54)

A few days later, Kissinger flew to Saigon to meet with Thieu, not knowing that Thieu had already received a ten-page document captured in Quang Tin province (in I Corps), a copy of the draft agreement that Kissinger had given to Le Duc Tho. Thieu had read the 'agreement' before Kissinger's arrival on October 19…and Thieu recalled that "I wanted to punch Kissinger in the mouth."(55) When Kissinger finally got around to passing a copy of the treaty, Thieu asked for a Vietnamese copy, with Kissinger replying that he did not have one. Eventually a copy in Vietnamese was passed to Thieu- who immediately recognized just how badly he and South Vietnam had been deceived. He and his close advisors noted 65 clarifications that needed to be addressed, 8 of which they considered very serious. Thieu said to Kissinger: "What does it matter to the United States to lose a small country like South Vietnam? We're scarcely more than a dot on the map of the world to you. If you want to give up the struggle, we will fight on alone until our resources are gone, and then we will die. The United States world policy dictates that you dance lightly with Moscow and Peking (sic), that you make different choices to follow your new strategies. But for us, the choice is between life and death. For us to put our signatures to an accord, which is tantamount to surrender would be accepting a death sentence, because without liberty is death. No, it's worse than death."(56) On October 23, Kissinger returned to seek Thieu's signature one last time. "He assured Thieu that there never have been talks or communications with the other side which have not been communicated to you. You have been apprised fully of every development as it has occurred and have been consulted on every move we have made with the single exception of the meeting on September 15 when we believed it necessary to move before we had heard from you. You must understand how it seems to an individual who has stood against 300,000 demonstrators, against bureaucratic and Congressional opposition, against public opinion and the press – this is why I took the liberty of speaking as I did."(57) Kissinger then repeated his request that Thieu not reveal to the Media the content and tone of their discussions and disagreements.

Kissinger's attempts to cover his own perfidiousness could not have been more obvious or shallow, but the pliant Media did not question or expose him.

At an October 26 press conference on the Vietnam negotiations, Kissinger concluded his presentation with ..."We believe that peace is at hand. We believe than an agreement is within sight, based upon the May 8 proposal of the President and some adaptation of our January 25th proposal, which is just to all parties."(58) On October 25, Kissinger had met with Max Frankel of the New York Times who in turn wrote a front page story to the effect that barring a supreme act of folly in Saigon or Hanoi a cease-fire was at hand. Frankel's story was followed by James Reston in an article entitled "The End of the Tunnel" and Newsweek described Kissinger's press conference candid and brilliant. Meanwhile, Le Duc Tho's concession regarding the lack of a guarantee that civilian prisoners would be released and that Thieu would be allowed to remain in control in Saigon was formally protested by the NLF to Lao Dong officials. Le Duc Tho and Huan Thuy and "the northerners assured Madame Binh (and the NLF) that the freedom of American POWs was tied to the liberation of civilians in the south and Thieu would be only a figurehead until the elections were held."(59

On November 20 Kissinger, Haig and Le Duc Tho met again in Paris, but after four days of talks both sides recognized that there was to be no agreement, primarily because Thieu was, understandably, unwilling to sign a settlement allowing North Vietnam forces to remain in South Vietnam.

"Nixon wrote in his diary on November 20 that Kissinger was more concerned with establishing a good record with Thieu, whereas he (Nixon) was more interested in saving South Vietnam. "As I told Henry when he began to rumble around to the effect that we have a very good record in this instance, I said, Henry, we're not concerned about being right on the record. What we are concerned about is to save South Vietnam"(60) And as Thieu remarked to Kissinger after reading the October "agreement" that Kissinger had reached with Le Duc Tho, "Why are you rushing to get the Nobel Prize?"(61) Le Duc Tho saw it another way. After Kissinger presented Thieu's 64 proposed text changes, Le Duc Tho remarked "We have been deceived by the French, the Japanese and the Americans, but the deception has never been so flagrant as now. You told us this (was a done deal) and you swallowed your words. What kind of a person must we think you to be?"(62) "The November talks ended as they had begun – stalemate. The December 12 meeting was merely a repeat. The 12 day heavy bombing by B-52s of Hanoi and Haiphong began on December 18 and continued until December 29, the sole exception being December 25. Fifteen B-52s were lost, as were 13 U.S. tactical aircraft."(63) The New York Times and the Washington Post, previously strong supporters of Kissinger, expressed outrage at the bombing.

"A point-by-point comparison of Hanoi's "ten-point program" of 1969 with the 1973 agreement shows that they are largely identical, even in wording. But there is one significant difference: the political provision that the Thieu government had to be replaced by a communist-approved coalition before there could be a cease-fire. Until October 1972, Hanoi insisted on this. In addition, the "Vietnamization" program meant that by 1973 Saigon's army was more able to defend itself than it was in 1969. But was it worth four more years of war in order to get a cease-fire that allowed Thieu to retain authority in Saigon? The deal would turn out to be costly: an additional 20,552 Americans dead, the near unraveling of America's social fabric, a breakdown in respect for government authority, the poisoning of America's reputation abroad and the spread of war to Cambodia and Laos."(64)

Amazingly, Nixon and Kissinger continued their secretive ways. They made secret pledges to Thieu: "You have my absolute assurance that if Hanoi refuses to abide by the terms of this agreement it is my intention to take swift and retaliatory action", read a Kissinger drafted letter from Nixon to Thieu on November 14. Another sent on January 5 pledged: "We will respond with full force should the settlement be violated by North Vietnam." And "These commitments were what finally convinced me to sign, Thieu later said."(65)

"On January 23 (1973), Le Duc Tho and Xuan Thuy met Kissinger (in Paris) with all their experts. The atmosphere was full of enthusiasm. Kissinger asked Le Duc Tho about the place for the return of the United States prisoners of war. Tho told him the United States pilots captured in North Vietnam would be handed over to the United States in Hanoi; those captured in South Vietnam would also be returned in North Vietnam, but for those captured in Laos, Vietnam would exchange views with its friends in Laos."(66)

President Nixon then declared ..."the treaty had achieved America's objectives of "peace with honor" and that the accord had "the full support" of South Vietnam's President Thieu. Nixon also told the American public, "Now that we have achieved an honorable agreement, let us be proud that America did not settle for a peace that would have betrayed our allies, that would have abandoned our prisoners of war, or that would have ended the war for us, but would have continued the war for the 50 million people of Indochina."(67) Astonishing! Had Nixon read the full "accord", or was he basing such statements on Kissinger's summary of it? Or was this just more political whitewash in action?

The Media was ecstatic and heaped extensive praise on both Nixon and Kissinger. In answer to a question asked during the self-congratulations, Kissinger confided to Ehrlichman that he estimated South Vietnam would last a year and a half, if lucky. He was not asked if this would qualify as the 'decent interval' he had earlier established as his goal.

In 1938, British Prime Minister Chamberlain returned from Munich to #10 Downing Street in London and decided to address the throng gathered outside. Mrs. Chamberlain told him to say that "you have brought back peace, but not just peace – peace with honor."(68) And so thirty five years later we again heard "Peace with Honor"being declared, with the 1973 version being as flawed as the 1938 statement by Chamberlain.

On September 22, 1973, Kissinger became Secretary of State, replacing William Rogers. Kissinger also continued his role as head of the National Security Council. On December 6, 1973, Gerald Ford became Vice-President, following Spiro Agnew's resignation. Nixon resigned as President on August 9, 1974 and Ford was sworn in as President at noon the same day. On April 30, 1975 the last Americans left Saigon by helicopter from the Embassy roof as North Vietnamese forces entered Saigon. The Media failed to mention or banner headlines reflecting Nixon's or Kissinger's "Peace With Honor".

Kissinger continued his "shuttle diplomacy" efforts throughout the remainder of the Ford years with varying degrees of highs and lows. He also continued to earn both praise and disdain for his propensity to engage in most, if not all forms, of deception. World leaders found him to be less than truthful and not trustworthy, but seemed willing to accept him because of his high position, coupled with his ingratiating ways. And so did the Media, which continued to woo him with flattery. His eagerness for favorable press play was all but legendary. Most of the Media – and the majority in Hollywood - were equally eager to provide him with the acclaim and approval he sought. He was still a powerful world figure, and accommodating to all who paid him public homage.

Within the Media, Kissinger had favorites and knew how to play them. Katherine Graham, Joseph Alsop, Joseph Kraft, Marilyn

Berger (Washington Post), Hugh Sidey (of Time and Life), Barbara Walters, Walter Lippmann, James Reston, Rowland Evans, Walter Cronkite, David Brinkley, Tom Braden, Marvin Kalb (CBS), Ted Koppel (ABC), Richard Valeriani (NBC), Max Frankel and James Reston (NY Times), Henry Grumwald (Time), etc., etc., were all but drooling in their praise. Kissinger was the conductor, and they were accomplished members of his musical ensemble.

The Soviet Union and China remained on center stage with Kissinger (and Ford), as did the Mid-East. However, one subject that would not go away was Vietnam. "All that the U.S. had left to show for the 58,022 dead were the shreds of credibility that came from having achieved a peace agreement that lasted long enough to disguise the American pullout. Neither the "decent interval" nor peace nor the honor that Kissinger had claimed in January 1973 proved to be long lasting. Kissinger now claims he felt that the final loss in Vietnam in 1975 represented a blow to America's credibility that in turn sapped the force of its threats and commitments around the world." "By our self-indulgence, we damaged the fabric of freedom everywhere. The surrender in Indochina ushered in a period of American humiliation that stretched from Angola to Ethiopia to Iran to Afghanistan."(69) Indeed, this was an accurate assessment by Kissinger, but, unfortunately, made only in hindsight.

Following the July 1971 meeting in Paris with Le Duc Tho, Kissinger reviewed with Nixon the status of negotiations: "A superb public relations record of genuine willingness to compromise difference and to let the South Vietnamese people decide their future freely. We have conceded everything remotely reasonable, short of a coup against Thieu – neutrality, limitation on military aid, a withdrawal deadline, *a large economic aid program* (emphasis added). Also, a willingness to take steps and make efforts greater than those demanded by our domestic opposition. A commitment by the other side stated even more clearly today by Le Duc Tho to release our prisoners in exchange for a date. Though this is not enough today we can return to it in the fall. I told them we were prepared in the next 5 years to *provide* $7.5 *billion* (emphasis added) in aid to Indochina, of which $2-2.5 billion could go to North Vietnam."(70)

In January 1972, Nixon told Zhou Enlai in Beijing: "We have

offered to withdraw all Americans, with "no tail" left behind – to use the Prime Minister's expression – and to have a cease-fire throughout Indochina provided we get our prisoner back. Nixon stressed that "the U.S. is prepared, in conclusion, to provide a very heavy economic assistance to Cambodia, Laos, and North Vietnam for *rehabilitation* (emphasis added) and to South Vietnam in the event a settlement is made. We don't want to leave a tail behind. We don't want bases. We would accept a neutralized area. On the other hand, it takes two to make a deal."(71)

In August 1972, Kissinger told Thieu that "I think they are keeping the prisoners as blackmail; in fact, I prefer that they don't return the prisoners of war and that there be no cease-fire before the election."(72) Is this part of his negotiating "plan", or what? Is Kissinger trying to say the U.S. is not concerned with battlefield casualties, or the prisoners, because we are primarily focused on the elections?

On October 9, Kissinger again met with Le Duct Tho. "It had taken over 18 months of talks and tens of thousands of lives to reach a point that might have been conceded from the beginning. The U.S. had neither the ability nor the will to force all Northern troops out of the South. Nor did Kissinger care all that much about the fate of the weak and myopic government of the South. He was there to get agreement on terms for a political battle between the Vietnamese, but the battlefield would be an uneven one, and the terms in contention were Le Du Tho's....(Kissinger) made it clear that no document would be signed in which the United States took responsibility for war *reparations*." ...but was willing to participate in a large-scale *reconstruction* plan throughout Indochina and to write this intention in a unilateral statement (emphasis added)."(73)

In response Le Duc Tho said "...to rapidly end the war and restore peace, to show our good will, we are proposing something new today both in content and in the practical and simple way to conduct the negotiations: an Agreement on Ending the War and Restoring Peace in Vietnam that would settle military questions involving troops withdrawal and the release of POWS, a cease-fire in Vietnam with international supervision and control and, what Le Duc Tho called healing *"the wounds of war and to rehabilitate the economy in North Vietnam.*(emphasis added). After the signing of this agreement, the cease-fire will become effective at once. (he added) "When the cease-fire is installed, all of these persons (cadres and workers of the PRG and NLF imprisoned in South Vietnam) must be released."..."Eventually he would be betrayed by Kissinger on this point."(74)

Kissinger and Le Duc Tho set October 31 as the date the treaty would be signed. "The Vietnamese delegation sent a letter of self-congratulations to the Politburo in Hanoi: We have achieved four objectives (end of U.S. military involvement, cease-fire in Vietnam, (and) withdrawal of U.S. forces; recognition of the existence of two administrations, two armies, two zones of control; recognition of the democratic liberties in SVN; – moreover U.S. acceptance of the *responsibility for post war reconstruction* (emphasis added)."(75)

Not all of this was well received in Hanoi. "The NLF formally protested to Lao Dong officials. Madame Binh met personally with Le Duc Tho and Huan Thuy, and "the northerners assured Madame Binh that the *freedom of American POWs* (emphasis added) was tied to the liberation of civilians in the south...and that Thieu would be only a figurehead until elections were held."(76) In early December 1972, "Kissinger cabled Nixon that the North Vietnamese were more intransigent that ever before. The NLF had been pressing publicly and privately for linking release of American POWs with political detainees, and Tho was beginning to raise the point. At a press conference on December 8, Madame Binh declared "that there is no reason we should free American prisoners while our compatriots remain in jail."(77)

Negotiations resumed on January 8 1973 in Paris (at the house in Gif-sur-Yvette) and continued into January 13. Le Duc Tho assured Kissinger that American prisoners in Vietnam, both military and civilian, were covered by Article 8a, and Kissinger accepted that there was no need for a formal understanding. However, Le Duc Tho suggested "that you should consider the question of political detainees in a very positive and in a very good way."(78) The final meeting was at Saint-Nom-La-Breteche...and reportedly covered the U.S. contribution to the reconstruction of North Vietnam plus two 'understandings' on the political prisoners being held in South Vietnam.

The final group of American POWs was released in Hanoi on

March 29, 1973, but "Le Duc Tho's warning to Kissinger on January 23, 1973, would prove prophetic: "Article 8c will be following you for a long time after the cease-fire."(79) (Article 8c of the Paris agreement stipulated that the two South Vietnamese parties would do their best to resolve the question of the return of the Vietnamese civilian (e.g., Viet Cong) personnel captured and detained in South Vietnam within 90 days after the cease fire. That Le Duc Tho singled out this particular article is of more than passing curiosity. A sister protocol further stipulated that each party would return all captured persons without denying or delaying their return for any reason. In my mind this highlights that Le Duc Tho strongly suspected, or knew, that Thieu intended to use the prisoners from the north as a bargaining chip, and in turn was letting Kissinger know that the north itself had bargaining chips that could be played, i.e., American POWs, and the U.S. commitments to provide reparations. That the North Vietnamese had all of Kissinger's promises, commitments, and concessions on tape (audio at least, if not both audio and video) throughout the secret "negotiating" sessions has to be assumed, for these meetings were held in houses or apartments in Paris controlled by the North Vietnamese. The opportunity was tailored made for them; not to have done so would have been incredibly foolish and unprofessional. The North Vietnamese representatives were neither.

Kissinger and Haig had earlier suggested to Thieu to hold prisoners as hostages to bargain for the removal of North Vietnamese troops from South Vietnam. "Indeed, Thieu's government reclassified many military prisoners as political detainees so as not to have to release them."... "Kissinger cabled Bunker (in Saigon) on February 24 that "the North Vietnamese have again complained to us on the question of civilian prisoners. They recall my promise during my visit to Hanoi that 5000 civilians would be released shortly, and note that not a single civilian has yet been returned by the GVN"...Bunker responded "I think we may find some reluctance on Thieu's part to granting amnesty to civilian prisoners before he has had an opportunity to bargain on the presence of NVA troops in South Vietnam. You will recall that we suggested to him that this was one of the ploys he might use."(80)

Thus the subject of 'Reparations', or the semantic equivalent, and that of the release of POWs held in all of Indochina – including U.S. POWs - were, inevitably, to meet head on. The American POW portion of the story remains particularly egregious and shameful.

"Garnett "Bill" Bell, Vietnam POW expert and then chief of the Pentagon's POW/MIA office in Hanoi, gave testimony in November 1991 to the U.S. Senate's Select Committee on POW/ MIA Affairs. In answer to the question posed by the Senate: "Was there hard evidence that any of the approximately 2,200 U.S. POW/ MIAs in Southeast Asia had been kept behind by the communists? "Yes", he said, becoming the first serving Pentagon official in memory to give that response. Bell said it was clear that at least 10 Americans had been kept by the Vietnamese."(81) Sadly, this was but a small part of the story of POWS being left behind.

"Alexander Solzhenitsyn, famed Russian witness to the communist prison system, put it this way: If the government of North Vietnam has difficulty explaining to you what happened to your brothers, your American POWs who have not yet been returned, I can explain this quite clearly on the basis of my own experience in the Gulag Archipelago. There is a law in the Archipelago that those who have been treated the most harshly and who have withstood the most bravely – who are the most honest, the most courageous, the most unbending – never again come out into the world. They are never again shown to the world because they will tell tales that the human mind can barely accept. Some of your returned POWs told you that they were tortured. This means that those who have remained were tortured even more, but did not yield an inch. These are your bravest people. These are your foremost heroes who, in a solitary combat, have stood the test."(82)

In the book "The Men We left Behind", there are dozens of tantalizing stories of multiple sightings of American POWs in Vietnam and Laos, and lists compiled of some 150 - 350 prisoners- or more - who were believed to have survived the war in captivity, but who were not released. There is considerable evidence of attempted cover-up by DIA personnel, the Pentagon, Congressional investigative committees, including Senators Kerry and McCain, Representative Sonny Montgomery, and the earlier 1977 Woodcock Commission, as well as plain disinterest on the part of members of previous administrations. Included are references to Kissinger's faulty recall in the early 1990s as to what commitments had been made. For example, "We were very careful throughout the course of the negotiation to ensure that there was never a linkage between actual release of our POWs and the actual delivery of reconstruction aid. – Henry Kissinger, November 2, 1992."(83) My, my - does that pass the smell test? Sounds more like Clinton's selected parsing of testimony!

"While the Montgomery Committee was in Hanoi (January 1976), Deputy Foreign Minister Phan Hien linked the POW/MIA issue to the Nixon offer of 3.25 billion in aid. On March 29, 1976, the DRV issued a press statement out of its Paris Embassy, demanding "that the United States fulfill its obligations under the Paris Agreement to 'contribute to healing the wounds of war' and declares itself ready for its part to undertake the necessary measures to look for Americans living in Vietnam.(84) In July 1976 in Manila Airport, Phan Hien said "The position taken by the Americans is the Paris Agreement no longer exists. That is why they refuse to implement Article 21 concerning the contribution to the healing of the wounds of war though the fact is that it was President Nixon himself [who] sent a letter to our prime minister. In this letter he committed that the U.S. would contribute $3,325,000* within the period of five years. But they are insisting in asking us to implement Article 8b (on POWs) while they are refusing to implement Article 2 (money) of the same agreement. I have never seen anything like this in international law and international practice." (85) This same book has many references to the North Vietnamese specifically seeking and citing Nixon/Kissinger commitments to provide funds to "heal the wounds of war", and disavowals by Kissinger, Ford, and Carter.

One of the more striking points the book raises is the question of why there were so few seriously injured prisoners returned, and especially why amputees were not returned, despite the North Vietnamese having more than adequate facilities to handle such cases. The Hanoi Hilton was the main warehouse for American POWs, but a dozen other prisoner camps in North Vietnam were identified by various sources, but they were not "investigated". These questions have never been answered satisfactorily, but neither has the Media asked. Why?

While I was in Laos (June 1973 – June 1976), the office received incomplete reports of American POW camps in eastern and northern Laos, including both military and civilian prisoners. Using the limited paramilitary resources at our disposal, we dutifully tried to track down these camps, but were not successful. Some, we were told, had "moved", or the "prisoners had been turned over to North Vietnamese", or simply the reported camp sites could not be "located". Photographs of some of the American POWs, including civilian air crews, showing them alive and well were included with some reports, but none of these persons were included in the groups that were released in Hanoi. I remember that (Col.) Bo Gritz later had copies of some of the same photographs that we had much earlier in Vientiane when he claimed to have "nearly" rescued two or three U.S. POWs in southeastern Laos, but failed when fired upon by Pathet Lao as the "rescuers" headed for Thailand across the Mekong River. Other than the old photographs he displayed, little else of his story appeared to be plausible or factual.

Our Media apparently did not judge the subject as worthy or opportune, for there was little pursuit of either the POW issue or "Vietnamese reparations". CBS, PBS, Time and the NY Times did not pursue them either. The Los Angeles Times did try once in October 1991. Ted Landreth, with CBS for years, concluded "The Media simply let us down on this one."

During the September 1991 Senate Select Committee on POW/MIA hearing, Sen. Bob Smith (R-NH) was the one senator who pushed the hardest on the question of POWs. "He challenged DIA on 14 of their 18 reports, including the information provided by Marine Robert Garwood, captured in South Vietnam in 1965, following his release to a visiting Finnish diplomat in Hanoi in 1978. D.I.A.'s bumbling, if not bizarre, responses to Smith's questions were not pursued by the Media, nor were subsequent, and often puzzling, efforts by some members of that committee to hinder gathering more specific answers. The questionable and unusual 'political' actions and remarks by Senators Kerry and McCain, both members of the committee, remain less than clear. The general topic was "old news" to the Senators, but not to the families of the POWs/MIAs; to them it was an ongoing tragedy. Yet it was largely ignored by the Media - and the politicians/bureaucrats. After all, think of the many red faces had a 'live' prisoner or two – or more - returned from Vietnam in the 1980s or 1990s. But Sen. Kerry pretty well took care of that possibility: "On 9 April 1992, the Chairman of the Senate Select Committee, Senator John Kerry of Massachusetts, in response to a protest by other members of the Select Committee, told the Select Committee members that 'all copies' (of the POW briefings) would be destroyed, "along with computer files of the briefing", McCreary (the committee's chief counsel) wrote on April 27 and May 3, 1992. The original briefing and 14 copies were reportedly shredded on April 10th."(86)

Further, "On July 14-15, 1992, Kerry's staff director traveled to Vietnam with Pentagon officials, who later reported on her meetings with Vietnamese officials. It was during a talk with Le Bang, Hanoi's top expert on the U.S., that Zwenig made a stunning statement concerning Kerry's expectations. "She said Senator Kerry believes the General Vessey cases (of 135 Americans who "were last known alive", or fell into the category of those for whom Vietnam should be able to account) can be resolved by the recovery and identification of *remains*, (emphasis added) through records, from witnesses of deaths, or some combination of these. She said that Senator Kerry believes SRV explanations of deaths could be based upon policies and inadequate records in the south", the Pentagon representative reported in a cable to the U.S."(87) Could Hanoi fail to get this less than subtle 'message'?

The fifteen month so-called "investigation" into the POW/MIA question finished in early 1993. "Even committee staffers were surprised, and disgusted, at the final report." "A third staffer was bitter about the report: "It's such an act of national betrayal. They made a promise they were going to try to get to the bottom of this and they didn't even try. Kerry and McCain clearly had their own agendas and misrepresented what the data showed. Did they accurately quote from the documents they had in front of them? No. When they talked about issues did they follow them through to completion to the extent that they brought out in the public all the things that we knew about the issues? Absolutely not, in no case, unless it served their own agenda, which was, I suppose, to kill this issue."…"A fourth Senate staffer, a highly experienced investigator, said: "It (the committee) had a shot at really doing something. And the chairman (Kerry) prevented it from happening. He blocked it every step of the way. It was the most frustrating and unbelievable thing I ever saw in my life."(88) Two other senate staff members are mentioned as also expressing dismay at the report.

To capsule key points in the shocking story told in *"The Men We Left Behind"*:

• "In 1983, National Security Adviser Robert "Bud" McFarlane assigned one of his top deputies to investigate how well the intelligence community was doing in its search for U.S. POW/MIAs. The intelligence expert was horrified at what he found. People working that issue could not find a POW if a POW dropped on their head."(89)

• "September, 1985: Admiral Thomas Brooks, a former DIA official states: "1. Follow-up actions have not been pursued. 2. A mindset to debunk has resulted in sloppy analysis. 3. The analysts do not have an open mind."(90)

• March, 1986: Colonel Kimball Gaines, a former DIA official, states: "1. the database is a wasteland. 2. There are unhealthy attitudes. 3. There is no coherent or disciplined collection effort. 4. There is not enough analysis. 5. There is almost a total lack of management. 6. There is intense effort initially focused on veracity of sources with a view to discrediting them. 7. The bias toward discrediting sources has resulted in reduced objectivity in the analysis."(91)

• September, 1986: General Eugene Tighe, former Director of DIA in his report states: "1. There is evidence that Americans remain alive in Vietnamese custody against their will, even in the limited sample of reports we reviewed. 2. An unresolved case, according to DIA, means we 'haven't yet proved these reports are false.' 3. The greatest problem associated with the POW/MIA issue is the lack of professional analysis of the available intelligence. 4. There is little evidence that any significant analysis of the intelligence received has been accomplished in recent years. 5. The routine approach to each source has been 'he's probably wrong'. 7. There is a dominant emphasis on the negative. 8. National technical means have never been levied proactively in an attempt to locate POWs. 9. The POW office only pursues leads that will help discredit the source. 10. The credibility of the DIA's judgment that a refugee account is a fabrication must be considered low. 11. DIA's use of the polygraph is wrong; it is not a substitute for analysis, and DIA does not seem to understand the frailties of the test. 12. We doubt every resolution of a case labeled as fabrication – we find most of this judgment without meaning, and untrustworthy. 13. There is an enormous volume of unanalyzed hearsay reports. Hearsay reports are impeached by DIA from the start – this is an irresponsible distortion of the intelligence process."(92)

• "The September 10, 1992, Defense Intelligence Agency report began: "Since the U.S. withdrawal from Vietnam in 1975, at least 18 sources, firsthand and hearsay, have provided DIA with details about an underground POW detention facility or facilities in Hanoi. Senator Bob Smith's, (R-NH) line of questioning during the Senate Select Committee on POW/MIA hearing on August 4th had backed the DIA into a corner. When the (September 10) hearings opened, Smith immediately went on the offensive: Sources reporting over a 15-year period have filled in details concerning the construction and operation of this underground U.S. POW detention facility as the Citadel next to Ho Chi Minh's Tomb. I will now summarize what the sources of these reports stated. And I ask my colleagues and the American people to listen for the common denominators – that is: locations, descriptions, and the type of people involved – listen for the patterns, and then you decide at the end, if you accept the official U.S. government position. I quote – "There is no evidence to sustain a belief that U.S. POWs were detained after Operation Homecoming in this area" You decide if you believe that statement. Senator Smith then proceeded to present a summary of 14 of the 18 DIA reports."(93)

• "The Defense Nuclear Agency debunked DIA's claim that the water table in the Hanoi area would not have allowed the building of such an underground structure, but a follow-up investigation by an investigative team never went to Hanoi."(94)

• "On October 17th, 1992, Presidential POW emissary John Vessey Jr. and Senator John McCain (R-AZ) jetted off to Hanoi for the "POW October Surprise" of the 1992 Presidential campaign. The men returned to Washington with promises that Vietnam had finally decided to release the mother-lode of POW/MIA information, including a huge storehouse of photos showing POWs and crash sites. Both men downplayed the fact that Vietnam had for years been denying having such files, and even claimed their records had been "eaten by termites." "The Bush (#41) loyalists also brushed past reports that the Vietnamese admitted possessing the files only after threats of Defense Secretary Dick Cheney that if Hanoi didn't admit having the files, the U.S. would break the news."(95) Unfortunately, the previously much touted photos and documents apparently proved to be of little value.

• "In early 1992, Dr. Steve Morris, a visiting professor at Harvard University, found a document in a Moscow archive that would stretch DIA's debunking skills to the maximum. It was a six page "Report of the Deputy Chief of the General Staff of the Vietnamese Peoples Army, presented by General Tran Van Quang "at a meeting of the Politburo of the Central Committee of the TV (Central Committee of the Workers party of Vietnam) on September 15, 1972. – According to the general, "The total number of American POWs captured to date on the fronts of Indochina, in other words in North Vietnam, South Vietnam, Laos and Cambodia totals 1,205 people. Of this number, 671 people were captured in North Vietnam and 143 pilots (were captured) in South Vietnam. "Sixty-five Americans had been captured in Cambodia, 43 in Laos, and 183 non-pilots in South Vietnam, for a total of 1,205. In an internal report, the DIA conceded: "It is completely plausible that a person of his (Tran Van Quang) distinguished command background, and eminent political standing, would be the person who could offer a political thesis to the politburo which involved political arguments advocated for the toughest deal to be made with the American negotiators."(96)

• "Vietnamese defector Dr. Dang Tan told the U.S. in 1971 that Hanoi's statements that it then held fewer than 400 U.S. POWs were "untrue". According to declassified CIA records, Tan reported "that already by mid-1967, when I departed NVN, over 800 American POWs were then in captivity in NVN."(97)"

• In 1978 a communist North Vietnamese intelligence analyst by the name of Le Dinh defected to the West. "Le Dinh heard at staff meetings that about 700 Americans still remained in Vietnam (after 'Homecoming'). The information was attributed to remarks by senior officers to the effect that the SRV retained a 'strategic asset' of over 700 American prisoners that could be used to force the U.S. to pay reparations," says the DIA's February 28, 1980 final debriefing report."(98)

• "Perhaps it was just a coincidence, but soon after the DIA found out about the Soviet POW report, Pentagon POW/MIA officials quietly ran thousands of POW files through a refrigerator-sized shredder at the U.S. Embassy in Bangkok, the military later admitted. By the time the U.S. Ambassador to Thailand learned of the shredding, which took place during March 25 through 27, 1993, it was too late. The original files of the Joint Casualty Resolution Center, the agency which looked for U.S. POWs during the war, were gone."(99)

• In March 1991, (Senator) Smith "began pushing for the Senate to establish a Select Committee to investigate the POW issue. Such a committee would respect no political sacred cows, employing a large staff and subpoena power to answer the most sensitive questions about the history of U.S. POWs from World War II to Vietnam. Senate Resolution 82 was a good idea, but despite Smith's fierce dedication, the proposal was going nowhere. Then, a hoax set the issue afire. A picture purporting to show three U.S. POWs alive in captivity years after the Vietnam War, dubbed the "Three Amigos" by the DIA, produced a media feeding frenzy on the POW issue. Into the sudden media spotlight butted the immaculately coiffed head of Senator John Forbes Kerry (D-Mass.), a man whose visage and ambitions bore no small resemblance to President John Kennedy. But Kerry's profile was not associated with courage by POW family members and activists. While he was a Vietnam vet, people active in the POW issue were deeply suspicious of Kerry, who had been a leader of the anti-war movement. During the spring of 1991, Kerry had pronounced that "It (the POW issue) has got to be put to bed – for the sake of those families, for the sake of this country and our politics, and in order to move on". He had then come out against the formation of a select committee."(100)

• "Kerry joined Smith in helping push the bill establishing a POW committee. By early August 1991, the Senate had voted to form the Select Committee on POW/MIA Affairs, and Kerry had been designated its chairman. One of John Kerry's first moves was to visit Vietnam, taking along his daughters for an eight-day trip in August which also included stops in Thailand and Cambodia. It is not known exactly what Kerry told Hanoi about his Senate committee. But after the trip, Kerry had dinner with Assistant Secretary of State Richard Solomon. According to a previously classified State Department report, Kerry told the State Department that "relaxing the (trade) embargo would increase U.S. leverage on the SRV (Vietnam) and improve cooperation on the POW/MIA matters." (101)

• "Kerry was pushing a soft line on Vietnam even before his views on where the POW issue should be heading – towards normalized relations with Hanoi. Smith, on the other hand, was not about to support Vietnam until it began real cooperation in accounting for live American POW/MIAs. On August 29, 1991, Kerry had his long-time aide Frances Zwenig contact the Bush (#41) Administration State Department. Zwenig asked for a favor. She explained that Kerry wanted to ensure that Bob Smith did not become co-chairman of the committee. Kerry's choice was Sen. John McCain (R-Ariz.), who, despite his experience as a Vietnam POW often served as an apologist for Republican inaction on the prisoner issue. McCain too had proposed softening the U.S. line on Hanoi, and he had also passed legislation, dubbed "the McCain Amendment," that appeared to open POW/MIA records but actually made it virtually impossible for the public to receive information on the fate of specific POW/MIAs. And in May 1991, "McCain referred repeatedly to the need to find ways to defuse the POW/MIA issue with the American public," the State Department reported. Perhaps most importantly, McCain was not regarded as the smartest member of the Senate, and had limited credibility on Capitol Hill because of his major involvement in the "Keating Five" scandal, in which he received perks linked to his support of corrupt savings and loan executives."(102) Zwenig's efforts on behalf of Kerry failed and Smith was made co-chairman. "Despite her failure at back-room political fixing, Zwenig was rewarded with a job as staff director of the Senate POW Committee, a role for which she had few qualifications other than loyalty to Kerry. A lawyer and former Peace Corps volunteer, Zwenig was unfamiliar with many military and intelligence topics vital to an understanding of the POW issue and the ability to penetrate stonewalling by the Pentagon and the intelligence agencies. Her role in the investigation would later spark bitter resentment among military and intelligence professional assigned to the committee staff."(103) Interestingly, Zwenig reportedly was rewarded for her work with Kerry on the Senate Committee by the Clinton Administration, which made her chief-of-staff for the U.S. Ambassador to the United Nations.

• "The (DIA/DOD) briefing was simply too politically hot for the Senate to handle. One investigator said: "Kerry got scared when he found out what the hell was going on, because this was not his agenda. His agenda was to open ties with Vietnam. I think he was stunned shitless on the 9th of April when we briefed that the intelligence indicated that there were guys alive through 1989."(104)

• "On April 9 1992, the Chairman of the Senate Select Committee, Senator John Kerry of Massachusetts, in response to a protest by other members of the Select Committee, told the Select Committee members that 'all copies' of the briefing. Would be destroyed, McCreary (John McCreary, intelligence officer with DIA) wrote in April 27 and May 3, 1992, Senate memos. The original briefing and 14 copies were reportedly shredded on April 10th"."When staffers later questioned the legality of destroying the records, according to McCreary, the committee's chief counsel, William Codinha, made light of the issue. "Mr. Codinha replied "Who's the injured party?" He was told, "The 2,494 families of the unaccounted for US servicemen, among others.' Mr. Codinha then said, "Who's gonna tell them? It's classified", recounted one of McCreary's memos."(105)

• "By the end of September's (1992) hearings on the Paris Peace Accord, and Kissinger's role, was still hotly debated. But a new consensus had developed that top U.S. officials at least suspected that U.S. POWs were being retained by the communists in 1973. Former CIA chief and Secretary of Defense James Schlesinger, supported by other Nixon-ear officials, told the committee: "I think that, as of now, that I can come to no other conclusion, Senator. That does not say that there are any alive today, mind you. But in 1973, some were left behind."(106)

• The report was released in January, 1993. "But instead of making clear findings, the Senate succumbed to mealy-mouthed "bipartisanship." As George Orwell wrote: "The great enemy of clear language is insincerity. "And perhaps no one should have been surprised that so little clear language appeared in the Senate's 1,223 page final report." "Concluded the Senate: "We acknowledge that there is no proof that the U.S. POWs survived, but neither is there proof that all those who did not return had died. There is evidence that indicates the possibility of survival, at least for a small number, after Operation Homecoming." "Even committee staffers were surprised, and disgusted, at the final report." Comments included "Kerry and McCain clearly had their own agendas and misrepresented what the data showed...Did they accurately quote from the documents they had in front of them? No."(107) "A fourth Senate staffer, a highly experienced investigator, said: "It [the committee} had a shot at really doing something. And the chairman [Kerry] prevented it from happening. He blocked it every step of the way. It was the most frustrating and unbelievable thing I ever saw in my life."(108)

- Let's fast forward to December 23, 2003 and an article by David Rogers in *The'Wall Street Journal* on presidential candidate John F. Kerry. Asked "What is your proudest accomplishment in politics? Kerry answered: "Working with John McCain on the POW/MIA investigation, to lift the embargo, and on the normalization of relations – to make peace with Vietnam."

There are many more disturbing details on the MIA/POW Senate Committee investigation, almost more than one can stomach. The arrogance, the politics involved, and disregard for the truth is spelled out in detail. That the Media ignored or failed to seize on this sorry charade is especially inexplicable and bothersome. Any investigative reporter worthy of the name could have had a career-high story, but none seized the opportunity, or focused on the deceits. How could the Media not ask questions? Or not expose the sham? Or not question why there had been earlier hearings? Or why and how had DIA had failed the country for so many years? Or why Kissinger had been so deceptive about reparations and insensitive regarding the POW/MIA questions? Or not question Kerry and McCain for their strange actions on the POW/MIA Committee? Can the Media justify its failure? Can Kissinger, McCain, and Kerry explain this web? They can, but won't.

Most of the Media barely made a ripple when Kissinger was named in 2003 to chair a panel formed to investigate "intelligence failures". When told he would be expected to identify his business relationships, Kissinger withdrew his name. Did the Media then dig into the 'associations' requirement that upset him? Not exactly...

"Kissinger Associates" and other Kissinger "for pay" activities require pages to relate, but the following is a partial list of major clients that had contracts or project arrangements with his firm in the (early) 1990s:

American Express and its subsidiary, Shearson Lehman Hutton
American International Group (insurance underwriter)
Anheuser-Busch
ASEA Brown Boveri, a Swedish Manufacturing firm.
Atlantic Richfield, the oil company

Banca Nazionale del Lavoro (BNL), a Rome Bank that made illegal loans to Iraq.

Bell Telephone Manufacturing of Belgium
The Chase Manhattan Bank

The Coca-Cola Company

Continental Grain (a privately held grain company.)

Daewoo, a Korean trading and construction conglomerate.

Ericsson, a Swedish telecommunications manufacturer.

Fiat, the Italian automobile company.

Fluor, a global engineering and construction company

Freeport-McMoRan, an oil, gas, and mineral company

GTE, the Connecticut-based telecommunications company

H.J. Heinz, the food-product conglomerate

Hollinger, Inc. a Toronto-based global newspaper company

Hunt Oil Co., a Texas-based firm

Merck and Co., the pharmaceutical company

Skandinaviska Enskilda Banken, a Stockholm-based bank, Trust Company of the West, (an investment management firm).

Union Carbide, the chemical and manufacturing conglomerate.

Volvo, the Swedish automobile company.

S.G. Warburg, the British investment bank.(109)

Interests also include Mexico, Turkey, the Ivory Coast, China, Zimbabwe, Japan. Indonesia, Europe, Venezuela, Chile, Seoul, Peru. Singapore, Malaysia, Panama, to name a few. He has made at least 15 trips to China on behalf of clients doing business there. In addition, he serves on various boards for which he is paid and receives substantial fees for speaking engagements, including his 'political expertise' appearances on various television networks. In the 1980s, Kissinger brought former Under Secretary Eagleburger into his "company" as an associate and contracted Gen. Brent Scowcroft, his long time deputy and former national security advisor, into his Washington office as vice chairman of the board."(110)

Perhaps because of "old times", or perhaps because of his sustained and extraordinary connections in the Liberal Media, Kissinger has largely escaped any unfavorable coverage, despite the extensive commercial interests he has had in countries which have not been 'friends' or supportive of U.S. interests. This was particularly true when thousands of Chinese students demonstrating in Tiananmen Square in June, 1989, were massacred by the PLA, and countless others were jailed – or simply disappeared. Kissinger defended Deng Xiaoping, and both he and Scowcroft hastened to use all their influences to keep the U.S. from exercising sanctions, arguing that "Whatever Americans feel personally about what happened, China is too important for America's national security to risk the relationship on the emotions of the moment."(111) The Media gave scant heed to such blatantly bourgeois statements, nor did it question the obvious self-serving economic reasons behind such pap. Hard to believe? Read on…

Kissinger has long promoted U.S. investments in China, creating special consultancies to do so. Billions of dollars have been channeled into the Chinese economy through Kissinger Associates. "Kissinger created the prestigious American-China Society, with himself and former Secretary of State Cyrus Vance as co-chairmen. Board members included ex-Presidents Gerald Ford, Jimmy Carter, and Richard Nixon, former National Security Advisers McGeorge Bundy, Robert McFarlane, and Zbigniew Brezinski, and former Secretaries of State Dean Rusk, Edward Muskie, Alexander Haig, and William D. Rogers." (112) Ask the Media, or Kissinger, or the above named, if there are any ethical problem involved. Quite a group, so double-check their answers for accuracy. And don't forget to ask about the pro-China Lobby in Congress...it's their ATM.*

"Kissinger Associates have persuaded clients to invest heavily in China. Chase Manhattan Bank had recently invested $270 million in the Daya Bay Nuclear Power Plant, which had an integral role in producing China's nuclear weapons. Another of the bank's multi-million-dollar investments underwrote leases for the purchase of aircraft by the Civil Aviation Administration of China."(113)*

"Kissinger had arranged for American Express, a company of which he was a board member, to invest $138 million in a thirteen-year loan to build an office complex in Beijing. At the same time he had arranged for another company of which he was a board member, the American International Group, to establish a joint venture with People's Insurance Company of China to handle insurance and reinsurance of Mainland China. The business was currently worth $25 million annually. Atlantic Richfield, another client of Kissinger Associates, had invested $170 million to develop the South China Sea natural gas field."(114)

"On the military front, Kissinger had continued to play what was called by Qiao Shi (head of China's secret intelligence service) his "China card," allowing the United States to help China to upgrade its satellites and purchase advanced missile guidance systems. One of these systems had been sold to the People's Liberation Army by the Israelis in 1987, with full U.S. approval. Currently, Grumman Aviation (has) a contract to provide advanced avionics for the Chinese Air Force's entire fleet of F8-2 interception jets, while Garret Aerospace was providing the engines for the C-8 fighters, the (Chinese) air force's short–range interceptors." "Against these genuinely massive investments, Qiao Shi knew that the question of (obtaining) his (requested) riot-control gear was a trifling matter. But Henry Kissinger was not a man to miss a deal. He had told China's spy master not only that he would receive the gear, but that US. investments would increase even further once George Bush (41) was in the Oval Office."(115)

"Qiao Shi believed the ability of Kissinger Associates to lobby even more effectively within the Bush administration would be enhanced by the presence of two top officials who had been members of the multimillion-dollar consulting company that bore Kissinger's name." "They were Lawrence Eagleburger, who would take the number two slot at the State Department and become one of President Bush's most trusted troubleshooters, and Brent Scowcroft, who had been director of the Washington office of "Kissinger Associates". He would be Bush's national security advisor. Together with James Baker, the two former employees of Kissinger Associates would form the policy-shaping "troika" around Bush."(116) Why was there be no Media "exposure" of this amazingly blatant effrontery?

Add the following to the above and the picture that emerges is more than a little disturbing. Li Ka-shing is described by China military specialist William Triplett as "the banker for the Chinese Army", with direct connections to the Chinese defense industry and its world wide interests. "Li serves with Gen. Alexander M. Haig, Jr., as a COSCO (China Ocean Shipping Company) advisor", per COSCO's Beijing website, June 2000. The former NATO Commander, Secretary of State, Presidential Chief of Staff, has been a paid domestic and overseas senior "honorary" advisor" to the China Overseas Shopping Company, the Red Chinese government's 600 merchant marine shipping arm of the People's Liberation Army. He lobbied to obtain the Naval Station in Long Beach for the Chinese. As owner of Worldwide Associates, Haig (also) advises United Technologies, manufacturer of jet engines, air conditioners, and elevators. The company has made billions in 17 joint ventures in China."(117) Diversified? Yes indeed! Mainstream Media coverage? Not yet! But then Kissinger and his liberal leaning colleagues have always been "helped" by the Liberal Media. Why should things be different now?

To return to Kissinger and Vietnam: his 'negotiations' record reflects little but shame, for he - and the U.S. - lost; the North Vietnam negotiators won, hands down. Thousands of U.S. military men died while Henry fiddled as a negotiator, as did many more thousands of South and North Vietnamese killed in the same time frame. And shamefully, unknown numbers of U.S. prisoners were left behind. There was no withdrawal of North Vietnamese forces from South Vietnam, only increases of NVN forces across all of Indochina. Vietnam, Laos and Cambodia fell to the Communists, and most of those that could not flee were subjected to the many horrors of reeducation camps and/or the vindictiveness of the 'victors'.

How can one "negotiate" when one steadily gives away his bargaining chips until there are few or none left? The mathematics involved are not difficult. Kissinger's performance (and his steady flow of erroneous reports of promise, expectations and progress) contributed substantially to the eventual alienation of Congress to the point that funds were denied to South Vietnam in the final weeks of their doomed effort to fight the North Vietnamese in 1975. All U.S. combat troops were already long gone from Vietnam, so Congress cutting funds no longer posed a threat to U.S. forces. However, its peevishly denying funds and equipment to the South Vietnamese effectively struck the final fatal blow to our previous ally. The Liberals simply have no conscience, and the U.S. was left without honor.

Words, not deeds, had become Kissinger's habit; promises and 'agreements' were simply broken; and deception became an art form. The domino theory played out not only in the three countries of Indochina, but subsequently in Africa and the Middle East as well. U.S. prestige was severely and deservedly damaged. The U.S. became widely identified as a "paper tiger".

Where was the Media during this shameful period? The Media failed dismally in meeting even its basic responsibilities regarding balanced reporting. While tried, the oft used excuse of "not knowing then" as opposed to "knowing now" wasn't then, and isn't now, good enough. Stated bluntly, Media did a lousy job of keeping the people of this country honestly informed. Were "Liberal Interests" simply at work, intent on covering the glaring mistakes and inadequacies noted during the period of time covered in this chapter?

Was the Liberal Media an integral part of the well organized anti-war movement? While "politicians" usually duck, and hide, and engage in double talk intended to obfuscate, the Media has an obligation to perform otherwise. This didn't happen, and the slow but steady downward spiral of our Republic continued unchecked

Kissinger Postscript

This addendum stems primarily from a program I watched on ESPN on September 9, 2003, featuring Henry Kissinger regaling the audience about his new book "The Anatomy of Two Major Foreign Policy Crises". His presentation prompted me to do more research, which in turn did not surface any redeeming attributes prompting a softening of the uncomplimentary conclusions I had reached earlier regarding the man for his painful 'Vietnam' performance.

In his inimitable studied and rambling style, Kissinger retold the saga of the 16 day conflict- and the period immediately after - that began on October 6, 1973, known as the Yom Kippur War (and as the War of Ramadan to Muslims) following Egypt's and Syria's well coordinated but unexpected attack on Israel. He featured his role, of course, and stated, inferred, drew word pictures, and gave statements that would lead to those listening that all but single-handedly he had saved the day by bringing Egypt and Israel to accept him as a fair minded arbitrator whose sole purpose was to restore peace in the area. During this process he noted how he personally had been able to persuade the Soviets to cooperate with the U.S. in this "tinderbox situation". Kissinger took credit for having arranged for a relief column to bring life sustaining supplies to the besieged 3rd Army of the Egyptians, akin to his claim of arranging for the impressive U.S. airlift of military supplies to Israel that Israel had desperately requested earlier. Others involved in this saga tell a markedly different story, complete with the deceits, posturing, and center stage desires – and disregard for life – that have emerged as the hallmarks of Kissinger.

In setting the stage, Kissinger said he was in New York City with his deputy (for the Mideast) Joe Sisco to attend the opening of the United Nations General Assembly and White House Chief of Staff Haig was with President Nixon in Key Biscayne. "A few hours after the war broke-out, Kissinger flew back to Washington to chair a meeting of the Washington Special Action Committee (WSAG), the N.S.C.'s crisis committee. Israel had already made an emergency request for military supplies, even though it was widely expected that it would beat back the Arab attack within a day or two. Deputy Secretary of State Kenneth Rush, representing the department while Kissinger wore his National Security Adviser's hat for that particular occasion, argued against sending supplies to Israel immediately. "They have no real shortage," he said. The Pentagon was even more strongly opposed. "Shipping any stuff into Israel blows any image we may have of an honest broker," said Defense Secretary James Schlesinger, whose incisive mind and prickly ego had made him Kissinger's new rival within the administration."

"Defense wants to turn against the Israelis", Kissinger reported by telephone that evening to White House Chief of Staff Gen. Haig. "Sounds like Clements", replied Haig, referring to Deputy Defense Secretary William Clements, a Texas oilman with pro-Arab sympathies."

"This began a critical dispute that would dominate Washington for a week: to what extent should the U.S. re-supply Israel."

"Even more controversial would be the ensuing historical dispute over the roles played by Kissinger, Schlesinger, and Clements in deciding the re-supply issue. Dozens of books and articles have tackled the question. Those sympathetic to Kissinger portray him as staunchly willing to help Israel as soon as its desperate needs became known by midweek, and they describe his efforts to overcome resistance from the Pentagon. Some critical accounts, on the other hand, accuse Kissinger of playing a devious game; they charge that he held back supplies from Israel in order to create a climate ripe for diplomacy, unfairly blamed the Pentagon for the delays when talking to Jewish and Israeli leaders, took credit for the delays when talking to the Soviets and the Arabs, and finally grabbed credit for the airlift of supplies when it began after a week."

"Because many of the participants recorded or transcribed their telephone conversations and kept notes at meetings – and because in the underhanded atmosphere of the Nixon administration such material tended to be copied and kept, sometimes shared – this historical dispute can now be greatly elucidated." (Nixon was already preoccupied with Watergate and was in Florida when the war began; Vice President Agnew would be forced to resign in the next two weeks; Nixon fired Attorney General Richardson and Watergate special prosecutor Cox shortly, leaving Kissinger – and Haig – freer than usual, including little consultation time with President Nixon.) "Kissinger assumed that Israel would win quickly, and he opposed giving it major support that could make its victory too one sided. "The best result", Kissinger told Schlesinger the first weekend, "would be if Israel came out a little ahead but got bloodied in the process, and if the U.S. stayed clean."(119)

"The strategy was to prevent Israel from humiliating Egypt again," Kissinger later explained." "American restraint, he hoped, might lead to an opening to Egypt, preserve, détente with the Soviets, and make Israel see the benefits of a negotiated settlement."

On the diplomatic front, "Kissinger wanted to delay efforts at the U.N. to impose a cease-fire-in-place for a few days until Israel had pushed Egypt and Syria back to their original lines. Otherwise, such a cease-fire would allow the Arabs to keep hold of the land captured in their surprise attack."

"Instead, Kissinger proposed a cease-fire status quo ante – in other words, a cease-fire that would require each side to return to where it was before the fighting began. Publicly, he argued that this was the only fair way to make sure that Egypt and Syria were not rewarded for starting a war."

Privately, his agenda was more complex. He realized that there was no way to get the Arabs to give up their own territory that they had finally retaken. His real goal was to get Israel to demand this outcome – so that if the battle turned and Israel captured new Arab territories, it would already be on record as favoring a return to the prewar lines. "There was no hope for a cease-fire status quo ante," Kissinger recalled, "but I wanted to get the Israelis to sign on to the principle so we could use it against them if they turned the war around."...On Sunday, October 7, the second day of the war, "Kissinger and Schlesinger agreed that the Pentagon would provide Israel with Sidewinder missiles and new ammunition, but it would require sending unmarked El Al planes to pick up these supplies secretly at a Virginia base." And both agreed that no U.S. aircraft would be needed.

"Late that Sunday night, Israel's Ambassador Simcha Dinitz

called to say that the Israeli planes had not been given permission by American military officials to land at the air base." "Oh, those goddamn idiots," Kissinger exclaimed, promising that he would try to clear-up the matter."(120)

"Kissinger, to whom linkage came naturally, immediately established some trade-offs with Dinitz and Soviet Ambassador Dobrynin. They centered on the most-favored nation (MFN) trade status that Washington had promised the Soviets in 1972. Congress, with Senator Henry Jackson in the lead, was threatening to block this new trade relationship unless Moscow lifted restriction on Jewish emigration. So Kissinger began to link the re-supply of Israel and the Soviet trade bill. Beginning on Monday he would continue to push for most-favored-nation status if the Soviets showed restraint in the Middle East Crisis".... "Early the next morning, Dobrynin called with a message from Brezhnev. "We feel we should act in cooperation with you," the note said. "I will include," he (Kissinger) promised, "some reference to MFN in my speech tonight."..."As for the Israeli side of the linkage equation, Kissinger told Dinitz that he expected Jewish leaders, in return for the U.S. efforts to re-supply Israel, to withdraw their support for Senator Jackson's amendment to the Soviet trade bill. He spelled out the deal explicitly to officials from the 'Conference of Presidents of Major 'American Jewish Organizations'; reluctantly, the group decided to go along." "Kissinger told Haig about this deal in a phone conversation on Sunday morning. Speaking about the re-supply of Israel, he explained: "If we support them, they would be willing to help with MFN." "The next day he called Ambassador Dobrynin and warned: "I'm going in this speech to mention our MFN position and I hope to God this is not a week what every Jewish league will start attacking me on this."(121)

"All these reveries were upset on Tuesday, October 9, the fourth day of fighting, when it became clear that Israel was in trouble. The Israeli cabinet had been meeting all night and had decided, among other things, to put its nuclear-armed Jericho missiles on alert. Early that morning Golda Meir called Ambassador Dinitz. Israel was being defeated, she said. He should call Kissinger immediately. Kissinger agreed to meet with him at eight-thirty in the Map Room of the White House. Dinitz's meeting with Kissinger was a tense one. Dinitz confided that Meir was willing to make a secret trip to Washington for a one-hour private session with Nixon to plead for more supplies. Kissinger dismissed that suggestion as unnecessary. Later, he would say that it smacked of "blackmail" because it was designed to put the administration on the spot."

According to both Kinitz and Kissinger, Dinitz did not mention – or threaten - that Israel was prepared to resort to nuclear weapons if its survival was at stake. "Kissinger later mentioned casually to Herman Eilts, who had been U.S. Ambassador to Egypt at the time, that Israel had given "intimation that if they didn't get military equipment, and quickly, they might go nuclear."

"After talking to Dinitz for more than an hour, Kissinger promised the ambassador that he would have a reply by late afternoon. Then Kissinger called a WSAG crisis committee meeting to prepare options for the president. The WSAG came up with a range of five options. Kissinger then met with Nixon privately in his Executive Office Building hideaway, where they agreed on the one that Kissinger preferred: a quiet and low-key re-supply of Israel with a modest amount of new planes and ammunition. In addition, Kissinger got Nixon to approve a crucial pledge: the U.S. would replace all of Israel's losses once the battle was over, thus the Israeli army would not have to hoard equipment. However, there would be no immediate American airlift. The re-supply operation would be kept "quiet" and Israel would have to make arrangements for picking up its new supplies."

"With that kind of movement, we won't be able to keep it quiet," Schlesinger told Kissinger that evening. "With all the Israeli planes flying around, it will be impossible for the Arabs not to find out."

"It is extremely important," said Kissinger, "to keep it as low-key as we possibly can." He then added that it would be wonderful "if we can get through this crisis without antagonizing the Arabs."

"By then – Wednesday, October 10 – the Soviets had begun their own re-supply operation of Syria. It was a modest airlift confined to ammunition and fuel (rather than tanks or planes), but it was still more overt than the U.S. effort. In addition, the Israelis were having problems collecting the U.S. equipment they needed using only their seven available transport planes."

"Kissinger then made a decision that turned out to be one of the most bothersome mistakes of the re-supply effort: instead of allowing American airplanes to transport the supplies, he decided that Israel should hire private charter companies to do it. This halfway solution, it turned out, neither helped the Israelis nor pleased the Arabs."

"The president, preoccupied with Agnew's resignation that Wednesday, had not been very involved in the decision-making about the war. The Soviet airlift finally engaged Nixon's fighting instincts. He was shocked to find out that five Phantom jets he had ordered sent to Israel two days earlier had not yet gone. "It should have been done," Nixon snapped at Kissinger. "Do it now!"(122)

Kissinger then called Scowcroft. "Look Brent, the Defense people are just going to have to stop dragging their feet," he said. The Israelis are going wild." Scowcroft promised that at least two of the Phantom fighters would go out the next day. Kissinger then called Haig. "Could you stiffen Schlesinger's back?" he said. "The guy is totally panicked. Clements is beating after him. If the Egyptians win, we will lose our position." On Thursday, October 11, newspapers carried stories to the effect that the U.S. was slow in providing Israel's needed supplies, and Schlesinger told Senator Jackson that he had no authority to requisition charters."

"Kissinger immediately called his chief Mid-East deputy , Joseph Cisco, to complain that the Israelis were raising hell about the difficulty of getting charter planes to pick up their supplies. The problem, Sisco explained, was that none of the charter companies was willing to come forward for such a dangerous and politically controversial job". So Kissinger ordered the military to charter twenty transport planes on its own for the Israelis to use." "Although Senator Jackson and others were blaming the Soviets for stirring up the crisis, Kissinger felt that their behavior had been restrained. The Soviets were being modest in the amount of supplies they were sending to Egypt and Syria, they were cooperating at the U.N., and they were pushing Egypt to accept a cease fire. In order to tout this triumph of détente, Kissinger called a press conference for the morning of Friday, October 12."

"At the press conference, Kissinger chose his words carefully. He said that he did "not consider the airlift of Soviet military equipment helpful," but he went on to label it "moderate". This was to be balanced against "the relative restraint that has been shown in the public media in the Soviet Union and in the conduct of their representatives at the Security Council. In short, he argued, détente was not dead."

In the meantime, the charter situation had still not been sorted out, and "Israeli leaders were in an uproar. Dinitz showed up at Kissinger's White House office that Friday night and explained that the situation was dire. Israel, he said, would run out of ammunition in three days."

"At this point an interesting shift occurred. Schlesinger became convinced that, as long as the U.S. was going to re-supply Israel, it should use its own military planes rather than continue to flop around trying to hire private charters. But Kissinger, who was hoping to begin a new diplomatic round, still wanted to avoid the use of U.S. military planes."

"Shortly after midnight, as soon as Ambassador Dinitz left his office, Kissinger talked to Haig on the telephone. "Jim told me it's suddenly critical," Haig said, referring to Schlesinger. "He's ready to move MAC (Military Assistance Command) aircraft in there immediately. I think that would be foolish."

"That would be disaster, Al," replied Kissinger. "How can he fuck everything up for a week – he can't now recoup on the day the diplomacy is supposed to start." He went on to express his disbelief at the Pentagon's claim that it could not hire any civilian charters. "You know goddamn well they didn't try."

"We do have the option of sending some American planes in there," Haig noted again. "I think that's a high risk for us."

"I think it's stupid," said Kissinger."

"As soon as he hung up, Kissinger telephoned Schlesinger, who was at home asleep." In his memoirs, Kissinger noted that he and the defense secretary were pretty much in agreement. But Schlesinger's recollection of the call was that Kissinger flew off the handle. "As Israel began to fall apart," Schlesinger recalled, "Henry began to fall apart."

"Kissinger began the conversation by saying that the lack of ammunition had caused the Israelis to stop pushing forward against Syria. This was "near disaster," he said, for America's diplomatic strategy, and it was all due to "massive sabotage" within the Pentagon. Yet when Schlesinger suggested abandoning the ill-fated private charter plan and using U.S. military transports instead,

Kissinger remained adamantly opposed. "One thing we cannot have now, given our relations with the Soviets, is American planes flying in there," he said.

"Schlesinger then got out of bed and went to the Pentagon. "There he reviewed the options....he came to the conclusion the re-supply could not be handled other than through an American military airlift all the way to Israel. The option that Kissinger was insisting on, the use of private charters, would not work. Schlesinger had available three C-5A transport planes, each able to carry up to eighty tons to Israel. As soon as the sun rose he called Haig to urge this course on Nixon."

"Nixon agreed and gave Kissinger the word as he was preparing to chair another WSAG crisis committee meeting that Saturday morning. "Do it now!" Nixon again told him. Kissinger did not object. In addition, he and Nixon decided that smaller transports would fly in other supplies, and fourteen new F-4 Phantom fighter jets would be sent to Israel immediately."

"Thus on Saturday, October 13, a week after the fighting began, a major American airlift finally got underway. A thousand tons of equipment a day began flowing, with flights landing almost every hour. More arrived on the first day than the Soviets had delivered to Egypt, Syria, and Iraq combined in the previous four days. After a week spent dithering over whether it was possible to get five F-4 Phantoms to Israel, forty were delivered over the next ten days.". "All of the delays at least produced a silver lining. The airlift was seen in Moscow and Cairo not as a major American provocation, but as a response to the one launched by the Soviets. (It) became clear that, if a cease- fire was timed right, the airlift would result in what Kissinger had hoped for at the start: a modest Israeli victory that encouraged Egyptian and Israeli flexibility while preserving the potential for American diplomacy."(123)

As for Egypt's beleaguered Third Army, "Anwar Sadat broke the impasse. He agreed to engage in direct Egyptian-Israeli talks at the military level in order to resolve the problem of access to his Third Army along the Cairo-Suez road. All that the Egyptian president asked was that one convoy be allowed through in the meantime to keep his men alive. Israel agreed. Shortly after midnight on the morning of Sunday, October 28, at a marker designating Kilometer 101 on the road between Cairo and Suez, Egypt's Lieutenant General Abdel Gamasy and Israel's Major General Aharon Yariv approached each other, offered awkward salutes, and then shook hands. Kilometer 101 was destined to become, so to speak, a milestone. The first direct peace talks between Israeli and Arab representatives in the quarter century since Israel had gained its independence had begun."(124

Kissinger met with Sadat in the Tahra Palace on 7 November. After the usual exchange of pleasantries, Sadat asked "And what about my Third Army"...What about the October 22 line?" "There were two options, Kissinger told him. Egypt and the U.S. could expend all their energies trying to force Israel to move back to the cease-fire line. Or, with the same amount of effort but a little more patience, they might be able to arrange a genuine disengagement of forces that would move Israel back from the Suez Canal. In the meantime, arrangements could be made to get a steady supply of non-military material to the Third Army. Sadat could choose, Kissinger said, he would do his best either way." Sadat said "that he would be willing to go for a full-fledged disengagement, as Kissinger preferred, rather than insisting on redressing the Israeli cease-fire violations that had ensnared his Third Army. In addition, he would begin the process of restoring full diplomatic relations with the U.S."

"With this sweeping gesture, Sadat moved Egypt from a reliance on Moscow to a reliance on Washington, and was in effect putting not only the Third Army issue but also Egypt's strategy of seeking a comprehensive peace on the line. In doing so, he paved the way for Kissinger's step-by-step approach, which became the basis for his shuttle diplomacy."(125)

"There was still one hitch in what was becoming a pattern. Kissinger had given short shrift to the ally on whose behalf he was nominally negotiating. Instead of consulting with Israel, or going there personally to present the deal, he sent aides Sisco and Hal Saunders."

"Golda Meir and her cabinet, however, were not ready to accept his plan for negotiations. Sadat might feel that the safety of his nation required bold strokes, but Israelis, understandably, felt that their nation required constant, excruciating vigilance."

"In what was also becoming a pattern, Kissinger's penchant for diplomatic ambiguity – creatively fudging issues, as he had done during the Vietnam negotiations, ran afoul of the Israelis desire to pin down each distinction with a hairsplitting rigor that would dazzle a Talmudic scholar."(125)

"In the meantime, the direct talks at Kilometer 101 were going better than Kissinger had imagined, disquietingly better. The problem Kissinger faced, it turned out, was not from the danger of disagreements there, but from the danger of a serious agreement, one that he would not control or get credit for. Israeli general Yariv and Egyptian general Camasy, sitting in that isolated tent in the desert, began to disprove the maxim that military men do not know how to make peace. Once they cleared up the rules of the road for the convoys to the Third Army, they turned to more ambitious proposals for a full-scale separation of forces."(126)

"Although still far from an agreement, the two generals were marching toward the type of disengagement accord that Kissinger hoped to negotiate himself. Rather than cheering them on, however, he began maneuvering to make them stop."

"As was often the case, his motives were a mix of valid policy concerns and personal vanity. On a more personal level, Kissinger quite simply liked to control important negotiations himself, and not incidentally garner the glory and potential peace prizes that came with them."

"At Kissinger's behest, both Sadat and Meir reined in their generals at the Kilometer 101 talks. The Israeli ambassador, although a Kissinger partisan, felt that it was largely a matter of ego. "Kissinger's view was that if any concessions were to be made, they should be made by him," Dinitz recalled. "He was very upset when he found that things were actually being settled by the generals at Kilometer 101. We had to make them stop. Ego was a weakness of his. But it was also the source of his greatness."(127)

That concept is a bit of a stretch. President Clinton's question regarding the meaning of "is" certainly encourages that we further parse the term "greatness" before bestowing it on Henry Kissinger. Opportunistic curio might serve more accurately. That he has not been pilloried by the press over the years – or by Congress - for his performance especially during the Vietnam era under presidents Nixon and Ford remains mystical. His having traded so successfully on his PRC contacts, as have the others associated with "Kissinger and Associates", with scant regard for the welfare of the

United States or for "human rights" issues is also more than a little unusual. While "Kissinger did express some shock at the "brutality" of the (Tiananmen) crackdown, his very brief statements along those lines were perfunctory. His analysis did not include the basic concept which all in China understood: the regime's method of dealing with unarmed student demonstrators was intentionally brutal and murderous; it was a planned means of eradicating once and for all any inclination the intellectuals might have to threaten the party's power or to cause public disorder. The general trend of his statements was to find excuses for the Chinese authorities. In late July, Kissinger wrote his column recommending against anti-Chinese sanctions. In November, he was in China, meeting with Chinese foreign minister Quian Qichen, who praised the former secretary of state for having "the courage of a statesman and the foresight of a strategist. "One of the people Kissinger and his group (he was accompanying a business delegation) met with on that trip was Deng Xiaoping...And after returning to the United States, Kissinger had dinner at the White House, where he reported back on his conversations at the highest level. In a single year, Kissinger may go to China several times, and the door is always open. In September 1990, he met with Li Peng, the man deemed most directly responsible for ordering the Tiananmen assault on students. The next day, Kissinger went to see Jiang Zemin, who noted that Kissinger was making his eighteenth visit to the People's Republic of China. And so it has continued."(128)

"In July 1995, Kissinger wrote in an article in the Washington *Post* " China is surrounded by countries whose strength is growing and who are not necessarily friendly to it, including Japan, India, and Russia, and it needs the United States to help prevent any country from achieving hegemony in Asia.". Kissinger also wrote, "Both America and China have their own reasons for opposing the domination of Asia by a single hegemonic power." Therefore, "China wants the United State to help balance its relationship with powerful neighbors, at least until it is strong enough to do so on its own." "Kissinger allows that human rights "must always constitute a central American concern," but that comment, in the absence of any specific recommendations about making that concern central, smacks of lip service."(129)

What Kissinger does not say as he expresses his views on American China policy is just how much he stands to profit himself from the very policies he urges the government to adopt. Kissinger's company (Kissinger Associates) represents many American corporations seeking to do business in China and who pay Kissinger large sums of money at least in part because of his unparalleled access to China's most powerful figures. "At the time of his 1989 statements, Kissinger had just formed a limited partnership called China Ventures to carry out joint ventures with 'China International Trust and Investment Corporation', headed by his frequent host in Beijing, Rong Yiren. In 1988, he formed the China-America Society with several former American presidents on its board. The society hosts banquets for Chinese leaders visiting the United States and sends delegations to China."(130)

"When John J. Fialka of *The Wall Street Journal* reported on these business interests, he gave Kissinger a chance to respond to the allegations of conflict of interest. It was "outrageous," he quoted Kissinger as saying, that "I would take a public position to curry favor with the Chinese government for clients." A few months later, complaining about criticism of his China dealings, "Kissinger called any linking of his views on China with his business interests a form of "McCarthyism." As Scott Thompson put it before the Senate Foreign Relations Committee, "The fact is that (Kissinger) does receive substantial remuneration from corporations with major investments in China, which stood to benefit from adoption of his positions." "The fact is also that the special way that China does business, and the willingness of former senior American officials to help it to do business, has resulted in a kind of new and powerful China lobby that perhaps only by coincidence, has declined to criticize Beijing's human rights record, has warned against "confronting" China or imposing sanctions, and has made substantial profits along the way."(131)

"Not long after the Tiananmen crackdown, President Bush sent his national security advisor, Brent Scowcroft, and his deputy secretary of state, Laurence Eagleburger, on a secret fence-mending trip to Beijing. Scowcroft had been deputy director of Kissinger Associates before going into the Bush administration. Eagleburger was its president before becoming Bush's deputy secretary of state. *The New York Times* reported that in 1988 Eagleburger made $674,000 from Kissinger Associates; Scowcroft earned $293,000 as a consultant to the group. As early as 1991, only two years after the Tiananmen massacre, Eagleburger was cautioning trade sanctions. "We firmly believe," he wrote in the *Washington Post* "that renewing China's MFN waiver – without conditions – provides our best interest for promoting positive change and U.S. interests in China." "A full "engagement" with China, he argued, is the best way to get the country back "on the path to freedom." Among the members of the board of Kissinger's America-China Society are Cyrus Vance, Haig, former secretary of state William Rogers, Robert McFarlane, and Zbigniew Brezinski . Haig is perhaps the figure who, after Kissinger, has most clearly mixed public positions of the sort that Beijing likes with business dealing in China on behalf of corporate clients. As secretary of state, Haig was the key Reagan administration official responsible for the 1982 Chinese-American Joint Communique that is one of the three key statements guiding relations between the two countries. (The other two are the 'Shanghai Communique of 1972' and the 'Joint Communique on the Establishment of Diplomatic Relations of 1978'.) In it, the United Sates agreed never to pursue a Two Chinas or a One China-One Taiwan policy, to keep arms sales to Taiwan not to exceed, "either in qualitative or in quantitative terms, the level of those supplied in recent years since the establishment of diplomatic relations between the United States and China," and "to reduce gradually its sales of arms to Taiwan, leading over a period of time to a final resolution." "Haig left the administration shortly before that agreement was signed. Beijing was grateful for his role in undermining Taiwan, and Haig was soon cashing in on that gratitude." "He (Haig) has been a senior consultant to United Technologies of Hartford, Connecticut, a manufacturer of jet engines and elevators, the world's sixteenth-largest industrial company, America's tenth-largest exporter, and a company that had seventeen joint ventures in China as of 1996." Meanwhile, Haig has written op-ed pieces and made statements in which he has opposed attempts to confront China because of it human rights record and urged the United States to find ever greater means of cooperation."(132)

"The comparison in this regard between Kissinger, Haig, and others like Eagleburger and Scowcroft on the one side, and James Lilley, the American Ambassador (to China) who did not attend that 1989 National Day rite, on the other is instructive. Lilley, a former senior official at the Central Intelligence Agency, has been a critic of China, a believer in the thesis that China's interests clash with those of the United States and that America does have an obligation to protect Taiwan from a forcible takeover by Beijing." "Does anybody think that, if Lilley decided to go into the consulting business, he would have much success trying to do the same thing?"(133) The answer is obviously no, but we do have William Jefferson Clinton, who – with Al Gore - likely vacuumed more money from China than 'Kissinger Associates' combined. And isn't it interesting that, with the exception of an occasional 'soft' article, the Media remains very quiet on the entire panoply of China – Kissinger – Haig - Human Rights - White House - China-America Association,- COSCO – Clinton/Gore, etc.?

Kissinger has said that his next book will be about the arts of diplomacy. Isn't that more than a bit Aeosopian? Or did he say "arts of negotiations"? Perhaps Al Haig will help him. The two of them have done very well for themselves - and their associates - in "negotiating" with the People's Republic of China. Perhaps they learned more from their negotiating efforts with the North Vietnamese than history credits them. How soundly the Mainstream Media sleeps. Amazing!

* Vietnam Voices, pg. 418

* Kissinger, Pg. 167, cited NY Times and Washington Post

* The reference was likely to the $3 *billion* figure Kissinger and Nixon had promised in 1973.

* Richard Bernstein and Ross H. Munro provide numerous examples of U.S. company's huge investment in China in their excellent book *The Coming Conflict With China.* Especially note pages 105-129., and 140-148.

CHAPTER 4

JAMES EARL "JIMMY" CARTER, JR.

("Self-delusion, Inc.")

Born on October 1, 1924, and raised near Plains, Georgia, the 39th President of the United States, "Jimmy" Carter, as he preferred to be called, came to the White House in 1977 by way of Georgia's Governor's mansion where he resided from 1971 to 1975. His path to the governorship was not spectacular, but followed his retirement from the navy after serving seven years (1946 – 1953), during which he became a senior-grade lieutenant. He first served on a battleship then requested and was assigned submarine duty, including attending a six month nuclear engineer course at Union College in Schenectady, New York. His decision to return to the Carter family peanut business in Plains was prompted by the illness of his father. His subsequent political efforts at the local and county levels reflected his ambition of becoming widely known and acceptable in Georgia politics, including serving as a state senator (1963-1967). In 1966 he failed in his effort to win the primary bid as the Democratic candidate for governor, but then spent the next four years campaigning hard for the 1970 governorship race, often joined by his wife Rosalynn (Smith) who he had married in 1946 upon his graduation from the U.S. Naval Academy.

"The Nation's News Media turned the spotlight on Carter in Georgia at his inauguration in January 1971, when he took over the governorship from the highly publicized segregationist Lester Maddox. In his eight-minute address, Carter said: "Our people are our most precious possession. We cannot afford to waste the talents and abilities given by God to one single person. .Every adult illiterate, every school drop-out, and every untrained retarded child is an indictment of us all. Out state pays a terrible and continuing human and financial price for these failures. It is time to end this waste... "At the end of a long campaign, I believe I know the people of this state as well as anyone. Based on this knowledge of Georgians north and south, rural and urban, liberal and conservative, I say to you quite frankly that the time for racial discrimination is over. Our people have already made this major and difficult decision, but we cannot underestimate the challenge of hundreds of minor decisions yet to be made. Our inherent human charity and our religious beliefs will be taxed to the limit. No poor, rural, weak, or black person should ever have to bear the additional burden of being deprived of the opportunity of an education, a job, or simple justice. (1)" It was likely one of the best speeches he ever made, although more pulpit in nature than memorable.

In his four years as Georgia's Governor, Carter's accomplishments were quite impressive. The number of black employees holding appointive jobs rose from three to fifty three; some three hundred separate state departments and agencies were consolidated into twenty-two; and he implemented "zero base budgeting" in each department of government resulting in a $200 million surplus during his administration.

Georgia law at the time did not allow Carter to succeed himself as governor, which prompted him to think in terms of running for President. His timing could not have been more fortuitous. In 1972 he had chaired the Democratic Governor's Campaign Committee and in 1974 he was chairman of the National Democratic Party Campaign Committee. With President Nixon's resignation, Carter, running as a "Progressive Southerner" against less than impressive or inspirational candidates, including a badly tarnished Ted Kennedy, easily won the nomination. Walter Mondale (of Minnesota) was his vice-president. In a close race against Gerald Ford, the scales tipped enough in Carter's favor when Ford inexcusably stated during the second debate that Poland and other eastern European countries were not dominated by the Soviet Union – a gaff from which he never recovered, thanks largely to the extensive replays by the Mainstream Media of Ford's mistake. Carter had run as an "outsider", was elected as an "outsider", arrived in Washington with a bevy of "outsiders", appointed more "outsiders", and remained a "preaching" outsider.

Carter said in his inaugural address "Ours was the first society openly to define itself in terms of both spirituality and human liberty. It is that unique self-definition which has given us exceptional appeal – but it also imposes on us a special obligation, to take on those moral duties which, when assumed, seem invariably to be in our own best interests. You have given me a great responsibility – to stay close to you, to be worthy of your and to exemplify what you are. Let us create together a new national spirit of unity and trust. Your strength can compensate for my weakness, and your wisdom can help to minimize my mistakes. Let us learn together and laugh together and work together and pray together, confident that in the end we will triumph together in the right." He concluded his address by saying that he hoped when his administration ended people might say about the nation: "That we had remembered the words of Micah• and renewed our search for humility, mercy, and justice; That we had torn down the barriers that separated those of different race and region and religion, and where there had been mistrust, built unity, with a respect for diversity; That we had found productive work for those able to perform it; That we had strengthened the American family, which is the basis of our society; That we had insured respect for the law, and equal treatment under the law, for the weak and powerful, for the rich and the poor; And that we had enabled our people to be proud of their own government once again; I would hope that the nations of the world might say that we had built a lasting peace, based not on weapons of war but on international policies which reflect our own most precious values. These are not just my goals. And they will not be my accomplishments, but the affirmation of our nation's continuing moral strength and our belief in an undiminished, ever-expanding American dream."(2)

This was not a speech that would ring any lasting bells in our country, or in foreign countries, friendly or other-wise, and certainly it raised more than one quizzical eyebrow. It was more a plaintive sermon than an inaugural address, apparently intended to rally what he perceived as a divided country and world. The reference to "Micah" was ill-chosen, and surely ruffled most Moslems around the world. On balance, the address sounded but a very faint ring for freedom. We had experienced communist triumphs in Vietnam, Laos, and Cambodia, the mid-east was at sword's points, Europe remained unsettled, Africa was not at rest, and the "Cold War" was still very much a threat. Carter's speech also mirrored an apology for the problems of the previous decade, or more, that our country had experienced both at home and abroad. Unwittingly, he was already on a gentle slope downward, both domestically and internationally, that would become increasingly steep in the days and months ahead. That he had run as an "outsider" and was, as he had oft loudly claimed, beholden to no one, would prove to be a two-edged sword. He had little support from either side of the aisle, having stepped on numerous toes in his successful rush to become the nominee of the Democrats, including defeating the Kennedy political machine in the process. The Republicans were not only in the minority in both houses of Congress, but still smarting from the very unsettling previous years under Johnson, Nixon and Ford. And the Media did not know which way to fire, so simply applauded – but not too loudly.

On his first day as President, Carter issued a blanket pardon to the approximately 10,000 men who had evaded the draft during the Vietnam War. He later made it possible for some 433,000 veterans who had been given a less that honorable discharge to have their cases reviewed, of which 29,000 did so. In a "fireside chat" delivered in the White House some two weeks after taking office, Carter, wearing a sweater, told Americans that the nation should lower their thermostats to save energy, and announced that all Federal buildings would do so immediately. (I remember sitting in my office in Langley (C.I.A.) wearing a top coat over my suit during that period, hoping for an early spring, and wondering just how the hell requiring government employees to work in very cold offices was expected to help solve the country's huge energy problems. Perhaps pneumonia and flu were to be considered patriotic duties, not illnesses.)

Carter's problems were not long in surfacing. His choice for

Director of Office of Management and Budget was Bert Lance, a 45 year old Georgia banker and long time supporter of Carter. However, the Justice Department had been investigating Lance for suspected illegal banking practices, including writing large overdrafts on his personal checking account. In late September, Lance resigned because of the controversy, although he was later exonerated. In the interim, Jody Powell, Carter's press secretary, had tried to influence Chicago newspapers to smear Sen. Charles Percy of Illinois, who, with Sen. Ribicoff of Connecticut, had gone to the White House to seek Lance's resignation. Lance had refused and Powell later admitted that his actions had been both inappropriate and "dumb". This reinvigorated the voracious appetite of the Media within the beltway, further weakening Carter's support in Congress. But more was to follow, including interest rates reaching 20%.

For example:

Carter brought civil rights activists and other advocates into his administration among the 500 appointees he made to key positions. This too did not necessarily set well with the 2.7 million other federal employees who were in place from previous administrations.

Carter also created two new Cabinet level departments: The Department of Energy, under James R. Schlesinger, and the Department of Education, under Shirley M. Hufstedler, a federal judge. This expansion of "federal government" was not a popular move, although most supported the idea of a need for an energy office.

Carter got a modest tax cut passed in May, 1977, but when the economy continued to falter he sought another, but more modest tax cut in 1978. However, double digit inflation had hit the country. Oil and gas prices soared, and gasoline shortages were not uncommon.

The Dow declined 25% in Carter's first year in office, reaching a low of 742 in early 1978.

Carter and Congress also disagreed over defense issues. Congress wanted to add a fifth nuclear powered aircraft carrier, but it was rejected by Carter. In the following year Carter wanted to increase defense spending, including a $33 billion project to build a new MX intercontinental missile system hidden in two thousand miles of tunnels.

Congress also rejected Carter's proposal to register young women for military draft, but approved reviving Selective Service registration of young men.*

In late 1980, Carter signed a law doubling the size of the Nation's national park and wildlife system, including 104 million acres of Alaska's wilderness areas. In August 2000, Alaska's Governor Knowles wrote Carter, who was in Alaska to commemorate the 20th anniversary of his Alaska National Interest Lands Conservative Act of 1980, that "Without any meaningful dialogue with the people of Alaska, you used our state as a media prop and platform to project your message to President Clinton"...and that "the refuge's future should be decided by wide debate rather than a last minute, unilateral action by an outgoing president at the midnight hour". Carter replied that he was merely expressing his private opinion; - which was touchingly modest of him.

Carter's sermons in the international arena focused largely on "Human Rights". Most nations simply thumbed their noses at him, including the Soviet Union, China, Argentina, Brazil, Uruguay, Germany (both East and West), and Iran. Only with Cuba were relations improved, primarily because Carter lifted previous restrictions on tourism there, and diplomats held discussions on fishing rights. Carter's "open heart and open arms" (3) statements led to 125,000 Cubans making their way to the United States, including thousands of criminals that Castro had released from prison so they could leave the country. While Carter's "humanitarianism" policies were a delight to the Liberals and to the naïve, they were amazingly selective, largely self- serving, and deceptively disingenuous. And none targeted China or the Soviet Union.

Carter "established full diplomatic relations with China on January 5, 1979, completing the process that had begun during the Nixon administration. At the same time, acceding to China's demands, Carter severed formal diplomatic relations with Taiwan and abrogated the mutual defense treaty with that country."...an action labeled as "cowardly" by conservative opponents. (4)

I guess "Jimmy" had just forgotten about the millions of Chinese over the years that had been killed and starved by Mao, etc., or the million plus Tibetans that Mao had his army kill, and that genocide was still being practiced by China in Tibet, as well as against the minorities living in western China. After all, "human rights" does not have an agreed or common definition. Perhaps "Jimmy" did not know where Tibet was located, or had not heard of Tibet's resistance against the People's Liberation Army for nearly 20 years, or heard of the escape of the Dalai Lama from Lhasa to India in 1959, or was concerned about the people behind the Iron Curtain.

Carter signed "new Panama Canal treaties on September 7, 1977 that would give control of the Canal and the Panama Canal Zone to Panama at noon on December 31, 1999. To overcome the objections and obtain ratification, Carter approved amending the treaties to affirm the right of the United States to intervene with military force to protect the Canal if necessary in the interests of national security."(5) This scenario had first begun under Ford and Kissinger (at the urging of the latter), but had remained in limbo because of strong opposition in both parties. But Carter signed a treaty with leftist dictator, Gen. Omar Torrijos, despite the polls in both Panama and the U.S. showing clearly that such a treaty was neither warranted nor wanted.

What, if anything. Carter received for this remains unclear, but in 1997 the ramifications and repercussions became crystal clear. In effect, "Panama sold the Chinese rights to two prime, American built port facilities which flank the Canal Zone both to the east and the west. The 50-year contract awarded Balboa, on the Pacific side, and Cristobal, on the Atlantic side, to a giant Hong Kong shipping firm, Hutchinson Whampoa, Ltd. By any analysis this company, headed by Li Kai-shing, is an interesting operation. Hutchinson had worked closely with the China Ocean Shipping Co, (COSCO) on shipping deals in Asia even before Hong Kong reverted to Beijing's control in 1997. COSCO, you may remember, is the PLA-controlled company that almost succeeded in gaining control of the abandoned naval station in Long Beach, California during the Clinton years. Li Kai-shing has served on the board of directors of China International Trust and Investment Corp., a PLA-affiliated giant run by Wang Jun whose name may ring a bell. Yes, the very same Wang Jun enjoyed coffee at the White House in exchange for a donation to the Clinton-Gore 1996 slush fund."(6)

Interestingly, "Shortly after Clinton announced the departure of the U.S. military from Panama, the Bank of China extended a 15-year $120 million loan to Panama at a 3 percent interest rate. This was during the 1996 U.S. presidential campaign; (and) veritable boatloads of Chinese money were adding to the coffers of the Democratic Party. The DNC chairman that year was Connecticut Sen. Christopher Dodd, who, along with Jimmy Carter's National Security Advisor, Robert Pastor, supported the Panama turnover as passionately now as he did in 1977."…"So when several American companies, including Bechtel International, tendered bids in the neighborhood of $2 million for leases on the strategic ports at Balboa and Cristobal, Panamanian President Ernesto "Toro" Balladeres secretly changed the rules of the bidding process and accepted the Hutchinson Whampoa offer for $22.2 million a year. Well, if the White House was for sale, why not the Panama Canal?"(7) Of course the Mainstream Media was quiet…would you expect otherwise?

On May 1977, the U.S. established relations with Vietnam and subsequently supported Vietnam's application for membership in the United Nations. Was this finally the "Decent Interval" Kissinger" had in mind? Or was it Carter's bargaining chip to avoid the question of POWs/MIAs and "reparations" that had not disappeared? For instance:

Prior to the above, North Vietnam's Phan Hien, "during a conversation with an Indonesian diplomat, commented on the upcoming visit of a U.S. delegation to Hanoi in mid-March 1977, saying that there were American prisoners of war, some of whom came from wealthy families, 6 still in the SRV, and they would not be released until U.S. financial aid for the SRV was forthcoming."(8) Interesting, yes?

"On March 8, 1977, at a "preliminary meeting", the Vietnamese (first) wanted to know if the Woodcock commission was going to be empowered to discuss the question of U.S. aid to Vietnam, negotiate and sign agreements. The talks ended without an agreement. The Vietnamese had clearly signaled that they intended to use the POWs they had kept as hostages in order to force the United State to fulfill its pledge to pay reparations. The Carter Administration countered by attempting to remove the POWs from the table with the Woodstock Commission findings. On June 2, Hien immediately went on the offensive, castigating the American side for making statements over the last month that would "harden opposition to aid with Congress." "He then went on to make the case that the Nixon letter offering $3.25 billion (plus another $1-1.5) was "a legal commitment which cannot be ignored."(9)

"Two months later the American Embassy in Paris received a call from the (DRV First Secretary Do Than) requesting an urgent meeting (and) saying they hoped the upcoming talks between the two governments "will achieve positive results in answering the interest of the two parties. "As a gesture of good faith, the Vietnamese (said) they were going to turn over the bodies of 12 American MIAs that they had "just discovered." "Do Than hoped "that the American party will respond in a positive manner to problems of interest to the Vietnamese party", i.e. reparations and recognition, in that order."(10)

"When Jimmy Carter (for whom Jesse Jackson had campaigned extensively) was elected, he immediately shut down an investigation into Jackson's finances. "Within months, Jackson saw friendly faces seeded throughout the new administration, and he began to dream a new scheme: Operation PUSH would transform itself into a service provider and cash-in on government welfare programs directly." "PUSH-Excel quickly racked up hundreds of thousands of dollars in donations from nonprofit organizations and received a $402,000 grant from the Los Angeles Board of Education. Carter's Health, Education and Welfare secretary Joseph Califano also approved the first of many large grants to PUSH-Excel in 1978. In total, four separate government departments, the Department of Education, the Department of Labor, the Department of Commerce, and the Department of Housing and Urban Development (HUD) gave a total of $6,677,704 dollars in grants to PUSH-Excel during the Carter years."(11) Wasn't that nice? Thanks to Jimmy, Jesse had found a new cash cow to milk!

The following relates another bureaucratic tale from the Carter years: By way of background, the five month reign of James Schlesinger as Director of Central Intelligence (from February 2 to July 2, 1973) proved to be less than positive for the Agency. He came to the CIA via the Atomic Energy Commission and the Department of Energy, and, like many of the other top management ping-pong ball like appointments in Washington, had also previously worked in the Office of Management and Budget. The selection process for most of these non-cabinet level positions was akin to a game of senior management "musical chairs", with the winners usually determined, of course, by political connections in and within succeeding administrations. Same rules for losers; "just wait your turn."

There was not a "wet eye" in Langley when Schlesinger left to become Secretary of Defense under President Nixon (and subsequently under President Ford as well.) On his first day at Langley, Schlesinger had assembled all available senior officers in the auditorium for a short address on "how-the-cow-was-to-eat-the-cabbage" under his watch. Making a grand entrance with his broken arm in a cast (suffered while bird watching in England was the explanation), he announced that henceforth the Agency would serve as an extension of the White House and whatever the White House wanted (from the C.I.A.) it would get. This reflected Nixon's and Kissinger's) less than enthusiastic support of the Agency over the years, and his still being miffed at the Agency for not covering for him in the lengthy Whitewater fiasco. More significantly, the D.C.I. had become a cabinet member, which theretofore had not been the practice, the previous DCIs usually having deliberately maintained an objective, professional distance from the policy makers. That Schlesinger, Kissinger, and Haldeman had had a strong hand in this realignment of musical chairs was accepted inside the "beltway", but less so in the C.I.A..

Schlesinger also acknowledged to the assembled officers that this change might not set well with all. He suggested that those that did not support his "new look" should feel free to retire. He added that there were to be extensive personnel reviews as well as a reduction in force. As for myself, I was scheduled for assignment to Laos, and thought myself particularly fortuitous in being assigned far from the dark clouds that were obviously gathering over the Langley flagpole. Laos simply was not on the preferred tourist stop list for senior government officials, members of congress, or most bureaucrats at any level. Schlesinger's management style proved to be more slash and burn than it was the result of a thoughtful review, and nearly a thousand career employees, including a large number of the Directorate of Operations (DDO) officers, left the Agency. Schlesinger was intent upon making his control known even though he came to the Agency as one from well outside the intelligence field, his closest previous experience having come from his time with the Office of Management and Budget. It was clear he was certain of his marching orders, and that he intended to follow them. Of course the Liberal Media was supportive and there was no Media protest of note, nor was there a meaningful whimper in congress. The bad taste of the Vietnam era had not gone away, and the blame game was still being played among the bureaucrats and politicians.

I cite the above "changing of the guard" only because it serves as a back-drop and mirror for what it foretold regarding Carter's appointment of Admiral Stansfield Turner to replace Schlesinger as D.C.I., a position he held from March, 1977 - January 1981. Turner was selected by Carter over a substantial number of more senior naval officers, at least some of whom had more experience in intelligence operations, but Turner was not selected for his qualifications to meet the many demands as the Director of Central Intelligences. He was to severely reign in the operational aspects of the Agency, reorganize it, or to completely restructure it. To do this, he brought an interesting assortment of equally unqualified senior staff persons with him to the job, including some from the navy, including a bit of nepotism, and other "friends" who had little or no idea as to what or how the intelligence community functioned. Almost immediately the DDO was given its marching orders in the form of what it could and was expected to do, as well as many more pages addressing what it would and could not do.* The Federal Bureau of Investigation was given similar restrictive instructions from Carter. (These extensive restrictions and "thou shall not(s)" remained in effect until lifted by Pres. Reagan under Executive Orders 12333 and 12334.)"

Turner also came armed with a sizeable sword which he intended to wield widely in further reducing the numbers of operations officers in the DDO, as well as in other Agency Directorates. His approach was to eliminate by GS Grades above GS 10 as many numbers within any particular grade level that might exceed the authorized or perceived strength of manning tables in the Agency, regardless of time-in-grade, expertise, potential, and accomplishments of the individual officers. To his credit, Turner at least listened to dissenting views of a number of senior officers, including myself, who pointed out the obvious fallacies of such an abrupt and unnecessary approach. Albeit reluctantly, he agreed to let the Agency's established promotion panel system address the perceived problem. However, word of his intent spread quickly throughout the Agency and within months a substantial number of already frustrated officers (mostly "wounded" veterans of the earlier Schlesinger swaggerings, and the extensive Colby "show and tell" testimonies to the Frank Church and Otis Pike committees in Congress) chose to retire or resign in less than a positive frame of mind. The end result was a much reduced number of experienced case officers, as well as the attendant loss of operational experience, foreign language(s) capability, counter-intelligence capabilities, and - to a certain extent - further generally decreased morale. Indeed, only in a very few grade categories were there "surplus" number of officers, most of whom had much left to offer. There was little compassion for most of those administratively released, even those who were nearing retirement age after providing many years of satisfactory effort, including many who had served overseas. Very little finesse was exercised.

Behind it all was the intent of the Carter administration to slice and dice the role of the C.I.A., or to restructure if not eliminate it. This was not a whisper, but a rather well-known goal, including identification of the anti-C.I.A. pot stirrers, and those who intended to blunt the effectiveness of the F.B.I., especially its Foreign Counter Intelligence efforts. Fortunately, there were some serving on Capitol Hill who opposed such draconian measures and their counsel slowed the ill-founded intentions. That the C.I.A. well in advance provided Carter with a detailed script of what the principals would agree to in the Camp David meeting(s) between Egypt's President Anwar Sadat and Israel's Prime Minister Menachem Begin in September, 1978, measurably reduced Carter's anti-C.I.A. attitude. Thanks to the intelligence provided him, Carter's role at Camp David required and was basically little more than his playing banker, water-boy, script prompter, and to pose for photos. Sadat had initiated the peace process by visiting Israel in 1977, setting the stage for the peace treaty that followed in 1979, for which both Sadat and Begin received the Nobel Peace Prize in 1978. Prior to that, Kissinger and Ford had also made extensive and 'promising' efforts to ease tensions in the mid-east. Unfortunately, the successful discussions between Sadat and Begin in September, 1978, was followed by a four day meeting in Iraq in November 1978 attended by 20 Arab League nations denouncing the Camp David agreements as usurping the rights of the Arab people. With the assassination of Sadat in Egypt in 1981 by extremist Moslems, peace in the area took numerous steps backward, and remains elusive today because of many of the same strident minority squabbles.

In January 1979, the Shah of Iran (Mohammed Reza Pahlavi) and his family left Iran because of the increasing demonstrations and strikes against his government, paving the way for the return of Ayatollah Khomeni after 15 years of exile in France. In November 1979, some 500 militant Iranian "revolutionaries" seized .the American Embassy in Teheran, taking 65 American hostages, a few of which were released by the end of the month. The remaining 52, less one who was later released because of poor health, were kept hostage until January 20, 1981. The "revolutionaries" demanded the return of the Shah (for his "crimes"), who upon leaving Iran had first sought medical treatment in the U.S. Carter then forced the Shah to find safe-haven outside the United States, an act some called shameful and certainly not in keeping with Carter's oft professed pursuit of "humanitarianism". The Shah died in Egypt on 27 July, 1980, in essence a stateless leader.

A real twist to intelligence gathering was that Turner and his staff brought with them the idea that the Agency did not really need to have unilateral sources of information (HUMINT) except in very unique circumstances. His concept called for case officers posted abroad relying on friendly and cooperative liaison relationships with host government intelligence services to meet the Agency's needs. Thereby this would not only save many overseas billets, but simplify the cumbersome collection process historically used involving development and use of unilateral sources. Why couldn't we simply gain the needed information from by our liaison counterparts? It took time, and many discussions, but this fool's errand approach was finally abandoned. Equally bad but actually experimented with was the idea of cross-training. For instance, a person who had been an intelligence analyst for a period of time would be selected to be a case officer overseas, or in charge of other case officers overseas. Or a case officer would be assigned to analyst duties. In theory it is plausible, but to expect one who is not experienced in the tradecraft arts to become a good case officer, or to lead good case officers, is more than a bit much…and not worth the increased risks involved, either to the foreign assets involved or to case officers. The case officer doing analyst duties is more plausible, but not exactly career enhancing or satisfying in most cases. Such proposals happen when there is unqualified leadership at the top. Turner was probably well intended, but on-the-job-training was, and is, not applicable to the intelligence business, nor is using it as political plum. With the exception of the very capable James Woolsey, from the 1990s to the present the CIA has been poorly led. The performances of the political appointments following Woolsey clearly underscore this point. But back to Carter…

"In April 1980, Carter ordered a secret military rescue mission, which (predictably) failed like a lead balloon. The operation was aborted after three of the eight helicopters broke down well before reaching Teheran. During the withdrawal, a chopper crashed into a transport plane, killing eight servicemen and dashing Carter's credibility."(12) Left unsaid was that the rescue attempt was botched from the get-go. First problem - it was a split military command (between the air and ground elements); second, - there were few if any rehearsals; third, - there were many unresolved unknowns, thus inviting the disaster that evolved; fourth, insufficient numbers of helicopters were committed to the demanding operation, including that the helicopters reportedly did not have screens to protect against the sand and dirt that the helicopters were to experience in the rescue effort;…and lastly, only the desert landing portion of the operation was achieved. (Agency assets in Tehran did their part such as being in place ready to block all roads near the embassy, etc., had the rescue actually been attempted, but many of those assets had to flee when subsequent investigations in Tehran by the Iranians threatened to identify them.) The actual helicopter rescue was to take place after the assault force had stormed the embassy and the hostages moved to a close-by soccer stadium, from which the helicopters would fly them to the desert rendezvous site and the waiting C-130. This was an operation not totally unlike the demanding but failed Son-Tay prison camp raid in November, 1970, during the Vietnam War (the camp no longer held U.S. POWs), but far less professionally pursued. The hostage rescue effort deserved, but did not get, the best. The helicopters had been exposed to salt water and the elements as they were being carried on ship deck to the launch point, adding to the ultimate SNAFU that resulted. So much for trying to go on the "cheap"…and in haste. As former DCI Richard Helms noted in his book, "A Look over My Shoulder" (Random House, 2003) although he had been posted to the American Embassy in Tehran from early 1973 until December, 1976, no one (civilian or military) contacted him in regards to the "rescue operation" despite his having a wealth of information and detail on the Embassy offices and grounds, local employees, likely ingress and egress routes, etc. But on second thought, that is not really surprising given the political dynamics and attitudes involved. To his credit on April 28, Secretary of State Cyrus Vance resigned in protest over Carter's decision to attempt the ill-fated "rescue" operation. Not surprising was that he was replaced by his deputy, Warren Christopher. How inspiring, to say the least! And the Mainstream Media? A few bleats and howls, mainly over the photos of the dead.

A final "operational" note that the reader may find of interest: In late 1976 and into 1977 the Agency developed a penetration of the Politburo of North Vietnam that led to uncovering a well funded and organized network of 'activist' radical, if not communist cells across the country, including college campuses. Working closely with the D.I.A. and its Director, Admiral Inman, the Agency determined that financial support for much of this network was provided by the Soviet Union, using Cuba's diplomatic pouch to the Cuban mission at the United Nations to bring large amounts of cash into the United States. The agency also learned that copies of sensitive State Department messages were being sent regularly in envelopes addressed to the North Vietnamese Embassy in Paris, France. Unfortunately, neither the Agency, nor the FBI, were allowed by Carter's Justice Department to move rapidly on either of these two activities. Given the restrictions that had been placed on both the Agency and the FBI by the Carter Administration, both were slowed to snail pace, each step along the way requiring specific approval in advance by the Justice Department, to the dismay if not chagrin, especially of the FBI, which had to resolve countless road blocks in their investigative effort.

Equally frustrating was the ruling by the Justice Department that even though the posted envelopes could clearly be seen (by rolling the flap a bit) to contain copies of secret State Department correspondence, the determination was made that it was not legal to intercept or interrupt the flow of U.S. Mail so long as it was part of the system. In other words, until delivered to France, these envelopes were simply part of the U.S. postal system. Fortunately, the French intelligence service proved to be cooperative and they passed to Agency officers much of the intercepted "mail" involved. In turn, subsequent investigations at State confirmed that their mailroom had indeed been penetrated and espionage was being committed. Eventually the guilty parties were appropriately charged in court. Equally interesting was the change in attitudes of some involved in this lengthy process. Ranging from initial expressions of disbelief and outrage that anyone could think that such things could be going on, the attitudes slowly changed to wry amazement and reluctant belief that such things could – and did - occur. One could have concluded that the "messengers" were held in lower regard by some in the administration than those who were blatantly working against our country...head shaking, to say the least!

"The problem, through Carter's and other liberals' eyes, was not the existence of a communist threat but our groundless paranoia about the threat. After all, if our fear of Communism was once "inordinate", surely it was because the Soviet Union and other Communist states were not particularly threatening, not particularly evil. Vietnam and its associated disillusionment had so soured liberals on America that they found it a terrible strain to condemn a communist nation. ...This was not so much isolationism as self-loathing...Post-Vietnam liberal isolationism saw the world as too good for us. Had ten new nations fallen into Communist hands between 1974 and 1980? Yes. South Vietnam, Cambodia, Laos,

South Yemen, Angola, Ethiopia, Mozambique, Grenada, Nicaragua and Afghanistan. Were nations from the horn of Africa to the Caribbean under siege by communist insurgents funded and supported by Moscow? Perhaps. But the Carter administration was untroubled by this."(13) As Secretary of State Cyrus Vance said: "The fact is", he warned, "that we can no more stop change than Canute* could still the waters". However, he failed to note that "No nation on the planet had ever chosen, through truly free elections, to be governed by Communists. And if they did hold phony elections, it was said to be "one man, one vote, one time". (14)

"In the same speech in which he "freed" Americans from fear of communism, President Carter sounded his call for a foreign policy based on human rights. What he did not say, but what the policy became in effect, was a program for lambasting America's noncommunist and anticommunist allies while permitting the monstrous human rights nightmare of the Communist world to go almost unmentioned."(15)... Jimmy Carter may not have been paying close attention, but during the 1970s the Soviets: threatened to enter the Yom Kippur War if Israel destroyed Egypt's attacking force; fomented revolution in a dozen countries; supported a network of provocateurs and spies within the United States; and circulated disinformation about America to the international press. They arrested Jews and Christians for the crimes of teaching religion or distributing Bibles, sentenced same dissidents to mental institutions, and armed and supplied the deadly North Vietnamese, North Korean, Somali, and Ethiopian regimes. The Soviets encouraged Castro to send his soldiers to several African nations to serve as a Communist foreign legion, and supplied communist insurgencies in El Salvador, Nicaragua, and Grenada, among other places." "The Soviets also gave funding and training to the Japanese Red Army, the Red Brigades, the Palestine Liberation Organization, the Baader-Meinhof gang, the French Action Directe, and the Popular Front for Liberation of Palestine — all of whom made murder of innocent civilians a part of their political agitation." "A glance at the rhetoric Carter and other liberals chose to use when speaking of the Soviet Union reveals, not the firm and consistent "containment" that so many would later claim, but instead a record of eager self-delusion and ready forgiveness of every

Soviet aggression."(17)

And "...it was the Carter administration that cut off aid to the Sandinista regime in 1980 due to its support for the FMLN guerrillas in El Salvador. Carter simultaneously resumed aid to the Salvadoran government. But it was easy for liberals to forget those facts when Reagan's anticommunism made them dizzy with outrage."(18)

"Though it would later be alleged (again) by liberals that American hostility drove the Sandinista regime into the arms of the Cubans and Soviets, the truth is that the United States helped to topple Somoza in the first place (imposing an arms embargo in 1978). With full U.S. participation, the Organization of American States, for the first time in history, issued a call for the replacement of the Somoza regime with a pluralistic, democratic opposition. The Sandinistas pledged to implement the reforms demanded by the OAS, including a promise to hold early elections. Taking them at their word, the U.S. offered generous aid and friendly relations with the new Sandinista Directorate when it first took power. The Carter administration provided $118 million in direct aid, negotiated with the international development banks for an addition $262 million, and arranged for refinancing of $500 million in private bank debt. This was more aid than any other nation provided to Nicaragua, and more than the U.S. had contributed to Somoza in the previous four years. Jimmy Carter also welcomed Daniel Ortega to the White House."(19)

"Deputy Secretary of State Warren Christopher (later secretary of state in the Clinton administration) testified before a House committee in September 1979 that "The (Nicaraguan) government's orientation, as revealed in its initial policies, has been generally moderate and pluralistic, and not Marxist or Cuban." Christopher further assured the Congress that the Sandinistas were desirous of "close and friendly relations" with the United States. Even after the new leaders had banned the opposition, begun to militarize the country, and visited Cuba, President Carter resented the suggestion that this should alarm us: It's a mistake for Americans to assume or to claim that every time an evolutionary change takes place in this hemisphere that somehow it's a result of secret, massive Cuban intervention. The fact in Nicaragua is that the

Somoza regime lost the confidence of the people. To bring about an orderly transition there, our effort was to let the people of Nicaragua ultimately make the decision on who would be their leader –what form of government they should have."(20)

"Carter was still battling Joe McCarthy. But the reality in 1979 was that Cuba was openly, not secretly, aiding the Sandinistas, though, to be sure, Castro advised the Nicaraguans to be cagey about their ideology. One of the "commandantes" said later "Our strategic allies (i.e. Cuba) tell us not to declare ourselves Marxist-Leninist. (Nicaragua) will be the first experience of building socialism with the dollars of capitalism." "It is difficult to understand how American liberals, having watched Communist regimes repeat the same pattern again and again for sixty years, could have failed to anticipate that the Sandinistas, once in power, would squeeze out all opposition, shut down the free press, and militarize the country. Yet time after time, liberals would follow their own well-worn pattern: first deny that the revolutionaries were Communists; second, assert that the no longer deniable repression was a response to U.S. actions; and third, laud the accomplishments of the revolution while offering perfunctory criticism (if that) of the human rights situation."(21)

Having served in Laos (1973-1976) and Burma (1977-1980) during the Ford (and Kissinger) and Carter administrations I can attest that the decline of U.S. prestige in Asia during that period was pronounced. In turn, that decline made the recruitments of well placed foreign nationals that much more difficult world-wide. There are basically three things that are key to a quality recruitment: one is the prestige of the U.S. with which the potential asset can identify; another is the personal rapport the case officer establishes with that individual; and third is the trust that the target has in both the U.S. and the case officer, including that he/she will not be exposed. With the absence of any one of the three, the prospective asset is usually going to decline, for his/her life may well depend on the trust and belief factors.

When the U.S. loses its luster, there are few words that can overcome that important factor. I should add that the Clinton-Gore administration was also considered very weak and less than impressive abroad. With friends I visited India, Nepal, Tibet, and Hong Kong in 1997 and we found there was little respect for the weakened U.S., with many questions as to what was happening in and to the U.S. In 1998 I visited Canada and the socialist-minded there thought it was great that we were emulating Canada's liberal slide, while the non-socialists despaired. In 1999 I again visited India and Nepal with two of my sons and a close friend, and we were often asked if the U.S. was going to survive the Clinton-Gore years. The beacon of freedom and strength had dimmed measurably in their eyes, and none thought that the U.S. recognized the threat the PRC posed throughout that area.

But there are more "Carterisms" that deserve to be noted: *

- "For the past many years, he (Carter) has been passionately anti-Israel, more or less embracing the PLO line. He has repeatedly been at the service of Yasser Arafat. After the Gulf War, the PLO chief was on the outs with Saudi Arabia because he had backed Saddam Hussein. So he asked Carter to fly to Riyadh to smooth things over and restore Saudi funding to him, - which he did. Arabs are also robust funders of the Carter Center."
- "As Brinkley writes in his book "the Unfinished Presidency", there was no world leader Jimmy Carter was more eager to know than Yasser Arafat. Shortly thereafter, Carter actually acted as PR adviser and speechwriter to Arafat." The entire composition is nauseating, but its flavor can be captured in a single line: "Our people, who face Israeli bullets, have not weapons: only a few stones remaining when our homes are destroyed by Israeli bulldozers." And Carter wrote "(Arafat) may well see the suicide attacks as one of the few ways to retaliate against his tormentors, to dramatize the suffering of his people, or as a means for him, vicariously, to be a martyr". That Arafat has spent his adult life as a "terrorist" without any regard for human life seems to have escaped the peanut farmer from Georgia…and the Mainstream Media.
- "A mission statement of the Carter (library) says, "Human rights' is a broad term, encompassing freedom from oppression and freedom of speech to the right to food and health." "As Jeanne Kirkpatrick – another Carter bogeyman – points out, it's odd how those with "political rights" also tend to have plenty of food, plenty of shelter, and plenty of health; it's further odd how those denied

"political rights" also tend to do without material comfort."

• "In 1997, Carter wrote an op-ed piece entitled "It's wrong to Demonize China." In it, he said "Westerners emphasize personal freedoms, while a stable government and a unified nation are paramount to the Chinese. This means that policies are shaped by fear of chaos from unrestrained dissidents or fear of China's fragmentation." "He also suggested that freedom of religion had come to China – causing activists in the field, who know the horrible truth, to gnash their teeth." (AND, tell that one to the Tibetans, and/or other minorities in China, Jimmy! You totally ignored their plight!)

• While in office Carter hailed Tito as a "man who believes in human rights". Since leaving office, Carter has praised Syria's late Assad (killer of at least 20,000 in Hama) and the Ethiopian tyrant Mengistu (killer of many more than that). In Haiti, he told the dictator Cedra that he was "ashamed of what my country has done to your country."

• While in North Korea, Carter lauded Kim IL Sung, one of the most complete and destructive dictators in history. Said Carter,"I find him to be vigorous and intelligent ...and in charge of the decisions about this country." "I don't see that they (the North Koreans) are an outlaw nation." Pyongyang, he observed, was a "bustling city" where shoppers pack the department stores, "reminding him of the "Wal-Mart in Americus, Georgia." Rip Van Winkle revisited?

• "During our preparations for the Gulf War, Carter actually wrote members of the U.N. Security Council persuading them to vote against the U.S. The Bush team discovered Carter's outrageous antics only when dumb-founded Canadian Prime Minister Brian Mulroney called Defense Secretary Cheney to ask what was going on. Some might call Carter's actions treasonous."(22)

• "In 1984 Carter visited the residence of Soviet ambassador Anatoly Dobrynin in Washington, D.C., to warn Soviet officials against President Reagan's re-election, saying that if Reagan were re-elected, "there would not be a single agreement on arms control, especially on nuclear arms, as long as Reagan remained in power." And "...in the waning days of the 1980 campaign, he sent the founder of Occidental Petroleum, and noted Soviet sympathizer, Armand Hammer to the Soviet Embassy in Washington for a secret meeting. Hammer asked if Moscow could help Carter, promising the president's gratitude in return." And "earlier still, during his 1976 campaign, Carter did something similar when he sent Averill Harriman to Moscow to assure the Soviets that Carter would be easier to deal with than Ford." (23) Such stirring qualities!

• Like his good friend Yasser Arafat did before him, Carter finally received his long sought Nobel Prize for peace in 2002. However, it was somewhat muted by remarks made by members of that selection committee to the effect that the award was more a criticism of the Bush administration by those who are anti-American (than a deserved tribute to Carter). Nor should we forget that on Clinton's behalf, Carter brokered an "agreement with North Korea involving several hundred million dollars in free fuel and food and two new light water reactors valued at a couple of billion dollars in exchange for North Korean promises to halt their nuclear weapons research programs. They promised and Carter claimed success."(24) Neither proved to be true.

Carter's trip to Cuba in May 2002, and his obeisant tributes to Castro and Cuba, including his supercilious attendance at a baseball game with the bearded dictator, complete with baseball cap (but looking more like a tired vendor than an "honored guest"), was not much different than his interesting ties to Arafat. Carter again managed to repeatedly attack both the U.S. and President Bush. While addressing the Cubans, Carter called for an end to the U.S. embargo on Cuba, praised Castro for all the wonderful benefits the latter had brought to the Cuban people, and urged the people to be politically active as "permitted" by the Cuban government (which shortly after Carter's departure, landed a large number of Cubans in jail as they attempted to follow his advice). And upon return to the U.S. he proudly reported that he and Rosalynn had found no sign of any weapons of mass destruction or indications of bio-terrorism supplies anywhere during their five day wanderings in Cuba. Given his extensive engineering background, superb investigative skills, and fearless record on pursuing human rights, one could conclude that Carter had readily demonstrated that he was now fully eligible to become a U.N. inspector.

What does one conclude about Jimmy Carter and his oft-repeated sermons on "human rights", and "caring", and "compassion", and "concern for our fellow man", and "being firm against those who do not share those same principles"? His deeds certainly do not match his vacuous words; his suit is empty. Amazingly, his "preacher's disguise remains intact.

One would think that Carter would have the good sense to simply be quiet as most former presidents, such as Ford and Bush, have wisely done. Carter was not a popular or effective president, and his record is, at best, very checkered. He came to Washington as a former governor of a small state, preached to an ever diminishing congregation, and lost badly because of his very inept performance. Doesn't he get it?

"The only two living former Democratic presidents publicly opposed the Bush administration's plan to invade Iraq: Jimmy Carter couldn't land a helicopter in a desert, but he seemed to imagine the public was hungry for his counsel on the war on terrorism. Carter is so often maligned for his stupidity, it tends to be forgotten that he is also self-righteous, vengeful, sneaky, and backstabbing. In a vituperative piece in the *Washington Post,* Carter peremptorily announced, "There is no current danger to the United States from Baghdad." The real threat to world peace and global human rights, he said, came from President Bush. The Bush administration, with its "belligerent and divisive voices," was becoming "similar to those of abusive regimes that historically have been condemned by American president."(25)

What prompts him to think people listen, or believe, or care now any more than when he was summarily voted out of office? In short, "Goodnight, Jimmy, Goodnight! Take the rest of the decade off, please!

And what does one conclude about the Liberal Media and its convictions, or lack of same, and its many failures to report objectively and accurately on the Carter administration? Most of us had worried and wondered how and why our Republic had regressed during these unsettling years... but little did we know that far worse was to follow. The braying Liberals, the Liberal Media, and the socialist Democrats in both houses of Congress made a large part of the final decade of the 20th century look as though a lip pursing

Jimmy - and Rosalyn - Carter had been "reborn" and returned to office as William and Hillary Clinton moved their act from Little Rock, Arkansas, to Washington, D.C., in January, 1993.

Yes, the physical differences are pronounced, yet much of the bull manure the Clintons found useful in Arkansas is simply repackaged liberal left refuse of the same low quality. And there was more hot air and heavy "baggage" hauled from Arkansas to the District of Columbia than had come from Georgia. Sadly, the administration of recessive forces began again in just four short years after Jimmy waved goodbye. Amazing, but such is life in these United States. Apparently Americans can't resist vaudeville, a glitzy carnival, and snake oil salesmen.

There are other less than admirable factoids on Jimmy Carter, both from his time in office and since, including such things as his performance on behalf of Pres. Clinton with (former) Haitian Pres. Aristide, his amazing simplistic and costly performance as he "hard-bargained" North Korea in the 1990s, his insightful visit to Cuba to assure us all is well there, and his truly insightful comments on the middle east to date. Had he learned to play the guitar and possessed a good sense of humor, Jimmy might have made it as an imitator of comedian "Lonesome" George Gobles, but alas, only his story about having once been pursued by a rabbit remains legendary. Seems fitting, somehow.

• A Hebrew prophet of the 8th century; a prophetic book of Canonical Jewish and Christian scripture.

• These figures are taken from "American Presidents", pgs. 398, 399

* 11th Century King of England, Norway and Denmark

* From the National Review, May 20, 2002 "There he Goes Again", Jay Nordlinger.

CHAPTER 5

WILLIAM JEFFERSON BLYTHE CLINTON

("Pretentious Bill &Hilarious Hillary")

There are some who have defiled the White House in various ways over the years, but none have done so much soiling in so many and crass ways as the 42nd president, William Clinton, and his wife, Hillary, during the eight years "Slick" Willie survived in office. The nation should have had at least an inkling of what was to come if only from their more than ten years of highly suspect and nefarious activities in Arkansas. Even as the Clintons left the Blair House on January 20th, 1993 to pick-up President and Mrs. Bush at the White House prior to the inauguration ceremony, they were crass. The reported exchange between Bill and Hillary could not be described as an endearing or promising one: "…it wasn't exactly all hugs at the Blair House, according to a U.S. Park police officer, who was standing in close proximity to the Clintons that morning, who overheard Bill as he came out of Blair House, saying "That fucking bitch!" When Hillary came charging out after him, she was saying, "You stupid motherfucker!"(1)

Almost equally shocking was that Harry Thomason and Linda Bloodsworth-Thomason, friends of the Clintons from Arkansas who had become Hollywood producers of prime-time entertainment, were in tow with the Clintons. President and Mrs. Bush had been kept waiting for the Clintons for 30 minutes, and did not know that the Thomasons were with the Clintons as their 'personal guests' to the inauguration. Thus we had, in essence, a preview of coming attractions. In this instance it was a gross violation of established protocol – but obviously that did not bother the Clintons, or the Thomasons.

Thus began the odyssey which continues even today. That the House of Representatives voted to impeach Clinton on two charges (perjury and obstruction of justice) on Saturday, December 19, 1998, followed by the Senate vote to acquit him on Friday, February 12, 1999 did not bother the whining couple; they simply continued to blame their problems on "right-wing" conspirators. The tactless "we're not leaving" departure of "Slick" Willie and "Carpet Bagger" Hillary from Washington on January 20, 2001 merely underscored their tainted reign.

Scores of books have been written about the Clintons. Most of them have been highly critical of them – deservedly so. This chapter will simply highlight some, but not all, of those deserved criticisms, leaving it to the reader to review the sordid details involved. If you are looking for praise of the Clintons or words supportive of their deeds, best you proceed to the next chapter, for you won't find such sentiments herein. Why? For the simple reason that such is not deserved. The Clintons have earned naught but enduring disdain, but this is getting ahead of the story.

William Jefferson Blythe (Clinton) was born in Hope, Arkansas, on August 19, 1946, three months after his father was killed in an auto accident. His mother, Virginia, remarried four year later to Roger Clinton and the family moved to Hot Springs, Arkansas, known as one of the better tourist playgrounds and "famous for its racetrack, its ornate bathhouses, its casinos, its prostitution, its sleight of hand and its epic defiance of Prohibition"(2) Another Clinton, Roger, was born there. Bill Blythe legally changed his name to Clinton at the age of 15. Reportedly a good student, he was active in non-athletic student activities, including debate, and won all-state honors as a tenor saxophone player. A doting and politically connected uncle who owned a Buick dealership in Hot Springs kept young Bill in cars and pocket money.

To summarize, Hot Springs was the spiritual capital of the "Dixie Mafia…formed (by) a ring of interlocking interests that covered Louisiana, Texas, Oklahoma, Kentucky, Tennessee and above all Arkansas. The coat-and-tie yuppies of the modern Dixie Mafia are the children and grandchildren of bootleggers, a provenance they share with Bill Clinton. The trade has evolved. Clinton's grandfather used to serve moonshine from behind the counter of his store in Hope. Now the business (in Arkansas) is a high-tech operation involving fleets of aircraft, off-shore banking, and deep reach into the U.S. federal government."(3) "During a ten year period from 1968 to 1978 when the Italian Americans were in the headlines for a spree of thirty murders, their redneck (Dixie Mafia) counterparts quietly dispatched 156 victims. There wasn't a well from Mississippi to West Texas that didn't have a dead body floating in it."(4)

• Clinton's high school achievements, plus the influences of his well-to-do uncle, his mother's many and varied social contacts, including Arkansas politicians, were equal to the task of gaining him admission to Georgetown University in Washington, D.C. There he helped cover his school costs by interning in the office of Senator William Fulbright, a leading and influential liberal Arkansas Democrat. In the summer of 1969, "Apparently Clinton promised to enter the University of Arkansas Law School, enrolling in the university's ROTC program if Colonel Holmes (a retired army ROTC recruiter at the University of Arkansas) would extend his (Georgetown) draft deferment. Holmes reported to the Wall Street Journal that subsequently Clinton was able to manipulate things so that he didn't have to go in."(5) Clinton denied the story by Col. Holmes, "...claiming that he had passed up enrollment in the University of Arkansas Law School to "put myself into the draft when I thought it was a one hundred percent certainty that I would be called", and added "I just didn't feel right about having a four-year deferment."(6) However, a letter that Clinton had written Holmes soon surfaced in which he thanked Holmes for "saving me from the draft" and explained why he broke his promise to enter the ROTC program at the University of Arkansas. "He boasted of his involvement in anti-war activities in Washington and in England. Then he reminded this World War II veteran, a survivor of the Bataan Death March who had languished in Japanese POW camps for over three years, of the student protesters' "loathing the military."(7) That he had chosen not to inform Col. Holmes, or the draft board, that he was making himself available for the draft, that he had kept quiet until he had received a high draft lottery number (all but insuring he would not be called), and the duplicity involved, seemed to escape the Media. "Reflecting back, one might well ask why only he (endured) so many scandals involving deceit, womanizing, petty corruption, and blatant lies. The scandals trailed back through all the years of his public life and into his earliest college days."(8) Perhaps the following examples help explain the real Clinton:

• Father Richard McSorley, a liberal Jesuit priest and activist at Georgetown University, wrote in 1977 (in his book *Peace Eyes*) that "...while I was waiting for the ceremony to begin (an anti-war demonstration outside of the U.S. Embassy in London in November, 1968, Bill Clinton, then studying at Oxford, came up and welcomed me. He was one of the organizers." These anti-American demonstrations were supported by the British Peace Council and the World Peace Council. Both organizations are documented as KGB front groups.

• The *Washington Times* reported that while in England, Clinton had been described by Lincoln Allison, a political lecturer at Oxford, as "being purely selfish...his protestation being based more on cowardice than ethically serious."

• To summarize: Clinton registered for the draft on September 8, 1964, and given Selective Service # 3 26 46 228; he was classified 2-S on November 17, 1964; on March 20, 1968, he was reclassified 1-A; Clinton was ordered to report for induction on July 28, 1969; but did not report; Clinton was reclassified 1-D after enlisting in the U.S. Army Reserved on August 7, 1969 under the authority of Col. Holmes; Clinton signed enlistment papers, and took the oath of enlistment, but failed to report to his "duty station" at the University of Arkansas ROTC in September 1969; Clinton was reclassified 1.A on October 30, 1969 following revocation by Col. Lewis of his "enlistment". According to Public Law 90-40 (2)(a), "registrant who has failed to report remains liable for induction." Clinton's birth date lottery number, 311, was drawn December 1, 1969, but anyone who had been previously ordered to report for induction was ineligible.* Did Clinton manage to mangle our laws to the point that Arkansas elected a felon as Attorney General and thence Governor? And did he take the oath of office as President as a felon? Or did Jimmy Carter's sweeping pardon of draft dodgers on January 21, 1977 include Clinton? I don't recall Media's defining – much less raising - any of these points

• On October 25, 1992, the *Washington Times* also reported a trip by Clinton in 1969 of some 40 days duration from England to "...Sweden, Finland, the Soviet Union and Czechoslovakia...and he stayed in one of Moscow's more expensive hotels, the National, much favoured at the time by the Soviet elite, after which he visited Prague, the capital of one of the Soviet Bloc's most repressive regimes. The trip to Moscow alone in those days could easily have cost $5,000." The Rhodes scholarship provided less than $2,800 for tuition, room and board; even with substantial help from his favorite uncle in Arkansas it would be hard to explain from whence the money came for these expenses. Clinton has also said that he visited Oslo and the International Peace Research Institute (another KGB sponsored front organization) located there. The most bothersome point in all this is that throughout that period of time *no* private foreign individual, especially an American, could simply "visit" Moscow – or Prague – without having been approved in advance, or invited, by the KGB as its guest. Neither the reason(s) for this trip nor the identification of his contacts in Moscow have been revealed by Clinton, nor has the KGB, so far, chosen to divulge any information on the subject. Perhaps some day.

• In Prague, Clinton stayed with Bedrich Kopold, a Marxist, one of the prominent leaders of the Communist Party in Czechoslovakia, a country wherein terrorist training was an accomplished art during the Cold War. This is not to infer that Clinton received any such training, for obviously he has no stomach for such physical things. His is more a record of leaving non-sexual physical activities to others.

• The answer to the questions of where the money came from for Clinton's travels is, according to Greg Huddleston of Brunley, Missouri, "the Vietnam Committee for the Solidarity with the American People". According to Mr. Huddleston's letter to the American Spectator (September, 1996), "It was entered in federal district court (Mace v Blunt, United States District Court, Western District of Missouri, Central Division, Case #92-4484-CV-C-9) that Clinton's "trip to Moscow and his extended stay at the home of his roommate from Oxford, Jan Kopold, was paid by the "Vietnam Committee for Solidarity with the American People." According to federal documents, the VCSAP was an organization established in 1967 by the National Liberation Front and its North Vietnam allies "to collect and organize pacifists and war dissenters to assist their efforts against the US imperialist war of aggression in Vietnam." The VCSAP did well under the direction of its appointed chairman, Hoang Minh Giam, who opened storefront offices in a number of cities, including Moscow, Hanoi, Algiers, Prague, etc. The North Vietnamese also had well funded and organized nets and cells across the United States, with heavy concentration - and cadres in place - on most college campuses. Where better to fan the flames of protest and discontent? Thus we appear to have the North Vietnamese, the KGB and Clinton on the same track. But why was there been no Media interest in this?

• Clinton's fellow Rhode Scholars at Oxford – and demonstrators - included similar minded socialist liberals Ira Magaziner, Robert Reich, and Strobe Talbott. Upon return to the U.S., Talbott found employment in New York at *Time.* Reich caught on at Dartmouth, then Harvard, and Magaziner at Brown University, where he successfully led the fight against the administration giving students failing grades. Why, the outrageous *audacity* of the Brown Administration to require passing grades!

• From Oxford, Clinton was off to Yale, more campus debates, further refinement of his gift of gab, and a law degree in 1973 – after working with Hillary Rodham, also a 'Yaley', on behalf of George McGovern's unsuccessful campaign for the presidency in 1972. No sense working when you can pose as a political organizer!

• When President Bush raised the subject of Clinton's patriotism during the first election debate in 1991, Clinton – and his entire retinue – heatedly accused Bush of *McCarthyism,*[1] and in turn accused the Bush administration of maintaining files on Clinton, Clinton's mother, and perhaps Hillary as well. That the State Department would routinely have had a record of passport data on the aspiring Rhodes Scholar(s) visiting England, or that the FBI would have had a file on Clinton because of his blatant anti-American activities – and his draft dodging antics - was not pursued by the Media. The topic was not raised again by Bush during the debates. Most of the Media were openly sympathetic to Clinton. Nor did the Media choose to investigate in any depth the many stories involving Clinton's venality, his womanizing, his use of narcotics, his (and Hillary's) unusual and suspect financial dealings while Governor, Clinton's Dixie Mafia contacts, the Mena airport story, his personal and questionable use of Arkansas's State Troopers, etc. Neither lying nor character, or even malfeasance, appeared to be of interest to the Mainstream Media, or of concern to the Clintons and their associates.

• At Oxford, famous for its Marxist bent, and its Rhodes Scholars, Clinton was one of six in his "class" of 32 who did not finish the school's requirements. His repeated and lengthy unexcused absence on "trips" has been offered as the explanation, although the Media generally ignored the entire subject. Simple question: Where did he go?

What had surfaced by this time merely foreshadowed that which would follow. The differences were primarily the backdrop for and scope of the Clintons' scandals both in Arkansas and during his eight years as our 42nd president. Unfortunately, the failure of the Mainstream Media to track the Clintons' guile accurately or in depth remained largely unchanged during the '80s and '90s.

To return to Bill Clinton efforts in Arkansas, he first taught law at the University of Arkansas and made an unsuccessful run for the House of Representatives in 1974. After playing house for about a year, Hillary and Bill were married in October, 1975 and both continued to teach law at the University. In 1976, Clinton, running unopposed, was elected state attorney general, and Hillary became employed by the Rose law firm. A quick study, Clinton came to know the various state offices, key players, and opportunities in Arkansas. In 1978 he was elected governor. However, he and Hillary managed to alienate the politically influential, the power brokers, and the voters by their indifferent and inattentive attitude, leading to his losing the governorship in 1980. The shock of defeat prompted Bill and Hillary to reassess their political practices and somewhat modify their cavalier ways, including Hillary deciding to use "Hillary Clinton", vice the "Rodham-Clinton", which had annoyed many Arkansans, and to revise her "hippy" image. Both apologized and sought forgiveness for their "mistakes" of '78 and '79. Their extended campaign effort to win the support requisite to regaining the governorship in 1982 was successful. As Governor of Arkansas, with Hillary's help, Clinton pushed through a uniform teacher testing requirement in Arkansas, the first state to do so in the country, which gave him national political recognition. But the day before the test was to be given, the previously sealed and secret test was leaked to the media and printed in a local newspaper. Despite protests that the test had obviously been compromised, it was administered. Most of the teachers passed. Surprise! Only in Arkansas.

Clinton quickly demonstrated a willingness to make accommodations in Arkansas beneficial to big money interests, powerful politicians, unions, and opportunists, thereby identifying himself as a leading contender for higher office. Stated more bluntly, while governor, he demonstrated he had few scruples, was adept at avoiding the truth, was a masterful 'yarn spinner', and had weaknesses that could be manipulated by the king makers. He proved he was malleable, able to play political hard-ball, and willing to do what was necessary to become the ultimate political candidate, with Hillary fully in lock-step with him. The Media were pliant and all but ignored what had transpired in Arkansas during the Clinton years, such as continued poverty, inadequate health care, poor academic achievements, and rampant corruption, including intra-state, interstate, and international schemes. One could say that Bill Clinton managed to make Huey Long look amateurish in comparison, with both having many similar traits. But in fairness, Huey did not have an equivalent sized cash register to the one that Bill had in Arkansas in the form of what Lasater, Tyson, 'bonds', and the many Mena tentacles represented.

The Inverted Pyramid Known As Mena

If memory serves correctly, Mena first came to the attention of the Media in the late '70s when Clinton agreed to Jimmy Carter's request that Arkansas "house" some of the 125,000 Cubans arriving from Cuba. This number included several thousand criminals that Castro had freed from Cuban jails, in the Castro-Carter approved exodus from Castro's "paradise" for those wanting to leave Cuba for the U.S. Clinton chose the Mena area in southwestern Arkansas because of its remoteness as the relocation site for the prison that was to be built there, primarily for the Cuban criminals, to the dismay of Mena's rural population. A new highway was included in the Mena package, as was an expanded airport, easing the local acceptance problems. Thus was established the setting for the ensuing years during which Mena would quietly prosper as a modern air head for various activities ranging from aircraft refitting to U.S. aerial support to Central and South American countries, to training of foreign air crews, to overlapping DEA and CIA covert operations, and opportunities for some Arkansas based entrepreneurs to illegally enrich themselves. None had to wait long for the multiple opportunities at hand, for Governor Bill Clinton was just the man to help all of them.

As noted in the chapter on Carter, in 1978 the U.S. chose to impose an arms embargo on President Somoza of Nicaragua and joined the Organization of American States (OAS) in calling for a replacement of the Somoza regime. Believing the Sandinista pledge to implement the reforms demanded by the OAS, including a promise to hold early elections, "...the U.S. offered generous and friendly relations with the new Sandinista Directorate when it first took power. The Carter administration provided $118 million in direct aid, negotiated with the international development banks for an additional $262 million, and arranged for refinancing of $500 million in private bank debts. As previously noted, this was more aid than any other nation provided to Nicaragua, and more than the U.S. had contributed to Somoza in the previous four years. Jimmy Carter also welcomed Daniel Ortega, the Sandinista president, to the White House. (9) Most of the Media chose not to headline, or challenge, this largesse or the reasons behind it, or the administrations' wooing of Ortega

"Though they lied about it, the Sandinistas began funneling money to the FMLN in El Salvador from their first days in power. In 1981, Tomas Borge proclaimed, "This revolution goes beyond our borders." "That same year a leader of the FMLN in El Salvador said, "We have the brilliant example of Nicaragua. The struggle in El Salvador is very advanced; the same in Guatemala, and Honduras is developing quickly. Very soon Central America will be one revolutionary entity."(10) "A record of Sandinista deliberations in 1979 was later smuggled out of the country. Called the "Seventy-two hour document," it revealed the classic Marxist terms in which the Sandinistas referred to themselves. The Frente Sandinista de Liberacion Nacional (FSLN) was the "vanguard" of the revolution, it said. Its alliance with more democratic anti-Somoza groups was merely tactical - an attempt to "neutralize" domestic rivals and wards of "Yankee intervention." "In fact, once in power, the Sandinistas intended to "crush" internal opposition and aid revolutionary movements abroad. Humberto Ortega, Nicaraguan defense minister and brother of Daniel, the Sandinista president, affirmed that "our revolution has a profoundly anti-imperialist character, profoundly revolutionary, profoundly classist (sic); we are anti-Yankee, we are against the bourgeoisie; we are guided by the scientific doctrine of the revolution, by Marxism-Leninism."(11) And "Sandinistas had enjoyed good relations with the Palestine Liberation Organization for a decade. Sandinista militants had received training in Fatah camps in Lebanon and Libya during the 1970s and fought with the PLO against Jordan's King Hussein."(12).

In September, 1981 Adolfo Calero became the political leader of the main resistance group opposing the Sandinistas. He invited other groups to join his Frente Democratico Nacional (FDN), and put all of them under the command of former National Guard Col. Bermudez. Joining the FDN were the 15th September Legion, MILPAS, and Brooklyn Miskitos. Bermudez went to Argentina and the U.S. in the fall of 1981 seeking military aid. Argentina provided military advisors in December 1981 and January 1982. A small number of U.S. advisors/trainers arrived between January and May 1982. The first organized raids and ambushes against the Sandinistas began in the April-May1982 time frame. After 1982, the "Contras" consisted of five separate armed forces: the FDN; Brooklyn Miskitos; Rivers Miskitos; Southern Indigenous Minority (SIM), consisting of Sumo and Rama Indians and the Creole blacks under the command of Commandante Samba (operating directly under Bermudez); and Southern Opposition Bloc (BOS), consisting of former ARDE (under Cruz and Alfonso Robello) and some who had earlier followed Pastora (once known as Commandante Zero) until he became widely discredited. Opposition to the Sandinista intent to communize Nicaragua was in place and growing.

Not only the usual far left/Socialist flacks such as Susan Sontag, William Sloan Coffin, Benjamin Spock, Noam Chomsky, Richard Falk, Peter Seeger, etc., and Hollywood wonders Ed Asner, Kris Kristofferson, Michael Douglas, Bianca Jagger, etc., but many U.S. politicians from the left joined in support for the Sandinistas. Vermont Democrat Patrick Leahy said on the Senate floor, "I embarked on my fourth trip to Central America in the last six years convinced the policies of the Regan Administration in Central

America are leading toward a major foreign disaster for out country. It is clearer than ever that the President's policy of military confrontation with Nicaragua is sinking the United States deeper and deeper into a quagmire."(13) Senator Ted Kennedy agreed with Leahy, as did Sen. Pell of Rhode Island and Sen. Dodd of Connecticut. In March of 1984, eight members of Congress signed a letter to Nicaraguan president Daniel Ortega: "Dear Commandante", they wrote, "We address this letter to you in a spirit of hopefulness and good will. As members of the U.S. House of representatives, we regret the fact that better relations do not exist between the United States and your country." "The implication, obviously, was that the Reagan administration was to blame for the poor state of relations between the two nations. If only the administration possessed the "hopefulness and goodwill" of congressional Democrats, things would presumably be different. The letter praised Ortega for *promising* (1) to hold free and fair elections, (2) to "reduce" press censorship, and (3) to permit "greater freedom of assembly for political parties They further praised Ortega for taking these "steps in the midst of ongoing military hostilities on the borders of Nicaragua. But the steps were not steps at all, merely "promises" The letter was signed by Democrats Jim Wright of Texas, Michael Barnes of Maryland, Bill Alexander of Arkansas, Matthew McHugh of New York, Robert Torricelli of New Jersey, Edward Boland of Massachusetts, Stephen Solarz of New York, David Obey of Wisconsin, Robert Garcia of New York, and Lee Hamilton of Indiana."(14) Most of the Mainstream Media praised the letter, ignoring the declared goals of the Sandinistas.

"In 1988 the Media Research Center (MRC) conducted a survey examining the coverage of television news. Monitoring the CBS Evening News, ABC's World News Tonight, NBC's Nightly News and CNN's Prime News, MRC looked at how the television networks reported on Nicaragua. They found that Sandinista promises to live up to agreements were lavishly covered while truculent or defiant statements by Sandinista leaders tended to be ignored. Five times as many stories covered Sandinista reforms as mentioned Sandinista violations."(15)

"The Reagan administration's policy toward Nicaragua evolved slowly. (President Reagan) authorized funding for a small force of Nicaraguans to take up arms against the Sandinistas. "Contras," cried the Sandinistas, meaning "counterrevolutionaries." But the label was never accurate. Most of those who fought against the Sandinistas were either former Sandinistas themselves, or ordinary peasants who had good and sufficient reasons for hating and fearing them."(16)" "In a prime-time televised speech in 1983, President Reagan made the case for aid to the Contras (meaning contra-Sandinista). The response from liberals and leftists was emphatically negative. Democrat congressman (eventually senator) Charles Schumer said: "It seems that from his speech last night, the president came out of a time warp. He arrived from 1967 and came to us today, with different names and different places, but the same exact arguments that we were hearing in the middle sixties about Vietnam."(17) Senator Dodd basically echoed Schumer's remarks, noting that "the insurgents were the remnants of the old Somoza regime – a regime whose corruption, graft, torture, and despotism made it universally despised in Nicaragua."(18) The Media? No change in the pro-Sandinista slants previously noted.

Factually, however, "One of the three leaders of the (Contra) movement, Alfonso Robello, was a former member of the Sandinista Junta. Arturo Cruz and Adolfo Calero were both opponents of Somoza, and Calero was imprisoned for his anti-Somoza agitation. Among the military leadership of the Contras, 53 percent were former civilians, 27 percent former guardsmen, and 20 percent were former Sandinistas. As the *New Republic* editorialized, "one doesn't raise an army of 15,000 peasants with promises of restoring a universally despised dictatorship."(19)

"While liberals blasted the Contras for being CIA tools, they never asked where the Salvadoran FMLN got its backing. In fact they argued again and again that the FMLN was purely homegrown and indigenous, despite its lavish backing by Cuba, Nicaragua, and the Soviet Union."(20) Nor did the liberals question who was funding the Sandinistas. Nor did the Media, whose sympathies and drum beats were, in the large majority, clearly with the Sandinistas and against the Contras.

"In April 1985, just after Congress had voted down non-military aid for the Contras in a party-line vote, Nicaraguan president Daniel Ortega jetted off to Moscow – presumably to celebrate. Democrats were furious – "politically-wise, he embarrassed us by his activities, to be truthful," Speaker O'Neill told the UPI wire service. Six weeks later, a sufficient number of Democrats having been shamed into second thoughts, the (non-military) aid package was approved."(21)

What had started out as a rather uncomplicated effort by the Reagan administration to quietly save Nicaragua – and El Salvador – from the communists gradually grew into a Mena based operation that experienced nearly as many pratfalls and problems as a slapstick comedy production, but given the politics, and politicians, and the associated corruption involved, there was little room for chuckles. In sum, under Carter, the Media in the United States had smiled along with delusional Jimmy in the belief that there was little or no reason to be concerned about the threat of Soviet sponsored Communism – anywhere. Never mind the extensive and constant Soviet sponsored disinformation about America in the international press, the extensive arrests of Jews and Christians in the Soviet Union, or the extensive jailing of dissidents there, the Soviet arming of North Korea, Somali, and Ethiopia, Soviet support to Castro as he provided troops to several insurrections in African nations, and their supplying of communist inspired insurgencies in El Salvador, Nicaragua, and Grenada, plus fomenting unrest in Panama. During this same period the Soviet Union was building its own military at an unprecedented rate, including weapons of mass destruction. The Soviets also supported avowed international terrorist organizations such as the Red Brigade(s), the Palestine Liberation Organization, the Baader-Meinhof gang, and the Japanese Red Army.* But the Media failed to connect the dots. Clearly, Reagan's policy was to confront, not appease, the communist threats.

The Sandinistas had begun their efforts to overthrow the Somoza government of Nicaragua in late 1978 and, strongly supported by Cuba, managed to gradually take control of Managua and the major population tract in Nicaragua. The establishment of the National Liberation Front by the Sandinistas was followed in 1980 by a distinct move to the left, including a combined effort with Cuba to wreak similar havoc and unrest in El Salvador. Even Jimmy Carter had considered this un-neighborly and publicly admonished the Sandinistas, threatening to suspend further U.S. aid. Of course, the United Nations simply stood - or sat - by and did nothing despite the ruthless treatment of rural and urban dwelling

Nicaraguans by the Sandinistas – or the increasing presence of the "Cuban volunteers" 'helping' them. Even the exodus of some 95,000 Nicaraguans from their homeland to nearby countries and the endless stream of reports of mistreatment of minority groups, such as the Moskito Indians, by the Sandinistas, or the influx of 3,000 Cubans – and hundreds of Soviet and eastern Europeans - in Nicaragua did not stir the United Nations. Soviet assistance was plentiful, including "350 T-55 tanks, 12 MI-8 HIP helicopters, six HIND attack helicopters, as well as artillery pieces, patrol boats, and airplanes. With Soviet and Cuban help, the Nicaraguans built a military air base at Puenta Hueta, with a runway 10,000 feet long – long enough to accommodate any plane in the Soviet arsenal."(22) The Media? Like the United Nations, little was said of atrocities, or the communization of Nicaragua by the Sandinistas, or the flow of refugees, or the large number (approximately 10,500) of "political prisoners" in Nicaraguan jails, or the presence of Cuban and Soviet military "advisors", etc. However, much Media praise was given for the "land reform" and other "promises" made by the Sandinista "Commandantes."

Given the situation facing the Reagan Administration at the time, especially the deteriorating situation in Nicaragua (and El Salvador), and a liberal congress opposing U.S. involvement or providing more help to the Contas, it didn't take rocket science to conclude that covert action was logically the best if not the only viable option. The question(s) then came down to how to get the needed supplies from A (the U.S.) to B (the Contras in Nicaragua), who could help in the process, and how much would it cost? Let's speculate a bit to see if the following pieces don't fit: (1.) CIA Director William Casey was a friend of Ronald Reagan and had been his campaign manager. (2.) George Bush, Reagan's Vice-President, and former CIA director, fit as the second leg of the triangle and could help Casey put together the requisite details. (3.) Mena, in a remote area of Arkansas, had already appeared on the Washington radar screen, thanks to the earlier Carter-Castro deal. (3.) John Paul Hammerschmidt, Arkansas's Congressman for the Third Congressional District – in which Mena is located – was one of George Bush's closest friends and his presidential campaign manager in both 1976 and 1980, and could broach using Mena with Clinton. Thus the remote location requirement and Arkansas politics problem were solved, and with some additional upgrading of the airport facilities and capabilities, the choice of Mena was a natural.

Continuing with the speculation, personnel selection should not pose a problem. On the NSC staff in the White House was a young Marine Lt. Colonel (Oliver North) who had previously worked with CIA officers in Vietnam, Thailand, and Washington, D.C. Also in the White House was a former CIA officer, Don Gregg, who had previously served in Vietnam, and was now on Vice-President Bush's staff. Nor would it be a difficult task to enlist DEA asset Barry Seal to handle the delivery challenges of the "A to B" operation. "At the age of 26, Seal had been the youngest 747 pilot in the nation. But his commercial aviation career with TWA was brought to an early and abrupt end in 1972 when he was arrested in New Orleans for alleged violations of the Mutual Security Act of 1954, which bars the export of explosives without the approval of the State Department. A DC-4 owned by Seal and loaded with 13,000 pounds of C-4 explosives, primer cord and blasting caps was seized in Shreveport, Louisiana. The government contended these were explosives earmarked for anti-Castro groups operating in Mexico. But the charges were never fully tried, and the case was later dropped without any real explanation."(23) Seal, after being charged with smuggling drugs, 'worked' with DEA to avoid jail-time. But as daring* and gutsy a free-spirit patriot that Seal was, his inclination to seize opportunities to make money, regardless of the risk, eventually led to his being killed, gangland style, by two Colombian hit-men on February 19, 1986 in Baton Rouge, Louisiana, in retaliation for Seal having crossed the drug cartels.

Terry Reed, a pilot and an engineer, had maintained irregular contacts with the FBI and CIA since his Vietnam and Thailand days, but he "was surprised when in March 1983 he met with John "Cathey" in Oklahoma City. There "Cathey" outlined to Reed what would be known as "Operation Donation"...and, after acknowledging that he and Reed had first met in Thailand (he as Ollie North then), said he would contact Reed when the plan was more complete."(24) Reed's pilot skills (and manufacturing knowledge) were anticipated as needed ingredients in the operation. "Cathey" kept his word, telling Reed that a "Barry Seal" would be contacting him. In December, 1983 Seal met with Reed and finalized Reed's association with the project. Seal needed Reed to fly drops to Nicaragua as well as to help train Nicaraguan air crews at Nella, a rural air-strip in western Arkansas not far from Mena, in twin-engine aircraft so that supplies could be flown to Nicaragua by Nicaraguan crews, thereby eliminating the need for U.S. crews to make the re-supply flights.

With the move of the Iver Johnson's arms-manufacturing company to Jacksonville, Arkansas from New Jersey, another needed support entity was now available to "Operation Donation". With minor modifications to the Iver Johnson manufactured weapons, they would be suitable for use by the Contras. The conversion process would be completed separately and away from the Jacksonville plant.

"To examine Mena is to examine the institutional condition of the United State. As for the president (Clinton), it exposes him as a remarkable counterfeit, willing to betray his liberal principles for self-advancement."(25)

Barry Seal was a key cog in the Nicaraguan "Donation Operation". "By late 1981 Barry Seal had made direct contact with the Colombian cocaine cartels. Over the next three years his fleet of aircraft would become the dominant smuggling conduit for the Medellin Cartel, which reportedly controlled 75 percent of all cocaine exported from Colombia. The Colombians learned to respect him. They were losing loads of cocaine at a fearful rate as the Reagan administration cranked up its war on drugs in the Caribbean. Seal knew how to get through the defenses. He had never lost a single shipment for the Cartel."(26)..."Seal told the President's Commission he had made about 100 smuggling flights without ever being intercepted by U.S. interdiction authorities. His average load was 300 kilograms, with a top smuggling rate of $5,000 a kilo. On one flight alone he earned a transport fee of $1.5 million. Using wholesale prices, he must have smuggled anywhere from $3 billion to $5 billion worth of drugs into the United States."(27) "Putatively, Barry Seal became a "snitch" for the DEA after he was caught conspiring to smuggle Quaaludes into Florida." "His lawyer was "the ubiquitous Richard Ben-Veniste, who later happened to be the chief counsel for the Democrats in the Senate Whitewater investigation, (who said) "I did my part by launching him into the arms of Vice President Bush, who embraced him as an undercover operative, Ben-Veniste later told *The Wall Street Journal.*"(28) (Ben-Veniste and Sen. John Glenn later combined to stone-wall the senate Whitewater hearings.)

The urgency pushing this scenario was a recalcitrant Congress that had not yet awakened to the true intent of the Sandinistas. The increasingly restrictive Boland amendments prohibiting any military assistance to the Contras prompted more expeditious actions – including training of Nicaraguan "Contra" air-crews - and an expanded search for finances to assist in supplying the Contras, the many contributors within the United States having already given all that they could.

On a second re-supply flight "that (Arkansas State Trooper and Clinton body-guard) L.D. Brown took with Seal on or around December 14, 1984, it was the same routine." "At the airport in Honduras Barry Seal picked up two duffel bags. Back at Mena, Seal opened one of the bags as they were getting into Brown's Datsun 310 hatchback. He reaches down and he pulls out a little wrapped package like I've seen kilos of coke packaged over the years and it had a number on it, number 2." When asked about Dan Lasater® , Brown said "I had met him a lot through the Governor. He came to the Mansion several times. There are not many people that can just come through the back gate and go in the kitchen door. We've been in his racing box at the races. We flew on his airplane. He was a fixture. I knew that he was involved in cocaine. I had been at Lasater's house one time where there was a silver platter of what I thought was cocaine and I got the Governor out of there. I said, "We need to go. Let's get out of here." "There were a lot of people there, a lot of girls there. He had to have seen it; I mean, it was obvious."(29)

Brown decided to withdraw his CIA application, which Clinton had helped him with, not certain that he wanted to be engaged in the type of activity he had been exposed to. "In January 1985 he (Brown) was visited at the Governor's Mansion by an emissary from the CIA, sent to placate him. The man seemed to know his way about the place. He came in through the little-known Spring Street access, and nonchalantly pulled up next to Hillary Clinton's parking spot. He was Felix Rodriguez, the legendary Cuban émigré who had tracked down Che Guevara in the mountains of Bolivia." (Rodriguez had served as a CIA contract officer under Donald Gregg in Vietnam.) "It was Gregg who brought him into the Central American operation.* Working under the code name of Max Gomez, he served as an adviser on aerial counterinsurgency against the FMLN guerrillas in El Salvador." (Rodriguez) "succeeded in reassuring Brown that "no more monkeying around with Seal would be involved. Don't worry about Barry," he said, "We're going to take care of him." "Brown's next assignment was to guard the shipments of the short-stock AK-47 rifles and other weapons that were arriving in metal canisters on a boat in the Bahamas, on their way to the Contras in Central America."(30)

While serving in Lima, Peru (1966-1969) and again in Vietnam in early 1971, I met and talked briefly with Felix Rodriguez, a survivor of the ill-fated Bay of Pigs fiasco. He had spent a short time in Lima in 1967 enroute to Bolivia where he was to accompany the Bolivian Army unit that had been formed to track-down the dedicated communist, Argentina born Ernesto "Che" Guevara. Supported by Castro, under whom he had served some nine years, Che had left Cuba to foster a peasant uprising in Bolivia, and, if successful there, then eventually in other countries in Latin America. From the multiple reports that were received from Peruvian sources, it was evident that Rodriguez had related in many of the bars and restaurants in Lima exactly what his "CIA mission" was. As it evolved, Che's small band of followers reportedly never exceeded 17 or so, with most of the Bolivians with him having been pressed at gun point into service. Despite the slow and cautious pursuit by the Bolivian army unit, Che was finally run to ground and captured in a remote area in the interior of Bolivia, thanks largely to a foot wound he had suffered. After being interrogated by his captors, he was summarily executed by multiple gun shots fired by one of the Bolivian army men. Rodriguez returned to Lima on his way back to Washington, and bragged that Che had given him his pipe shortly before being shot. His departure performance in Lima was reportedly much like his arrival one. I was surprised to see Rodriguez in Vietnam, for at least to me it was not a logical posting for a contract Cuban, although the Chief of Station in Saigon at the time had enduring connections from his and Rodriguez's earlier "Bay of Pigs" days. Rodriguez again impressed me as not having lost any of his obvious, if not brash, self-confidence or his compulsion to gab. That he had been very involved in Central America activities and with most aspects of the Mena project was not surprising.*

Among the more troubling things that evolved from the Mena pursuits were the drug and money laundering efforts, both of which continued for an extended period of time. These were apparently well known in Arkansas by a wide range of entities, including Bill Clinton, many Arkansas State Troopers, at least some Arkansas bankers and bondsmen, contract air crews, the Arkansas Media, as well as DEA and FBI personnel assigned there, etc. Certainly there were multiple ancillary developments, but the Media paid little attention. Mainstream Media? Hello out there, anyone home?

• "There was no need for Clinton to create the Arkansas Development and Finance Administration (ADFA) in 1985. The state already had the Arkansas Industrial Development Corporation. These two bodies had semi-independent boards. ADFA gave Clinton a patronage machine that answered to the governor alone, and allowed Clinton to tap into the huge reserves of the Arkansas Teachers Retirement System. At the same time, Clinton steered bond business to Lasater, and low interest industrial loans to the others in the Arkansas group – Seth Ward, for instance, the father-in-law of Webster Hubbell – frequently without due diligence and over the objections of the agency staff." "They were giving money away like candy to the insiders" said Mark Swaney (lead advocate of the Arkansas Committee investigating the alleged nexus of drug-running, money laundering, and covert activities linked to the Mena airport)."(31) Media coverage? Try somnambulistic.

• "Funds had been flowing offshore. ADFA had done at least $250 million worth of business with the Fuji Bank of Japan. In December 1988, for instance, ADFA raised $50 million for the purpose of building and buying houses for the needy of Arkansas, but instead ADFA wired the $50 million to the Fuji Bank, Grand Caymans Branch, account number 63119809. It was a nice piece of arbitrage profiteering. ADFA raised the money at 7 percent interest with tax exempt status, and loaned the money to Fuji at an interest rate of 9.37 percent, generating a profit of about $1 million on a ten-month investment."(32) Media pursuit? No? Strange…

• "Lasater & Company was deeply involved in the demise of thrifts in Illinois. One of them, First American Savings & Loan of Oak Brook, Illinois, fired back with a lawsuit against the firm in October 1985, alleging that Lasater had transferred "unprofitable investments from his personal account to the account of the Plaintiff after trading was closed. The suit was taken over by the government's Federal Savings and Loan Insurance Corporation in April 1986, after First American was seized by regulators. It hired the Rose Law Firm to handle the case. The attorneys were Hillary Clinton and Vincent Foster. They settled for a modest $200,000 in a sealed agreement that drew a veil over the reasons for the collapse of the S&L. *The Chicago Tribune* ran a blistering series on Hillary Clinton's role. It accused her of a "glaring conflict of interest" for negotiating a "secret, out of court settlement" that ended a suit"against a family friend and influential benefactor of her husband."(33) "Patsy Thomasson handled the case for Lasater, who was by then in prison. She was more than just an executive vice-president. She was also a "financial and operational principal" of the brokerage firm, with direct liability for criminal misuse of client's accounts. According to the Arkansas Securities Commission, she held the key 24, 27, and 53 series brokerage licenses, making her one of the most highly qualified broking principals in the country."(34) Most of the Media simply harmonized "Silent Night".

• "When the *American Spectator* in the late summer of 1995 published reports that arms and drugs had illegally passed through Arkansas's Mena airport with Clinton's knowledge, the entire media, with the exception of the *Wall Street Journal* and the *Arkansas Democrat Gazette* ignored the story. This was mainstream press at its most perverse. When it had ballyhooed the *Spectator's* reports of Clinton's promiscuity, the press said the reports were not news. Criminal conduct would constitute news. So, here were reports of criminal conduct, and the mainstream press went limp. Ignored were the *Spectator's* reports of drug trafficking, money laundering, and illegal arms shipments, all witnessed by a member of the governor's security detail and corroborated by a half dozen other witnesses. The European press was more attentive. In the summer of 1984 the *Economist* of London warned of "a peculiar pattern of suicides and violence surrounds people connected to the Clintons or their associates. From early 1964 onward Ambrose Evans-

Pritchard wrote a series of reports to the *Sunday Telegraph* linking Don Tyson to drug trafficking, Clinton to drug use, and questioning the unsolved murder of a Clinton security expert who supposedly kept damaging files on him."(35) Our Mainstream Media? Surely you jest!

• "All the reports of slippery bank dealings in Arkansas began with (Jeff Gerth's) March 8, 1992, story in the *New York Times*, where Gerth reported that the Clintons went into partnership with James B. McDougal in 1978 to buy land along the White River and establish the Whitewater Development Corporation. Gerth reported that "at times money from Mr. McDougal's savings and loan was used to subsidize it". "And Gerth describes the Savings and Loan, Madison Guaranty, as a "failing savings and loan association," regulated by an agency of Clinton's government."(36) Other U.S. Media? Hardly a squeak.

• The damaging financial reports continued. On July 13, 1994, the Associated Press reported that as much as $400,000 in unsecured loans was made to Governor Clinton during the 1980s to promote his political agenda, though at the time his annual salary was only $35,000."(37) The Media snored on.

• "During Governor Clinton's 1990 campaign he garnered nearly $200,000 in unsecured loans from the Bank of Perry County. Some of these loans were neither reported in campaign finance records nor repaid. One was the $50,000 loan that Clinton called on the Stephenses and their associates to guarantee just before the November election. The owner of the Bank of Perry, Herb Branscum, a former Democratic party chairman and generous supporter, was thereafter rewarded with an appointment to the state highway commission."(38) Interesting, but not to the Media.

• "By the time Clinton announced his presidential candidacy he had appointed every member on every state regulatory panel. The result was vast corruption. In the area of bond trading alone, the incidence of securities fraud in Arkansas was probably the highest n the nation. Little Rock's so-called "bond daddies" were a national scandal. Arkansas also joined the savings and loan follies with extreme abandon, and it accumulated one of the highest rates of thrift failure in the country."(39) All quiet on the Media Front.

• "The national press never looked very carefully at Clinton's record as governor. For example, in December 1992, when the president-elect was snaring headlines worldwide with the Economic Summit he had convened in Little Rock to conjure up a more robust American economy, Governor Tucker was just a few blocks away, having called a special session to bail out the state's Medicaid program. Owing to Clinton's negligence, it was bankrupt."(40)

There is more, but the examples given certainly should underscore the temerity of the Clintons. Before continuing with other events during the reign of "King" Bill in Arkansas, the reader may recall or want to know that the UNO ("Contra") candidate, Mrs. Violeta Chamorro, on February 25, 1990, won handily over Sandinista candidate Daniel Ortega, to the surprise - and dismay - of the Sandinistas. The liberal left media was in Nicaragua in force, as was Ed Asner, and as were both Jimmy Carter and Sen. Dodd, the former as an observer, and the latter reportedly more interested in Bianca, who was there as a strong supporter of Ortega. The Liberal Media and other "observers" were there to see Ortega win, but were surpassed in amazement by his defeat only by the embarrassment of the *Washington Post* which had predicted an overwhelming victory for the Sandinista candidate. The freedom loving Nicaraguans had refused to be intimidated by the Sandinistas, despite the latter's extensive threats and violence. The final Contra victory had fittingly come at the voting booth. Now back to Arkansas—

• In January 1985, Clinton announced "My brother has apparently become involved with drugs, a curse which has reached epidemic proportions and plagued the lives of millions of families, including many in out state." "In the spring of 1984, Detective Travis Bunn, a highly decorated Army Special forces sergeant, had recorded Roger Clinton saying: "I've got to get some for my brother, he's got a nose like a Hoover vacuum cleaner." (41) Bunn's investigation was taken over by the State Police at the request of the governor's mansion.

• Roger Clinton later moved into a corporate suite at Vantage Point. "From time to time the Governor would appear, usually in the middle of the afternoon. "They were sharing joints of marijuana. (The management) could hear Roger saying what it was, where he got it from, what it was like. The (manager) could hear the Governor bleating his approval: "This is really good shit!" "It was not just marijuana either. Two or three times a week the governor was buoying his spirits with a snort of Kid Brother's Colombian rock. The repartee was coming through the vents."(42)

• "Roger Clinton's band used to play at a Little Rock nightclub. "Big Brother would come by from time to time with one or two of his State Troopers. Roger had all the pretty girls and drugs and the fast life, and Bill was pretty envious of this. On one occasion "Roger the Dodger" came back to the bar and said he needed two grams of cocaine right away. They carried out the deal near the ladies room. The Dodger then borrowed her (waitress Sharlene Wilson's) "tooter, her "one-hitter", as she called it and handed it to the governor...."I watched Bill Clinton lean up against a brick wall. He must have had an adenoid problem because he casually put my tooter up his nose," she said. "He was so messed up that night he slid down wall into a garbage can and just sat there like a complete idiot."(43)

• "Sharlene Wilson was surprisingly frank about her job at the Mena Airport in the mid-1980s. The cocaine was flown in on twin-engine Cessnas, sometimes as often as every day. "I'd pick up the pallets and make the run down to Texas. The drop-off was at the Cowboys Stadium. A lot of the cocaine that came into Mena was taken up to Springdale in northwest Arkansas, she said, where it was stuffed into chickens for reshipment to the rest of the country. She had another job, she used to pick up cocaine deliveries on the railway tracks near the little town of Alexander, thirty miles south of Little Rock. "Every two weeks, for years, I'd go to the tracks, I'd pick up the package, and I'd deliver it to Dan Harmon, either straight to his office or at my house. Sometime it was flown in by air, sometimes it would be kicked out of the train. A big bundle, two feet by one and half feet, like a bale of hay. Roger the Dodger picked it up a few times."(44)

• "But in the summer of 1987 one of the drops disappeared. Furious, Harmon brought out some of his men to watch the delivery on the night of August 22. They were expecting a delivery of 3 to 4 pounds of cocaine and 5 pounds of "weed". Harmon and his men told me to wait in the car, which was parked off Quarry Road. "I heard two trains. Then I heard some screams, loud screams. When Harmon came back, it looked like there was blood all down his legs. She learned later that a group of boys had been intercepted at the drop sight. According to Sharlene some of them had managed to get away, but Kevin Ives, 17, and Don Henry, 16, were captured. Harmon's men interrogated them as they were lying on the ground, face down, hands tied behind their backs. They were kicked and beaten, and finally executed. One of the boys was stabbed to death with a "survival knife". The bodies were wrapped in a tarpaulin, carried to a different spot on the line, and placed across the railway tracks so the bodies would be mangled by the next train."(45)

• "At 4:25 AM the three drivers of a Union Pacific train coming up from Shreveport caught site of an obstruction on the line. They jammed on the breaks but there was no chance of stopping the immense freight train in time. The Arkansas medical examiner, Fahmy Malak, ruled the deaths an accident. He said the boys had smoked twenty marijuana joints and fallen into a trance of the railway tracks, side by side." "That Malak survived in Arkansas is a testament to Clinton's power", wrote Meredith Oakley in her dispassionate Clinton biography *On the Make*, "He repeatedly lied about his credentials, misconstrued his findings, and misinterpreted autopsy procedures. In the lab, he misplaced bodies and destroyed evidence. On the witness stand, he was a prosecutor's dream."(46)

• "In April 1988 a second autopsy was conducted by the Atlanta medical examiner Dr. Joseph Burton. It showed an enhanced photograph he had taken (to six other confirming examiners) of the "penetrating wound" into the thoracic and left lower chest cavity of Don Henry was a "stab" wound, consistent with it having been inflicted by something as a large cutting edge knife." "He also found that Kevin Ives had been smashed in the head with a rifle butt, probably Don Henry's .22 caliber hunting rifle. There was "considerable reaction with the lungs of both boys" indicating that they had not died immediately. The level of marijuana in Kevin's blood was 97.9 nanograms per milliliter, consistent with having smoked two cigarettes over the previous few hours. Don Henry's level was slightly higher, but not enough to induce collapse." "The preponderance of evidence in this case indicates that Kevin Ives and Don Henry sustained injuries prior to impact with the train, that these injuries were inflicted on them by another individual or individuals, that their bodies were placed on the track."(47)

• "It was at this stage that the Clinton administration in Little Rock began to exhibit the body language of alarm. In May 1988

Governor Clinton's chief of staff, Betsey Wright, deflected an attempt by the grand jury to subpoena two outside pathologists who had looked at the train deaths during a review of the Arkansas crime lab. Wright responded with an affidavit asserting that the doctors had not been contracted "to provide second opinions on specific cases." "It was gratuitous obstruction. The grand jury, highly irritated, then issued a subpoena for Betsy Wright herself. For weeks she defied the order."(48)

• "Dan Harmon had managed to take over the case, first as a concerned private attorney and then as a special deputy prosecutor appointed by his friend, Judge John Cole (a close friend and supporter of the Clintons)." "He promised the grand jury to turn over every stone until the fiendish killers were caught and brought to justice. Linda (Ives) believed him. In December 1988 the grand jury reached the end of its natural life and was disbanded. It was only later that Linda Ives would be told by two frightened jurors that Harmon had prevented the grand jury from calling witnesses."(49)

• Keith Coney, who told his mother he knew too much about the railway deaths and feared for his life, died in a motorcycle accident after a high speed chase. Coney had been with the two boys a few hours before their deaths. Linda now believes that they met up again at the tracks. "I'm sure now that there were three of them out there, at least, and he was the one who got away," she said." "Boonie Bearden, a friend of the boys, disappeared. His body was never found." "Jeff Rhodes, another friend, was killed with a gunshot to the head in April 1989."(50)

• "Aircraft with no lights were observed flying very low over the tracks at night. One informant staked out the area and observed a twin engine plane coming in at approximately 3:00 A.M. at least once a week. "It would fly in extremely low over the field, reduce speed, before throttling up again. By the field is a children's colony that is lit up each night like a 'Christmas Tree'. That was the beacon."(51)

• "It was Jon Brown who finally broke Sharlene Wilson and extracted her confession. He then discovered a fresh witness, a lad who had been out with two friends that night looking for a marijuana patch. The witness had been about sixty feet away, hidden below the bank, watching a group of men talking on the tracks.

"One of them I definitely recognized as Dan Harmon. Then I noticed two more people, Kevin and Don, walking down the railroad tracks." "At first it looked as if Harmon was just talking to the boys, but then a shot rang out. The witness turned and ran." "It was at this point that the FBI took charge. Phyllis Cournan, an athletic, single-minded agent from Philadelphia, had recently arrived in Little Rock on a routine assignment. She immediately gave the boy a polygraph test, which he passed, and placed him in the witness protection program. Cournan was now being accompanied by an FBI agent from the Hot Springs office, Floyd Hayes, who was assigned to be her partner."...."then, after eighteen months, the probe suddenly collapsed. In November 1995 Linda and Larry Ives went to see Special Agent Bill Temple, the number two man in the FBI office in Little Rock, and given a taste of the bullying insolence of the FBI." ": He was so arrogant and smug" said Linda." "He said, 'Maybe in light of the fact that there was no physical evidence, maybe it's time for you all to realize that no crime occurred.' I slammed down my notebook and said, "I don't have to listen to this bullshit' and walked out."(52)

• "She went public, accusing the FBI of working to cover-up the murder of her son. The chief of the FBI's Little Rock office, I.C. Smith, countered in the local newspaper, *The Benton Courier*, saying that the Bureau had a "very real problem" establishing federal jurisdiction in the case, and anyway it was not clear that the boys had been murdered. He said that Linda Ives had "badly misquoted" Agent Temple's remarks." "Linda Ives telephoned Phyllis Cournan, who had been present at the meeting. With the tape recorder running, Linda Ives extracted from Cournan an acknowledgement that Temple had been quoted "verbatim". "Linda had learned that there was no use giving quarter to Louis Freeh's FBI."(53)* "Linda Ives, the housewife from Benton, had outmaneuvered the Bureau. But she still did not understand what it was about her son's death that had caused a federal grand jury probe to be shut down in early 1991, or why the FBI had backed away in November 1995, or indeed why the Justice Department's prosecution of Dan Harmon in June 1997 was confined to racketeering, when they knew perfectly well – or so she had to assume – that he had murdered her son. Linda Ives, Jean Duffey, and John Brown all came to the same conclusion. They were pitted against Dan

Lasater – the Dixie Godfather, and the friend and provider for the Clinton brothers."(54) The Clintons had eagerly "kissed the ring" of Sen. Fulbright, the original Arkansas Godfather, but then readily kissed Dan Lasater's as well throughout the 1980s and early '90s. The Mainstream Media? Hear no evil, see no evil, etc.

The debris created during the Clintons' years in Arkansas, ranging from illegal money laundering schemes, to illegal trafficking of drugs, to illegal banking practices to extortion and common graft have been succinctly documented by authors of the books cited in the end notes. Also included in the debris should be mentioned the names of others who were subjected to the quirks of the Dixie Mafia. The body count, while not complete, is impressive. For example:*

- Jerry Parks, head of Clinton's gubernatorial security team in Little Rock, was gunned down in his car at a deserted intersection outside of Little Rock after being forced to stop. The dossier he had compiled on Clinton was stolen. Reportedly, he had been a confidant of and was working with Vince Foster on the dossier before Foster's sudden "death".
- Bill Shelton, Arkansas state trooper and fiancée of Kathy Ferguson, ex wife of Trooper Danny Ferguson. Shelton allegedly committed suicide at the grave site of Kathy, who allegedly had shot herself in her home, although her packed bags were next to her body. Kathy was reported as a possible corroborating witness for Paula Jones. Danny Ferguson was the State trooper who was alleged to have escorted Jones to Clinton's hotel room. Both died in mid-'94.
- Gandy Bugh, attorney for Dan Lasater in a case concerning alleged financial misconduct; died in an alleged suicide by jumping out the window of a multi-story building, in May,'94.
- Dr. Ronald Rogers, Arkansas dentist killed in a plane crash as he was enroute to an interview about Clinton with the London "Sunday Telegraph".
- Stanley Huggins, partner in a Memphis law firm investigating Madison Guaranty. His 300 page report remains "hidden".
- Suzanne Coleman had an affair with Clinton when he was attorney general. Died of "suicide" with gunshot wound to the back of her head. No autopsy performed. She was pregnant at time of her death; rumored it was Clinton's.

• Paul Wilcher, Washington attorney investigating government corruption. Died June, '93. Was found dead on a toilet seat in his Washington apartment. No cause of death was determined. He was investigating the Mena drug and gun-running stories as well as the ATF assault on the Branch Davidians at Waco. He had delivered a 99 page affidavit to Janet Reno three weeks prior to his death.

• Florence Martin, an accountant related to the Barry Seal case. Dead of three gunshot wounds to the head.

• Vince Foster, White House Counsel, died July 20, 1993. He was found dead in Ft. Marcy Park outside of Washington, a supposed suicide, but reports of a cover-up persist. This "mystery" and highly suspect "suicide", obviously mishandled by the Heinz mixture of investigators, centered on Foster being reported dead two hours before his body was found in a remote area of a Virginia park. How he got there has yet to be determined, nor how he was able to drive without the keys to the car in his possession -or in the car - or able to walk approximately 700 feet from where his car was parked without a trace of the parking lot or park on his clothes or shoes. The gun that he used to kill himself did not have his prints; there was no exit wound in his head; there was little blood; and his glasses were many feet from the body. Photographs taken at the scene disappeared and medical reports conflict. The keys appeared only after Craig Livingston, patron of Hillary Clinton, and Associate White House Counsel William H. Kennedy III visited the Fairfax County Hospital morgue "to identify the body". The FBI and the Park Police antics soon came to resemble the old TV Comedy "Hee-Haw." Special Counsels Robert Fiske and Kenneth Starr rubber-stamped the "suicide" report. Miguel Rodriguez, a member of Starr's staff, had pursued his own investigation and disagreed with Starr, but resigned on March 23, 1998. What drove Foster to "suicide" remains unclear, although reportedly he was preparing to reveal details of some of the Arkansas scandals, about which there were many. For example:

• Mary Mahoney, former White House Intern for Bill Clinton. She was killed at a Starbucks Coffee shop in Georgetown. Reportedly, she was about to go public with her story of sexual harassment by Clinton as part of Paula Jones lawsuit. Mary was shot five times and her two associates were also shot and killed while cleaning up after closing. No money was taken.

• James McDougal, died of an apparent heart attack in an Arkansas Federal Penitentiary while in solitary confinement, during which his prescription medicines were denied him. He had been Clinton's right hand man in Arkansas and the banker who financed Whitewater. (His wife, Susan, chose prison time rather than to cooperate with a Special Counsel and give testimony against Clinton.)

• Keith McKaskie had information of the Ives/Henry deaths. He was stabbed to death in November 1988.

• Gregory Collins had information on the Ives/Henry deaths. He died from a gunshot to the face in January 1989. Jeff Rhodes had information on the deaths of Ives, Henry, and McKaskie. His burned body was found in a trash dump in April 1989, but had died of a gunshot wound to the head.

• James Milam had information on the Ives/Henry deaths. He was decapitated. The state medical examiner, Fahmy Malak, initially ruled death due to natural causes.

• Jordan Kettleson had information of the Ives/Henry deaths. He was found shot to death in the front seat of his truck in June 1990.

Three others were attacked and badly beaten, plus had other attempts made on their lives. Gary Johnson had videotapes of Clinton at Gennifer Flowers apartment. His tapes were stolen; Dennis Patrick, a court clerk, had millions of dollars laundered through his account with Lasater & Co. without his knowledge; Larry Nichols, a former official at ADFA and author of "The Clinton Chronicles", which broke many of the Clinton scandal stories was beaten; Reporter L.J. Davis, a reporter investigating Clinton scandals, was attacked in his hotel room in Little Rock. His notes were stolen.*

Before we leave Arkansas, it is worth reading what Meredith Oakley wrote in the *Arkansas Democrat Gazette* about Bill Clinton the day he announced his candidacy for the White House. "His word is dirt. Not a statesman is he, but a common, run-of-the-mill, dime-a-dozen politician. A mere opportunist. A man whose word is fallow ground not because it is unwanted but because it is barren, bereft of the clean-smelling goodness that nurtures wholesome things. Those of us who cling to the precepts of another age, a time in which a man's word was his bond, and, morally, bailing out was not an option, cannot join the madding crowd in celebrating what is for some Bill Clinton's finest hour. We cannot rejoice in treachery. The bleaters who care more for celebrity than veracity are basking in a false and empty light. They trumpet the basest form of political expediency, for they revel amid the debris of a broken promise. Clinton will never accept that assessment of his actions or his following. He subscribes to the credo that the anointed must rule the empire, and he has anointed himself. In his ambition-blinded eyes, one released from a promise has not broken any promise. He ignored the fact that he had given his own pardon." Prophetic words – and accurate - to say the least; and, considering the multiple shames of Bill's and Hillary's eight years in the White House, less damning that they might or should have been.

So it was off to 2400 Pennsylvania Avenue for the variegated collection of Arkansans that the Clintons had knighted to join them, including an assemblage of lawyers chosen to keep the problems of the Clintons well under control. The basic plan had been formed and now it remained only to appoint other similar minded leftists/liberals to key positions. The control apparatus that had proven effective for them in Arkansas would serve as the formula to be implemented in the Executive and Judicial branches. They were confident their fellow Democrats in Congress could be counted on to do the rest.

Bill's and Hillary's cabinet choices were interesting: few had experience in or were qualified for the positions they were to fill; few had made a positive mark in their chosen professions to date; their political agendas were remarkably similar; Hillary had selected or approved all of the nominations, with Bill, of course; and each nominee readily swore allegiance first and foremost to the Clintons. The White House staffing quickly resembled a pre-school gathering or a high school recess. White House dress codes became jeans, tennis shoes, or 'whatever'. Clearance requirements were ignored for months. The attitude of the Clintonites was obviously disdainful not only of the military assigned to the White House, but of most other outsiders as well. James Carville and Paul Begala were role models for Clinton followers. They were the "attack dog" enthusiasts, as well as incessant distorters of facts.

Like all Clintonites, they were strong proponents of a much "expanded government" (socialist/communist) programs and controls.*

Clinton's choices did not seem to have a high survivability rate. Secretary of State Christopher again proved ineffective and was finally replaced. (However, he reappeared to advise Al Gore on his 2000 run for the roses, but again failed.) Lloyd Bentsen, previously known for his 'special' $10,000 "Breakfast Club" sessions while a Senator from Texas, soon resigned his Treasury post. Secretary of Agriculture Mike Espy left because of legal problems. Secretary of Defense Les Aspin also left in shame after the Somalia debacle. Bruce Babbitt, Secretary of the Interior, was in hot water from the 'git-go' for his many highly questionable "land" deals with BLM and/or developers. Secretary of Commerce Ron Brown, former DNC Chairman, already under investigation and rumored to be on the verge of exposing some of the administration's illegalities, died in a still mysterious plane crash. He was found in the plane with a hole in the top of his head very much resembling a gun shot wound. The video tape on Ron Brown's "death" is a bit grisly but intriguing. So is the fact that no autopsy or investigation followed. A stewardess on his plane was reported as having survived the puzzling crash, but reportedly suffered a broken neck and/or a slit throat during her subsequent evacuation.*

Then we had quixotic but otherwise less than qualified Rhodes Scholar Robert Reich as the talkative Secretary of Labor. Jocelyn Elders was Secretary of Health and Human Services but soon became too controversial because of her suggested sexual recommendations for public schools, such as encouraging masturbation and handouts of condoms as routine fare for nearly all ages. Henry Cisneros did not last long as Secretary of Housing and Urban Development because of mistress and 'perjured' deposition problems. Hazel O'Leary had to leave her Secretary of Energy position for conduct in office bordering the extreme edge of the illegal. Richard Riley, Secretary of Education, while totally pliant to the Clintons, proved to be both ineffective and largely unproductive. Same for Sandy Berger. Clearly, "The Right Stuff" the cabinet was not.

Webster Hubbell had hoped to be named Attorney General, but did not meet all of Hillary's requirements. Too, the Clintons' concern was that the Rose Law Firm connection might be too obvious in the selection, so Hubbell was named Assistant Attorney General. Attorney General Janet Reno, herself a liberal 'Judge" from Florida and supportive of homosexuals, was named after two other liberal nominees had to be abandoned by the Clintons for having failed to pay Social Security benefits to their hired help. But with Hubbell in place and a very appreciative Janet Reno, the Clintons were confident they would be as protected as they had been in Arkansas. After all, Reno had the F.B.I., the D.E.A., the A.T.F., the I.R.S., I&N, the Federal Marshals, etc., and, equally important, all the Federal Attorneys across the country under her jurisdiction. Underscoring this control, Hubbell ordered all the previous Federal Attorneys to resign following Clinton's January 20, 1993 inauguration. The Mainstream Media? Hail to the Chief!

The first major policy effort by Clinton was a near disaster. He had pandered to the homosexuals during his campaign and promised them much. He attempted to effect a change in the long-standing practice of not allowing homosexuals into the military services, but had not accounted for the overwhelming resistance to such a change by the general public and the Pentagon. Clinton hurriedly backed-off, and finally agreed to the "Don't ask, don't tell" dictum. Then it was Hillary's turn, having been touted during the campaign as the "two-for-one" partner that Bill would bring to the White House. Assisted by Ira Magaziner, her incredibly large and costly health care proposal was a work-in-progress that took months of preparation by several hundred government workers dragooned from many government offices. The cost of this plan to the tax payer was estimated to be $13.8 billion dollars, as opposed to the $100,000 cost figure claimed by the Clintons.* The bottom line? It failed muster and was finally dropped by Congress as an un-winnable, failed pursuit. But such small set-backs as that seldom bother Socialists and/or Liberal Leftists; they just bide their time for another opportunity.

The Clinton record is less than clear on meaningful initiatives regarding or interest in matters of "common welfare" by the White House. There was constant talk, of course, but few deeds of real merit. However, there were more than a few significant and worrisome events that deserve review, for they reveal the tawdry baseness of the Clinton Administration at work.

• Waco: "The FBI assault on the Branch Davidian community in Waco, Texas, on April 19, 1993, incinerating at least 76 people, most of them women and children, after FBI tanks went smashing through the walls of Mount Caramel, stands as the worst tragedy precipitated by government action on American soil this century. You have to go back to the slaughter of 200 Sioux Indians at Wounded Knee in 1890 to find an abuse of power on this scale. Just like Waco, Wounded Knee was designated a "battle" by officials; and just like Waco, the victims were demonized as sexual deviants. Infants too young to use masks were subjected to six hours exposure to CS gas, a weapon banned in warfare under the Paris Convention on Chemical Weapons. The Justice Department plan, in effect, was to torture the children to force the surrender of the parents. There is expert opinion that some of the babies would have died from the gas alone, without the fire. So I bristled when Attorney General Janet Reno testified to Congress that the raid would have been a success if the FBI had increased the dosage of chemical poison. I do not believe that the Branch Davidians were stockpiling weapons for offensive action, or any action. I do not believe that they "ambushed" the ATF as federal agents were "peacefully" trying to deliver a search warrant. I do not find the FBI account of the final conflagration to be remotely credible. Bill Clinton remarked at a press conference two days after the fire, "I do not think the United States government is responsible for the fact that a bunch of fanatics decided to kill themselves:".(55) And hiding behind Reno's combat boots was cowardly Bill, who had left the Waco problem in "Ranger" Reno's soiled hands. Shame on all of them, and especially the Media for not exposing the farce.

• It seems strange that Louis Freeh and his mental midgets had not learned a damn thing from the clumsy and sorry Ruby Ridge experience of the previous August. The same collection of inept organizations (ATF, Federal Marshals, FBI) had gathered to arrest Randy Weaver…and ended up killing Randy's wife, his son, his son's dog, and wounding Randy and his friend Kevin Harris. The claim was that they had subdued an "armed separatist brigade", as the *N.Y. Times* chose to label it, obviously accepting FBI handouts to the media. And of course helicopters had also played a role. Who needs "negotiators" when you have snipers available? The ensuing law-suits went against the government and for the Weavers, thus providing some sense of justice. One would think that Clinton, Reno, and Freeh – and the Media - would have recalled Ruby Ridge when the Waco plan was being drawn. Or was it intended that there would be no Davidian survivors?

• "To compound the injustice, the Waco survivors were subjected to vindictive prosecution. Some of the Davidians were sentence to 40 years in prison, even though the jury found them innocent of conspiracy to murder. To the horror of the jurors, convictions on minor charges were misconstrued by the judge to vastly increase their penalties. As jury foreman Sarah Bain said later: "the federal government was absolutely out of control here. We spoke in the jury room about the fact the wrong people were on trial, that it should have been the ones who planned the raid and orchestrated it and insisted on carrying out this plan who should have been on trial."(56)

I remember watching the Waco fiasco unfold day-by-day on TV and telling my sons that if ever they wanted to learn how *not* to mount an operation they should pay attention to what was happening at the Davidian compound. It was patch-work planning at best and pathetically amateurish - and deadly - at worst. For example, "negotiations" by ridiculous, crude intimidation as practiced by some of the FBI "wizards" that were doltish in concept. The one or two "on-track" negotiator(s) were simply excised from the process in the rush to finish the operation. Prior to the final effort by the FBI manned tanks, with their hastily trained crews, to breach the exterior walls of the compound, there had been assaults by ATF personnel, firing from roof-tops, and strafing by helicopter gun-ships as well. Hostage negotiators failed to convince others more senior not to use the tanks and the CS gas. It was obvious that command and control elements were confused and frustrated, and being pressured by Washington FBI desk jockeys, and the top levels of the Justice Department. In video(s) I've seen on Waco, one AFT agent manages to shoot himself in the foot after clambering onto a roof; another part shows two ATF agents climbing through a window into a room on the second floor, followed very closely by two other ATF agents on the roof who then fire into the same room that had just been entered by the first two. It was more an attitude of damning the Davidians than anything else, underscored by the FBI's failure to have a fire department standing by at the site. This remains unforgivable, for it merely insured that there would be little or no chance of survival for those inside. And the immediate bull-dozing of the "crime scene" was obviously intended to obfuscate or obliterate any evidence contrary to "official" claims, such as the denials by the FBI of having used "flash-bang" grenades – or other incendiary devices. The "cleansing" of the area after the fire did not take long; for unlike the fire trucks, bull dozers were ready to go. And the shameless cover-up of Waco continued in Washington by Justice, the White House, Congress, and most of the Media – but history can not be revised that easily.

The Clintons also treated us to "Travelgate", the Hillary inspired firing of the career employees in the White House Travel Office on trumped-up charges, which were eventually found groundless. It later surfaced that documents vital to Billy Dale's defense had been destroyed with the knowledge of the Justice Department.

"Whitewater" (and Casa Grande) hearings by Congressional committees surfaced illegal real estate and banking ventures by the Clintons in Arkansas. Aided by Janet Reno, who hampered the investigations to the extent she could, and the collusion of Liberal Democrat Senators Glenn[2]-, Leahy, Reid, Daschle, Levin, Kennedy, etc., and Representatives Bonior, Waxman, Rangel, Dingle, Obey, Schumer, Nadler, etc., the investigations faltered and failed. Only a few carefully selected witnesses were allowed by the Democrats to testify; all of whom had convenient memory loss or were less than truthful. Months elapsed before some of the long sought-after Whitewater associated records were 'found' in the Clinton living quarters area of the White House. Surprise!

• White House aide Cheryl Mills told the House Government Reform and Oversight Committee that she and White House Counsel Jack Quinn had withheld documents from investigators for 15 months, including a memo suggesting Clinton wanted the $1.7 million White House Data Base shared with the DNC.•

• Fox New's Carl Cameron revealed "that FBI surveillance observed Charlie Trie's employees destroying evidence in the campaign fundraising investigation. On the eve of Senate hearings into campaign finance abuse, the Justice Department pulled back on the warrants and the search of Trie's office leaving frustrated FBI agents to watch as more documents were destroyed."•

• Thanks to the Clintons and Janet Reno, these 'investigations' became charades on the 'Hill'. FBI Gate magically disappeared as did Trooper Gate. Again, Reno and her "Justice" Department obfuscated or hindered meaningful hearings, including restricting and/or denying vital subpoenas. Clinton's Arkansas 'Storm' Troopers went largely untouched, including key State Troopers who had previously and effectively done Clinton's (and Lasater's) biddings. Craig Livingston, the individual "responsible" for the FBI files (reportedly as many as 2,000*) being in and used by the White House for political purposes, was identified as having been hired at Hillary's behest, but that investigation was also effectively blocked by Reno's Justice Department. Livingston, and Marceca - who had been hired by Livingstone - escaped unscathed by claiming faulty memories and repeatedly pleading the Fifth Amendment.

• Buddy Young, former head of Clinton's State Trooper security detail in Arkansas, was promoted to a senior position in FEMA paying $92,000 a year. He moved from Washington, D.C. to Denton, Texas the day after Vince Foster was found dead.

• Among those receiving unexpected audits by the IRS were Juanita Broaddrick, Paula Jones, Kathleen Wiley, Linda Tripp, Dolly Kyle Browning, Katherine Prudhomme, Billy Dale, Bill O'Reilly, Kent Masterson Brown (who brought the lawsuit compelling Hillary's task force members to be revealed), the National Review, American Spectator, Christian Coalition, Citizens for a Sound Economy, Heritage Foundation, National Rifle Association, Freedom Alliance, Western Journalism Center, National Center for Public Policy Research, Fortress America, and Citizens Against Government Waste. All had incurred the wrath of the Clintons. Subsequently many were defended by *Judicial Watch*, or the charges finally dropped by the IRS.

• Mark Middleton, friend of Clinton and an Arkansas lawyer, invoked his Fifth Amendment rights 28 times in refusing to testify at a House committee hearing on campaign finances abuses during the 1996 presidential election when asked if he conspired with government officials in China - or elsewhere - to illegally funnel contributions to the DNC or the Clinton-Gore re-election campaign.

• President Clinton, according to the *Washington Times*, indicated that his memory had failed him 267 times in regards to questions asked about the Paula Jones case. For instance: "I don't remember" was given 71 times, "I don't know" 62 times, "I'm not sure" 17 times, and "I have no idea" 10 times. The Media did not question this incredible performance.

• Former White House Counsel Charles Ruff used "I don't know", "I don't remember", "I don't recall", or "I have no specific recollection" 12 times within the first 30 minutes of questioning before a congressional committee investigating White House 'E-mails'. The Media again evidenced little interest.

• In FBI interviews, Vice President Al Gore changed his answers when confronted with documents in a fund-raising investigation. In one FBI interview, Gore said 23 times that he was unable to recall aspects of the November 21, 1995 meeting and other fund-raising issues brought up by the FBI. The Media? No follow-on stories.

Bill Clinton was clearly wagging the wrong finger when he blatantly lied on television to the American public about his relationship with Monica Lewinsky. Despite his oft repeated protestations to the contrary, he finally admitted to having had an "improper relationship" and "apologized" for his behavior. He later paid for his lies. In addition to being impeached by the House of Representatives for perjury and obstruction of justice, Clinton had earlier been held in contempt by a federal judge and been fined $90,000 for perjury, in addition to settling the lawsuit by Paula Jones for $850,000. He was subsequently suspended from the practice of law in Arkansas and disbarred. Despite the 'Clinton Legal Defense Fund' that Democrats established for him, his multiple legal fees were likely well into the seven figures. Some suits against the Clintons remain. As an aside, a measure of the Clintons as vengeful persons can be made from the "releasing" by Northwest Airlines of Steve Jones, Paula Jones' husband, from his 16 year job as a ticket counter person five days after the Clinton settlement. The IRS had earlier audited them, and now it was Northwest's turn to implement another nasty "Arkansas twist" to the Clinton-Jones script. No prompting involved, of course. And little Media coverage, of course.

Clinton's Real Legacy:

The People's Republic of China should have invited Bill and Hillary Clinton to the recent successful launch and recovery of their first space craft. After all, it was he who made possible this quantum leap forward by the PRC into the space age. It did not happen overnight, nor did it result solely from covert attempts by the Chinese to acquire space technology secrets from various locations in the U.S. over an extended period of time. Nor did the success result from Chinese students having come to the United State to study space technology and space engineering, then returning to China with advanced knowledge. Some may have come from a combination of these, but most of it came from the Clinton administration that blandly "gave" incredible amounts of advanced technology to the Chinese in return for millions of dollars in cash for political purposes. Just as the Mena operation in Arkansas had started modestly, the greed factors began to grow rapidly and became out of hand. The root of this problem, like Mena, had it's origin in Arkansas, thus we have the very long shadow of a China-Gate factor. Unfortunately, this one will continue to haunt the United States and our allies for decades to come. "® Indeed, look at what China is doing with the huge trade surplus that it has accumulated at American expense:

- It has built what will soon be the world's largest exchange reserve, a future source of political as well as economic power.
- It is using billions of American dollars to buy state-of-the-art weapons systems from Russia and Western Europe.
- It is subsidizing its own trade deficits with countries that are not as open to Chinese exports as the United States.

To remedy this, we must start by taking steps that will reduce the trade deficit with China. "The United States should allow China to export all it can sell in the United States but on one condition: that the trade deficit ratio be gradually reduced by an increase in American exports, brought about by the elimination of China's unilateral trade barriers."(57)

The Clinton administration gave China what it wanted: Most Favored Nation status and WTO membership. That the U.S. did little beyond posturing in regards to Human Rights, e.g., Tibet, was also obvious, as was China's conclusion that the United States was still but a "*paper tiger*", a determination China had reached and preached beginning in the early 1960s.

How did we get from the years of uneasy and uneven relations between the U.S. and China in the '60 and '70s to the U.S. aiding and abetting China's role on the world stage in the '90s? A partial answer is the establishment of diplomatic relations with China by Nixon and Kissinger in 1972, and subsequently the incredible investment opportunities that 'Kissinger & Associates' developed with and in China after Kissinger left the government. But that in and of itself pales in comparison to what Clinton first pursued in Arkansas with a Chinese agent, and subsequently what he and Al Gore engaged in throughout most of the 1990s. For example:

• "a Communist Chinese banker (agent) rescued the Clinton-Gore team not once but twice in 1992". "within five-and-a-half years, the Clinton-Gore administration would be the subject of worldwide ridicule and a wave of congressional investigations for having maintained itself in power with a flood of tainted money...(that) in return for money, Bill Clinton and Al Gore would literally sit down to supper with a procession of agents from the Communist Chinese army, Chinese intelligence, the Macau criminal syndicate, and Cambodia's biggest drug trafficker. Who would have guessed that Clinton and Gore would use the power of the White House itself to lobby for foreign powers?"(58) "The Clinton-Gore record of campaign finance abuses is staggering:

• Chinese agents ensured victory for the Clinton-Gore team in the 1992 general election with a massive cascade of illegally laundered foreign funds into key states.

• More than one hundred potential witnesses of illegal foreign campaign contribution to the Clinton-Gore team have fled the country, taken the Fifth Amendment, or refused to be interviewed by investigative bodies

• Chinese agents helped secure the 1992 Democratic presidential nomination for Clinton with a multimillion-dollar loan from an Arkansas bank under their influence.

• Chinese agents became the number one donors to Clinton and Gore in 1992.

• The Clinton Justice Department allowed the statute of limitation to run out so that illegal donations from Chinese agents could escape prosecution.

• Leading donors to the DNC are business associates of a major

Tiananmen massacre war criminal that even the Chinese jailed.

• Chinese agents tried to use their donations to Democratic senators to pressure Taiwan's banking authorities.

• Chinese agents gave $100,000 to convicted felon Webster Hubbell at the very moment the independent counsel's office was seeking his cooperation in its investigations of the Clintons.

• Chinese agents helped fund Dick Morris's brilliant stealth advertising campaign against the Republicans in 1995-1996.

• A Macau criminal syndicate figure who exploits women for prostitution laundered more than $1,000,000 in illegal donations to the DNC; he met Clinton on a number of occasions, including visits to the White House; and he may have funneled more than $300,000 in illegal cash donations to the Democrats and/or hush money for Hubbell.

• A Chinese agent who is also in the business of exploiting women for prostitution donated hundreds of thousands of dollars to Bill Clinton and Al Gore, and he sat beside both the president and vice president at intimate fund raisers.

• A Chinese agent/DNC donor with personal connections to Clinton and Gore is a business associate of Cambodia's largest narcotic trafficker.

• A Chinese agent who specializes in illegally laundering funds for the Democrats has an eight-year business relationship with Gore.

• Leaders of a Thailand conglomerate that is in business with Middle Eastern terrorists and with China's biggest arms smugglers had a White House meeting at which they were illegally solicited for campaign donations.

• A Chinese PLA (People's Liberation Army) spy laundered illegal campaign funds through Johnny Chung and met Clinton twice at fund-raisers.

• A Macau gambling figure and a Hong Kong billionaire in business with the PLA donated hundreds of thousands of dollars to Clinton's favorite "charities".

• One of the officers of an American defense contractor in business with China's missile builders became the number one contributors to the Clinton-Gore campaign in 1995-1996.

• After illegal campaign funds began to flow to the DNC from a PLA spy, the Clinton White House and his Navy Secretary helped the PLA's arms delivery boys secure the port at Long

Beach".(59)

"The bottom line: Clinton and Gore got away with it. At this writing, not one single person on the recipient side of the ledger has been charged with any crime. Some of the conduits for foreign money, such as Charlie Trie and Maria Hsia, have been indicted, but only for minor offenses. Clinton's Executive Order Number 1 has become an obscene joke."(60) "During the 1992 campaign candidate Bill Clinton bitterly attacked George Bush's China policy. Clinton said, "In the three years since the massacre in Tiananmen square, the Bush administration has pursued an ill-advised and failed policy of constructive engagement with the aging leaders in China."(61) Play it again, Bill!

Add to the above these few snippets:

• China Ocean Shipping Company (*COSCO*): The PLA's arms delivery team has a new home in California thanks to high-level intervention from the White House and Clinton's Secretary of the Navy.

• *National Security Adviser Sandy Berger*: Berger was, according to Representative Nancy Pelosi, the point man for the trade office of the Chinese government. Surprise?

• *COCOM*: As soon as President Clinton came to power, the administration began dismantling the export control system, paving the way for American defense production equipment to be made available to the PLA.

• *Satellites:* Under the Clinton administration the PLA launchers have learned to make their missiles more reliable; the satellites are used for forbidden purposes; and the operators are simply members of the PLA in civvies. (Thank you, Bernard Schwartz!)

• *Machine tools:* Machine tools we use to make ICBMs, strategic bombers, front-line fighters, and advanced air lifters went to China.

• *Supercomputers:* The Commerce Department doesn't even know how many supercomputers the United States has sold to the PRC, under whose control they are, or for what purpose they are being operated."(62)

"When national defense specialists gather to look over the debris of the Clinton-Gore years, we normally talk about the losses – the lost army division, the lost air wings, the lost carrier battle groups – from even Bush's (Bush senior) low post-Cold War levels.

Where once we had 15 carrier battle groups, we now struggle to keep 10. In 1990 we had 18 army divisions; now we have 8, and they are lighter than their predecessors. We have half the number of Air Force and Navy Tactical Wings we had under Bush Sr. Active duty military personnel are down by a third ."(63) All of these points should be highlighted in the Clinton library. Our Media can certainly take pride in the excellent coverage they provided us during the Clinton years - or didn't the Mainstream Media recognize these numbers before?

"The Clinton-Gore foreign policy, which places trade and commerce ahead of national security, blinds the administration to the rampant proliferation reported by the allied intelligence agencies and (some of) America's own press corps."(64) "Communist China sees but one major roadblock to achieving regional hegemony – the Pacific stretch of the United States. Consequently, the Party's military buildup has been tailored to counter America's military capabilities – from its nuclear missiles, to the Sovremenny destroyers, to its information warfare program, to its satellite program, American is now a target. Faced with this threat, what was the response of the Clinton-Gore administration? It:

• permitted the sale of hundred of supercomputers to the Communist Chinese, exposing America to the threat of destructive information warfare;

• failed to enforce American anti-proliferation laws, so that it is only a matter of time before some PLA weapon is used by a rogue state in the Middle East or South Asia, raising the spectre (sic) of germ or nuclear warfare.

• allowed PLA rockets to be made more reliable by the technical assistance of Loral, an American corporation whose president, Bernard Schwartz, was the 1996 Clinton-Gore campaign's leading donor;

• referred to Taiwan as "the turd in the punch bowl" of U.S.–China relations, essentially selling out a democratic government to a tyrant;

• watched (author's note: and abetted) as a Communist Chinese company gained control of the Panama Canal;

• exposed the United States to derision and distrust by the way the Communist Chinese agents have so easily been allowed to buy influence at the White House."(65)

"By ignoring the PRC's fifty-year history of naked aggression and her overt threats to the United States and its democratic allies, Clinton and Gore abetted Communist China's military ambitions. Perhaps most odious has been how that administration rehabilitated the image of the PLA, a brutal army responsible for tens of millions of dead civilians in China, Korea, Tibet (*), and elsewhere. Under Clinton and Gore, six of the seven active-duty uniformed generals directly responsible for the June 1989 Tiananmen Square massacre were welcomed to Washington."(66) Did the Mainstream Media cover these points in depth or question these absurd "appeasement" policies? Of course not!

In September (1996), the first newspaper stories appeared about large campaign contributions to the Democratic National Committee. A number of contributions were connected to the Lippo Group, the Indonesian conglomerate owned by the Riady family. "The Riadys were an Indonesian family worth at least a billion dollars. They controlled the Lippo Group, a conglomerate with interests in the United States, Asia and China. James T. Riady, the son of the family patriarch, Mochtar Riady, was only 20 in 1977 when he went to Little Rock, where he met and became friends with the young Arkansas attorney general, Bill Clinton. In the 1980s James Riady ran the Worthen Bank in Little Rock and continued the close association with the governor. Hubbell had been the attorney for both James Riady and the Worthen Bank."(67) "James Riady made about 20 visits to the White House after Clinton was elected and continued to pursue ties with his old friend. Sherburne and Fabiani (lawyers at the White House) knew for more than a year that *The New York Times* was asking some questions about $100,000 that Riady had given to Webster Hubbell. The payment had been made to Hubbell after he resigned from Justice but before he had plead (sic) guilty to bilking his former partners, including Hillary, and clients of $400,000. A $100,000 payment to Hubbell by Riady had overtones of hush money – money paid for Hubbell's silence, Sherburne realized."(68)

Shortly after the first presidential debate in October, some reporters posed direct questions about Riady and the relationship between Clinton and Riady, and if campaign contributions had been made. "Bruce Lindsey, Clinton's friend and deputy counsel, said he would go to Clinton himself and get the answers about

Riady. He reported back (to Sherburne) that there had been some meetings. Riady had said he was glad China was being given favorable trade status. He also had stressed the importance of good relations between the two countries. But Lindsey said the relationship between the president and the Riadys was routine and basically social. There is nothing there about Hubbell, he said."(69)•

• On the plane to San Diego for the next debate, "Sherburne sat next to Sandy Berger, then the deputy national security adviser. She asked him for a short course on Indonesia and the Riadys. She read through her stack of papers. Included was a March 9, 1993, letter to the president from Mochtar Riady, the head of the family. It was four pages, single-spaced, urging Clinton to normalize relations with Vietnam and pursue better economic ties with China. There was also a reference to seeing the president at his inauguration, and a request to allow the Indonesian president to attend a G-7 economic meeting. This certainly was not an exclusively social relationship."(69) Lindsey's response to Sherburne's protest that this could not be described as simply a social contact was that "he didn't care what the press believed." Most of the Media has yet to play the Riadys, Charlie Trie, John Huang, Johnny Chung, Ng Lap Seng, Ted Sioeng, or Maria Hsia cards, or the multiple Chinese cards involved either; perhaps they are waiting for a story with impact. That 20 of 21 witnesses left the country (only Charlie Trie returned) and 18 foreign witnesses refused to be interviewed, or to make statements, would seem to be worth further coverage. The China connection stories have all but disappeared. But there is more in the unlikely event the Media is interested:

• "...at her confirmation, Secretary of State Albright conceded to a senator that PLA companies were smuggling germ warfare equipment to Iran, but that the United States had applied no sanctions to punish this illegal act. In other words, the Clinton administration let PLA arms dealers believe that it wasn't serious about stopping this kind of trade."(70)

• "To the best of my knowledge, no one has said anything to me about any espionage which occurred by the Chinese against the labs during my presidency" said Clinton in early 1999. Yet Sandy Berger, Clinton's National Security Advisor, was informed at least three years earlier of the problem. Excuse me, but doesn't there seem to be a substantial disconnect here that needs to be explored and explained? Liars Inc., perhaps?

• "In recent years – with the World Trade Center bombing in New York(®), the U.S. embassy bombings in the Middle East, and in Africa, and so forth – we have seen how America has become a target for terrorist attacks. Eventually terrorists will try to target the United States of Americans abroad with WMDs. This is what is at stake when Clinton lifts sanctions against PLA companies to please his leading contributor" (71) The pictures of the president and the vice president of the United States surrounded by Communist Chinese agents made the front pages from Tokyo to Tel Aviv. The message(s) Bill Clinton and Al Gore to the world were clear.

Our Commander-in-Chief, despite his "loathing of the military", finally got to play 'King of the Hill'. In contrast to the United States deploying military troops in eight foreign campaigns from 1946 to 1991, "During the Clinton years, we deployed troops to at least forty separate foreign locations. The Clinton administration called these "unanticipated missions." My peers, fellow pilots, and aircrew members called them "channel surfing." Somalia, Rwanda, Haiti, Macedonia, Bosnia, Ecuador, East Timor, Kuwait, Liberia, Albania, Congo, Gabon, Sierra Leone, Afghanistan, Sudan, and Iraq were countries that saw direct U.S. military intervention of one sort or another. The results were ruinous, the specific engagements troubling. The Somalia debacle of 1993 was Clinton's first exercise in military deployment and the first case in point. Under Operation Restore Hope, then-president George Bush deployed 28,000 U.S. forces in late 1990 to provide and protect the distribution of food and medicine in war-torn Somalia. I flew into the country in mid-1993, when our mission was still a humanitarian effort to feed and clothe the citizens of Mogadishu. It was a very pro-American, pro-UN environment. The next two or three times I went in, I noticed that we seemed to be stepping up the war-fighting aspect, taking in troops and equipment. The mission seemed to be evolving. .It was evident that the situation was sliding toward chaos. The degree of intensity was increasing dramatically. The operation had grown into a nation-building commitment under Clinton. The critical juncture came in May 1993, when President Clinton declared the mission accomplished. But rather than bringing our forces at home, he left 4,500 American men and women in place as part of his vague policy of "assertive multilateralism." "The Clinton administration had developed a loose national strategy of "engagement and enlargement"; foreign intervention would be primarily peace operations, and potential military interventions would be through international institutions like the United Nations. On June 5, 1993, twenty-five Pakistani soldiers contributing to the U.N. effort were attacked and killed by Mohammed Farah Aidid's militia. UN bases came under mortar attack. Missiles were launched at U.S. helicopters. Snipers tried to pick off UN and American troops. Because the Clinton administration provided insufficient support, our forces had to depend on UN units. Our commanders weren't happy with the situation, nor that their request for armored vehicles and tanks to protect the troops, and other support, had been denied by Secretary of Defense Les Aspin. This was an obvious recipe for disaster, and the price America paid was the lives of eighteen American soldiers, the bodies of two desecrated and dragged through the streets. Seventy-three other soldiers were injured. Help was pinned down only a mile away, all because the request for armored vehicles and tanks had been denied. Clinton (then) announced he was sending in an additional 1,700 troops and 104 armored vehicles to "do the job right".(73) I wonder if his lip was quivering.

"In September 1994, we began our final pullout from Mogadishu and the UN operation in Somalia from the historic successes in the Persian Gulf to being beaten by a ragtag group of thugs in Somalia, and all because of an administration that had no real, definable idea of what it was doing there and none of the military and foreign policy expertise requisite to complete a mission of this complexity. More significant, an Arab extremist whose name few Americans had ever heard – Osama bin Laden – had achieved a victory, and his international al-Quaeda terrorists, who were fighting alongside Aidid's thugs were emboldened to strike again. It became a common joke among soldiers, sailors and airmen: "We're doing so much more with so much less that we should be able to do everything with nothing."...PDD's-25 (Presidential Directive) nice words, but they were ignored. More notably, perhaps, but not receiving as much press, PDD-25 permitted U.S. forces to serve under foreign commanders. This tenet would come back to haunt us."(74) "At the same time we brought our forces out of Rwanda, we put them into Haiti. On September 15, 1994, President Clinton declared that "our mission in Haiti, as it was in Panama and Grenada, will be limited and specific." "The Clinton administration had already stumbled in Haiti once. In the same month as the tragedy in Mogadishu, the administration had sent the USS *Harlan County* to the docks of Port-au-Prince as a show of force. But an angry mob of Haitians was enough to force the U.S. Navy vessel to turn back. This time, with Operation Uphold Democracy, the Clinton administration deployed twenty thousand troops at an expense of at least $2 billion into the poorest country in our hemisphere – but again without a coherent strategy. President Clinton's declared intent was to restore democracy in the form of elected president Jean-Bertrand Aristide and to rebuild a nation that was disintegrating. Ironically, Aristide, the man backed by the Clinton administration, (and Carter) was a committed socialist who had made a career of loathing the United States. The Clinton administration kept American troops deployed in Haiti for five years. We spent $3 billion in Haiti, but the economy only worsened, unemployment climbed above 70 percent, and drug trafficking increased."(75)

Nor should we forget Bosnia, where there were, and are, many European interests, but very few, if any, American security interests. That unauthorized adventure has been referred to as 'Albright's 'war', but Clinton was her boss. Most of the Media chose not to criticize the administration, or expose the accrued dollar costs to tax-payer's, or note the many failures involved, not only in Bosnia, but in the other actions cited above as well – including participating U.S. military resources being placed under disjointed and confused UN command. These oversights by the Media remain incomprehensible. The dissemination lids were kept on tightly by the Clintons, just as they were during the mid-1990s when Osama bin Laden took center stage. Like twin ostriches, the Clinton-Gore appeasement, cover-up, and "spin" era was maintained in high gear in the U.S., and from Europe to the Mid-East to Asia. The Mainstream Media simply chose to ignore the downhill ride in America, the failed patchwork operations, and the escalating threats abroad.

In a September 7, 1999 news letter from David Horowitz (President, Center for the Study of Popular Culture), "Seven months ago, a bipartisan committee of Congress concluded its investigation of the most destructive episode of espionage in the nation's history. Two months ago, the committee's report was finally released by the Clinton White House after about a third of its finding was purged "for security reasons". The *Cox Report* revealed that Bill Clinton's administration has presided over the worst catastrophe that has ever befallen the security establishment of the United States."

The *Wall Street Journal* summed up its conclusion in this way: "The espionage inquiry found Beijing has stolen U.S. design data for nearly all elements needed for a major nuclear attack on the U.S., such as advanced warheads, missiles and guidance systems. Targets of the spying ranged from an Army anti-tank weapon to nearly all modern fighter jets. Most wasn't done by professionals, but by visitors or front companies. Lax security by the Clinton administration is blamed in part, and satellite makers Hughes and Loral are criticized." "It was Clinton's personal intervention that secured a top security clearance, a high post in the Commerce Department and a key fundraising spot at the Democratic National Committee for Chinese agent John Huang. It was Huang who orchestrated Clinton's major fund-raising campaigns, both legal and illegal, including the famous affair at the Hsi Lai Buddhist Temple where a photo of Vice President Al Gore shows him sandwiched between a high-ranking Chinese intelligence official and a gangster from an Asian "Triad" with ties to the Chinese dictatorship. Before being recruited by Clinton, John Huang was an Arkansas based employee of James Riady, scion of the Riady family empire, a multi-billion dollar octopus interlocked with China's military and intelligence agencies. The Riadys, who are ethnic Chinese, are avidly pro-Beijing. They are also co-owners of the Worthen Bank, which mysteriously burned some years ago with Hillary's Castle Grande billing receipts, and which launched John Huang's career." "The massive loss of American security assets during the Clinton years can best be understood as a Theft, a Giveaway, and a Cover-up. The success of these operations depended on a China Connection – an orchestrated penetration of the American executive by Chinese military and intelligence asset, abetted by the President and Vice President of the United States." "Among the factors enabling this conspiracy to complete its transfer of advanced military technologies from American laboratories to the Chinese dictatorship was the systematic destruction of security controls by Clinton's former Energy Secretary, Hazel O'Leary,* and the reckless declassification and release of America's nuclear secrets including the release of eleven million pages of previously restricted documents. But the dismantling of America's security controls was not confined to the nuclear labs. While O'Leary unlocked the gates to Los Alamos and Livermore, the White House was systematically removing security controls on American missile, satellite and computer technologies crucial to China's development of a modern military force. To their credit, some officials in government resisted these treasonous efforts, but the Media failed to publicize the problem. When the "thefts" came to the attention of the appropriate security agencies, it was necessary for the abettors to take countervailing measures, including obtaining White House help. Throughout this process, the White House directed a massive cover-up of its own misdeeds, deliberately failed to clamp a tight lid on security when the most damaging leaks were first discovered in 1995, then continued the transfer of sensitive technology secrets to China even after the hostile agendas and practices of the Beijing regime were undeniable." Kudos to the *Wall Street Journal* for breaking the story, including exposing the Clinton Administration, and its trying to "hide" the Cox report." But Janet Reno and the FBI survived with only minor scratches.

The article concluded "These facts are distressing enough. More ominous is the deafening silence that has greeted their release. One would not know from reading the nation's press or watching the television news services that America had suffered its worst "defeat" in peacetime ever. You would not know that the twenty-year advantage the U.S. held in military technology – an edge that has been the only real guarantee of America's security for half a century - had been lost virtually overnight; and you would no know about the reckless, corrupt and lawless acts of the Clinton White House in respect to China, or if they raise the serious question of whether treason has been committed."

The entire Horowitz news letter is not quoted above, but its major points are, including that the Media again was/is inexcusably derelict, especially in not trumpeting that the White House had effectively concealed key parts of the *Cox Report*. This in itself is both unbelievable and shamefully suspect. In effect, most of the Media – and Congress - simply acted as co-conspirators! Hopefully one of these days the full report will be available and justice eventually served.

Thus China evolves as the real Clinton-Gore legacy; not the impeachment of Clinton because of his zipper problems and a White House intern. Clinton's sustained actions on behalf of China clearly borders on treason against the United States. On November 15, 2003, a brief article from Beijing states: "China sent a communication satellite into orbit today in its third launch since its fist manned space mission last month, the government said. The Zhongxing-20 satellite went up from the Xichang Satellite Launch Center in the southwestern province of Sichuan, the official Zinhua News Agency said. It was carried up by a Long March 3-A rocket, Xinhua said. The launch was declared a success from the Xi'an Satellite Monitor and Control Center 25 minutes later, when the satellite entered orbit, Xinhua said." Beijing was plain rude not to publicly thank Clinton, Gore, Berger, O'Leary, Richardson, Loral, Hughes, and Kissinger & Company, etc.

The oaths of office of the president and vice president – "to protect and defend" - have not been upheld, and the U.S. is – and will be - far less secure because of the deliberate duplicity and sickening greed of the Clinton-Gore administration, aided and abetted of course by the some members of Congress and the inattentive Mainstream Media. And thank you too, Loral (and Loral president, Bernard Schwartz, who reportedly donated amounts in excess of six figures to the Democrats) and Hughes, and all others who climbed aboard the lucrative China dollar train and helped create this threat. And thank you Bill and Al, for ignoring the growing Muslim terrorist threat first identified early during your administration as originating from Osama bin Laden and his al-Qaeda followers. These dual threats to the U.S. will be with us – and the free world - for many years to come. But there are other examples that should be cited to underscore our thanks, such as:

- According to Accuracy in Media (August 20, 2002), Dale L. Watson, formerly with the FBI, "In 1993 told Congress that the FBI had identified "a significant and growing organizational presence" of foreign terrorists in the U.S. but that the bureau had them under control. He said the FBI had an "enhanced capability" to track their activities. He testified that the FBI was aware of a terrorist plan called BOIJINKA that was found in 1995 in the computer of the mastermind of the World Trade Center bombing, Ramzi Yousef. Watson said BOJINKA was a plot to explode bombs on U.S. airliners. True, but it also included plans to hijack airliner and use them as missiles to attack important U.S. buildings. Watson didn't recall that when Special Agent Kenneth William in Phoenix sent a memo to Headquarters last year about the many Middle Easterners taking flight training. He didn't help get a warrant to investigate one who had aroused suspicion, Zacarias Moussaoui."

• The National Review, November 25, 2002: "an agent from the FBI's New York office told Congress how his Washington bosses had ordered him not to track suspected terrorist Khalid al-Midhar. Precisely because the CIA had told the FBI that Midhar was an al-Qaeda operative, the FBI could do nothing: Data obtained through intelligence channels could not be used to launch a criminal investigation. Midhar was let be. Thirteen days later he helped hijack the plane that struck the Pentagon." "In May 1994, FBI director Louis Freeh told Congress that criminal cartels are now richer and stronger than many nations. The *New York Times* echoed that judgment: "The threats (today) are less nations than gangs. The harm they can do the United States is a law enforcement problem." "The doctrine was refined by former Carter defense secretary Harold Brown, chosen by to head the Commission on the Roles and Capabilities of the United States Intelligence Community. Brown redefined international terrorism, traditionally considered a form of political warfare, as a form of "global crime." Terrorism joined narcotics transshipment and trafficking in weapons of mass destruction in that category." "Presidential Decision Directive 24 (May 3, 1994), which gave the Bureau control of counter-espionage, was just the start." "And on June 21, 1995, Clinton issued Presidential Decision Directive 39, which made the Bureau the pointed end of the counter-terrorist program. "Clinton named the FBI the lead: agency in that effort, and PDD-39 itself specified: "Lead agencies are those that have the most direct role in and responsibility for implementation of U.S. counterterrorism policy." Now the FBI and the Justice Department would play major policy roles. The Clinton administration deployed various fogs and euphemisms for this *coup de bureau.* Janet Reno used words like "partnership" and "coordination". National security adviser Sandy Berger invoked a corporate-merger model. Former CIA officer Robert Baer, expressing a more accurate viewpoint in his 2002 book, was less delicate: "Over at the FBI, director Louis Freeh couldn't wipe the smile off his face. He was dismantling the CIA piece by piece." By 2000, Freeh was speaking openly of "FBI leadership in national security." "Eleanor Hill, the intelligence committee's chief investigator testified that before 9/11, "An overall assessment of the risk to America was not prepared." (She) noted, "the United States continued to rely on what was primarily a law-enforcement approach to terrorism. As a result, while prosecutions succeeded in taking (an occasional) individual terrorist off the streets, the masterminds of past and future attacks often remained beyond the reach of justice." "In May, 2002, President Bush actually expanded the FBI's counterintelligence mandate".

• In an October 10, 1999 Washington Times (weekly edition) article, noted "There's a real concern that the Justice Department is not doing its job." And "complained that Justice is providing President Clinton with political cover on issues ranging from the campaign finance scandal to his granting of clemency to Puerto Rican terrorists, (and) also would look into the department's probe of nuclear secrets stolen from U.S. weapons laboratories."

• In the same issue of the Washington Times, an article by Jerry Seper noted "FBI agents assigned to the campaign-finance task force told a Senate committee on Sept. 22 that Justice Department officials blocked their efforts to pursue key investigative targets – including information that Charles Yah Lin Trie was bringing in "duffel bags full of cash" to the Democratic Party. Sen. Fred Thompson (said) "During the 1996 presidential campaign, Charlie Trie contributed $220,000 to the Democratic National Committee and $789,000 to the Presidential Legal Expense fund. Most of the money was not Trie's, much of it was from sources in Macau, Hong Kong and the People's Republic of China. Mr. Smith, along with agents Daniel Wehr, Roberta Parker and Kevin Sheridan, told the committee that their Justice Department supervisor, Laura Ingersoll, who eventually was replaced as the campaign-finance probe's lead attorney, prevented them from executing search warrants they sought to stop the destruction of evidence (at Trie's Little Rock office). Mr. Wehr also testified that "Miss Ingersoll told the agents they should "not pursue any matter related to solicitation of funds for access to the president, adding that the reason given was: "That's the way the American political process works." And the fallout? Ho Hum.

• According to July 2002/NewsMax.com, "The TWA Flight 800 investigation of the destruction on July 17, 1996 was 'wrested' by the FBI "from the National Transportation Safety Board (NTSB), the agency that typically handles crash investigation. NTSB investigators later complained that "overbearing" FBI agents "immediately took control, and hampered a lot of things we did." The bureau stalled the investigation so as not to interfere with Clinton's election. But evidence exists that FBI agents badgered dozens of witnesses – who said they saw a missile hit the plane – into changing their story".

• From the same issue of NewsMax.com: "Mueller steadfastly maintained that what the uninitiated might perceive as timidity was really the careful navigation of turf problems. Manhattan District Attorney Robert Morgenthau complained that the feds were refusing to share important evidence with state prosecutors who had their own prosecutions brewing. During a 1991 Senate hearing on the BCCI, Mueller (then with the FBI) tossed off the tempest with one of his classic minimizing retorts: "The criticisms of the (FBI) are fundamentally unfair. To the extent the allegations could be pursued, they were pursued." Eschewing any blame, he embraced only the credit for helping to untangle a complex international investigation and winning the forfeiture of $1.3 billion of the bank's U.S. asset. The heaviest casualties remained in the ranks of the victims." "Unperturbed by the imperturbable prosecutor, 10 years later the *Wall Street Journal* was still citing the BCCI as proof that Mueller was a "peculiar choice" to direct the FBI. After leaving the criminal division in 1993 for private practice, "Mueller calls good friend Eric H. Holder Jr., the U.S. attorney for the District of Columbia, and asked to assume the relatively mundane duties of a trench prosecutor fighting the tidal wave of homicides in the nation's capital. It was Holder who in 1998 recruited Mueller to take over the ailing U.S. attorney's office in Sam Francisco. However, as those who debate his suitability as FBI director point out, there is a universe of difference between a U.S. attorney's office and the sprawling FBI empire with 27, 400 special agents, professional staffers and support personnel occupying 56 field offices, 400 satellite branches and 40 offices overseas." "Mueller, with just days at the FBI-helm pre-9/11, had an easy pass on the 'who-knew-what-when' hit list. That is, until he blew it with ill-rehearsed damage control tap dancing that came off in the end like so much inept waffling, covering up, - worst of all, unfathomable lack of knowledge regarding the most vital investigation in U.S. history. Mueller's old nemesis from nomination days, the *Wall Street Journal*, grimly noted that only after FBI whistleblower Coleen Rowley's tell-all memo was released did Mueller suddenly become privy to embarrassing field intelligence such as the "Phoenix Memo" with dire warnings about U.S. flight schools where Arab men with ties to terrorist groups were training."

• From the same issue of NewsMax.com, "If Mueller's benefactor was Eric Holder, Teflon Tenet's patron is the senior President George Bush. The retention of Clinton appointee Tenet at the CIA after the Bush victory is largely due to advice from the father to son. Tenet wisely courted the senior Bush, a former director of central intelligence. In 1998, the fawning Tenet even renamed CIA headquarters the George Bush Center for Intelligence. Reportedly, the elder Bush's advice ran counter to the consensus of the current president's transition team, who pegged the CIA a disaster area and recommended Tenet's dismissal" But Bush #43 didn't listen. Lastly, "During all of 1998, the same year FBI counterintelligence briefed Congress on the al Qaeda hijack plot – Clinton met with his Cabinet exactly twice: once in January to lie to them about his relationship with Monica Lewinsky, and again in August to admit the affair."

• According to March 2003 NewsMax.com (and as eventually covered in detail by the O'Reilly Factor, Fox News, including interviews of the two FBI agents) "The Clinton-Reno justice Department refused to allow two veteran FBI agents assigned to the anti-terrorist probe to investigate a key figure tied to Osama bin Laden. FBI special agents Robert Wright and John Vincent told Brian Ross of ABC news that they were called off criminal investigation of suspected terrorists linked to the deadly bombings of two U.S. embassies in Africa. U.S. officials say al-Qaeda was responsible for both the embassy attacks and the Sept. 11, 2001 attack in the US. Wright told Ross that when he pressed for authorization to open a criminal investigation into the money trail, his supervisor stopped him. Wright and Vincent said that some of the money for the attacks led back to the people they had been tracking in Chicago and to a powerful Saudi Arabian businessman, Yassin al-Kadi, one of 12 Saudi businessmen suspected of funneling million of dollars to al-Qaeda (with) extensive business and financial ties in Chicago. Chicago federal prosecutor Mark Flessner was assigned to the case despite efforts Wright and Vincent said were made by superiors to block the probe. Flessner told Ross that the two agents were helping him build a strong criminal case against al-Qaeda and others. "There were powers bigger than I was in the Justice Department and within the FBI that simply were not going to let it happen. And it didn't happen, Flessner said."

Further reflecting the incorrigibility of the Clinton-Gore administration:

• Enumerated in the September, 2000 issue of INTERNET Vortex, "Clinton's administration set the following records: "the most convictions and guilty pleas; the most cabinet officials to come under criminal investigation; most witnesses to flee country or refuse to testify; (he) is the only president ever impeached strictly on grounds of personal malfeasance; (and) the greatest amount of illegal campaign contributions from abroad;

• Crime Stats: Number of individuals and businesses associated with the Clinton machine convicted of or pleading guilty to crimes: 47. Number of convictions during Clinton's presidency: 33; Number of indictments/misdemeanor charges: 61; Number of imprisonments: 14; Number of congressional witnesses who have pleaded the Fifth Amendment, fled the country to avoid testifying, or (in the case of foreign witnesses) refused to be interviewed (as of September 1999): 122.;

• Number of times Hillary Clinton said "I don't recall" or its equivalent in a statement to a House investigating committee: 50. Number of times Bill Clinton said "I don't recall" or its equivalent in the released portions of his testimony on Paula Jones: 271. Total number of facts or events not recalled before official bodies by (Clinton administration persons) Bill Kennedy, Harold Ickes, Ricki Seidman, Bruce Lindsey, Bill Burton, Mark Gearan, Mack McLarty,

Neil Eggleston, John Podesta, Jennifer O'Connor, Dwight Holton, Patsy Thomasson, Jeff Eller, Beth Nolan, Cliff Sloan, Bernard Nussbaum, George Stephanopoulous, Roy Neel, Rahm Emanuel, Maggie Williams, David Tarbell, Susan Thomases, Webster Hubbell, Roger Altman, Hillary Clinton, Bill Clinton: 6,121. Number of campaign contributors who got overnights at the White House in the two years before the 1996 election: 577.

In an April, 1999, (19-25) National Weekly Washington Times, Phyllis Schlafly noted that "Executive Orders (EO) originate with the constitutional text that says the president shall "take care that the laws be faithfully executed." However, "laws must mean laws that are already passed, not laws that the executive order purports to create. Some of Mr. Clinton's executive orders (now up to 279, and not all numbered) are federal land grabs over property that belongs either to the states or to private landowners. Land use and zoning are quintessentially matters of state or local, not federal, jurisdiction. Mr. Clinton's surprise grab of 7 million acres of Utah land for a national park in 1996 just happened to include a trillion dollars worth of clean-burning, low-sulfur hard coal. Mr. Clinton's removal of this huge natural resource from commercial availability tremendously enhanced the value of the second largest source of environmentally safe coal, which just happened to be owned by Mr. Clinton's Indonesian friends, the Riadys, who gave millions of dollars to Mr. Clinton's presidential campaigns in 1992 and 1996. For the Mexican and Brazilian bailouts, Mr. Clinton used executive authority to raid a U.S. treasury Department fund set up in the 1930s for the specific purpose of being available to stabilize the U.S. dollar. The president certainly was not authorized to give this money to foreign governments so they could make their loan payments due to Treasury Secretary Robert Rubin's old firm, Goldman Sachs. (You don't really suspect, do you?) Clinton's EO 13919 of June 3, 1994, titled "National Defense Industrial Resources Preparedness" asserts plenary and dictatorial authority over citizens, food, transportation, energy, health, contracts, materials, and resources, to be exercised by the National Security Council and FEMA (Federal Emergency Management Authority Agency.) Many still wonder if the real purpose of that Executive Order was to grab emergency powers had we been bitten by the Y2K bug, or in case another calamity might encourage similar suspensions of normal rights. Maybe the Media might tell us.

(Author's note: While serving on a local industrial environment advisory committee in Henderson, Nevada in the 1990s, I read the unclassified portions of the FEMA plan while attending a FEMA briefing and was shocked by the wording that clearly made FEMA a supreme entity in the event a national emergency of any kind was declared by Washington. Clinton could, using the old Emergency Powers Act, assume dictatorial powers and FEMA would carry his flag, including FEMA being able to direct local, state and national police and armed forces. I called the alarming phrasing of the FEMA "doctrine" to the attention of some of my fellow committee members. The FEMA representative making the presentation attempted to assure us that the document applied only to "sensitive" industrial entities, despite the printed words clearly indicating otherwise. Knowing that Clinton's former Arkansas State Trooper "friend" had been promoted from his Texas FEMA regional position to become #2 in FEMA was considerably less than reassuring to most of us.

The *Washington Times* April 19, 1999 noted, "Mr. Clinton's EO 13083 on federalism, which was a transparent attempt to rescind the 10th Amendment, did give Congress a wake-up call. Congress and the American people must call a halt to any assault on our 'separation - of – powers' form of government. James Madison called the accumulation of executive and legislative powers in the same branch "the very definition of tyranny." That most of the media totally ignored and did not challenge Clinton's abuses of Executive Orders should not come as surprise. That it did not recall Madison's warning is also not surprising.

In *The American Spectator* (November 1999), Byron York wrote "The Clinton State Department declined to help the Senate reach those who were abroad and the Clinton Justice Department declined to help them reach those (witnesses) who were resisting. But the most extraordinary thing about their (the witnesses) performances was the across-the-board memory failure that struck witnesses when confronted with tough questions. Susan Thomases, Hillary Clinton's New York confidante, was unable to recall significant facts and events a total of 41 times. Margaret Williams, the first lady's chief of staff, said she couldn't recall at least 30 times. And Nussbaum said he could not remember at least 71 times. So that is perhaps the ultimate, corrosive legacy of Bill Clinton. His tactics reduced the natural conflicts between the White House and Congress to their rawest, state-of-nature level. "I don't have to give you that information and you can't make me give you that information," Bernard Nussbaum said at the beginning of the long, ugly process. Things went downhill from there, and might sink further still.

In an October 2003 article in *Judicial Watch*, entitled Congressional Investigation Uncovers Clinton-Enron Connection, "While the Bush administration has been criticized for its ties to energy interests, a recent Congressional investigation suggests it was the Clinton administration that first turned a blind eye to corruption at Enron Corp. In 1999, the IRS made a criminal referral to the Reno Justice Department to investigate possible bribes paid by Enron officials to Guatemalans close to the country's former president in order to win a lucrative electric power contract. On May 21, 1999, the IRS district director in Houston sent a letter to Janet Reno indicating that Enron may have violated the Foreign Corrupt Practices Act, which forbids U.S. companies operating overseas to use bribes to win contracts. Reno declined to investigate. Now the Senate Finance Committee is attempting to find out why."

A similar article by Alexander Cockburn, *Creator's Syndicate,* noted that "the true binge (of corporate purchases of shares) got going between 1994 and 1998, when non-financial companies sank themselves in debt by either repurchasing their own shares or acquiring shares as a result of mergers. The annual value of repurchases quadrupled, testimony to the most sustained orgy of self-aggrandizement by an executive class in the history of capitalism. Top corporate officers stood to make vast killings from their options, and by the unstinting efforts of legislators such as Sen. Joe Lieberman, they were spared the inconvenience of having to report to stockholders the cost of these same options. Enlightened legislators had also been thoughtful enough to rewrite the tax laws in such a manner that the costs of issuing stock options could be deducted from company income...Democrats in the U.S. Senate in early July rallied in defense of the accounting "principles" that permit the present deceptive treatment of stock options. Not just Joe Lieberman, the corporations' favored errand boy, but Tom Daschle of the northern plains. These are the Augean stables, and who aside from Ralph Nader has the credentials to talk seriously about cleaning them out?"

In a *Washington Times* article in December, 1999 by Sean Scully, entitled "DOE Loses Track of Its $120 Billion Budget", "The Department of Education has admitted, eight months late, that it cannot account for parts of its $120 billion budget for last year…The department is not the only federal agency to fail its audit this year – Agriculture, Defense, Transportation and Justice were unable to balance their books – but Education is the last agency to submit the annual audit report which was due March 1. Secretary of Education Richard W. Riley defended his department's record and said Republican remarks were "recycled old charges" He called the charges "simply another attempt to obscure the success of this administration in advancing the quality of public education at all levels." Mainstream Media? "Do Not Disturb".

In the week of June 3, 2003 edition of *Human Events*, an article by Marc Morano entitled 'Enron and the Clinton Administration': "Enron saw the opportunity to exploit the newly minted Clinton Administration before Clinton was sworn into office, and saw 1992 campaign issues including investment tax credits and restriction on carbon dioxide as beneficial to the company…By 1995, Enron was able to successfully secure financing for the Dabhol Power plant in India with loans from the U.S Export-Import Bank totaling $298 million to cover about 32% of the costs. Enron's ownership stake was 80% in the Indian power plant. Enron was able to secure another $100 million in investment money from the U.S. federal agency, Overseas Private Investment Corporation (OPIC), In 1996, the investment corporation provided another $200 million in "political risk insurance" for the India project, according to OPIC documents In an Oct. `15, 1996, memo from John Palmisano, senior director for environmental policy and compliance at Enron, it was noted the Clinton Administration sought the company's help in gaining support from China and India for proposed climate change regulations. .In fact, in 1997, Enron joined forces with the Clinton Administration to defend OPIC from budget cuts in Congress. The Kyoto Protocol on climate change represented an opportunity for Enron to make a profit, because the carbon monoxide requirements of the treaty dovetailed with numerous corporate projects that emitted less carbon monoxide and other greenhouse gases. Vice President Al Gore championed the Kyoto Protocol, which would have required 40 industrial nations to drastically reduce greenhouse gas emissions by the year 2022. Zond Energy Systems partnered with DOE to develop a (highly touted) wind power technology facility in Storm Lake, Iowa. A Sept.17, 1999, press release by the DOE boasts of the "public-private partnership" between Zond and the department." (Enron acquired Zond Systems Inc., a California company, in 1997.)

In *Human Events*, August 13, 1999, Baker Spring wrote that "President Clinton announced on June 20, 1999, while in Cologne, Germany, that he and Russian President Boris Yeltsin had agreed to negotiate amendments to the Anti-Ballistic Missile (ABM) Treaty...But there is one significant problem with this decision: Russia is not, and never was, a party to the ABM treaty, which the United States signed in 1972 with the Soviet Union As legal experts recently testified before Congress, the ABM Treaty became obsolete in 1991 when the Soviet Union dissolved. The President apparently believes that if he just pretended that the ABM treaty is in force with Russia, then Congress will acquiesce and the treaty will return to force. This approach not only misrepresents the facts, but it also keeps Americans vulnerable to missile attack as the administration honors the terms of a defunct Cold War relic."®

In a November 10, 2003 review of the book *Legacy*, (by Rich Lowry), *National Review* writer Mona Charen noted in her November 10, 2003 article that "Bill Clinton spent more money conducting polls than all other president in U.S. history combined. Polling substituted for leadership, and public relations commonly trumped substance. In November 1995, a bomb exploded at the offices of the Saudi National Guard, killing five Americans. Seven months later, the Khobar Towers complex, killing 19 Americans. Lowry writes, "For the Clinton team, desperate times called for the usual measures – a poll." Having responded to the dual bombings in Africa with a pin prick, he (Clinton) didn't respond at all, choosing to view it not as an attack upon the United States but an effort to "derail" the peace process between Israel and the Palestinians By demonstrating the United States was so weak and indecisive throughout the 1990s, Clinton convinced our enemies that we were, in the words of Osama bin Laden, "a paper tiger."" Interestingly, this same term was used widely by the People's Liberation

Army in Tibet in the late 1950s and 1960s in assuring the Tibetans the U.S. would not help them simply because the U.S. was only a "paper tiger". Did the "Mainstream Media" jump on any of the above examples of great statesmanship by Clinton? If so, only in the classified ads.

"As the left's new pet, Henry "No longer a War Criminal" Kissinger said, "Regime change as a goal for military intervention challenges the international system established by the 1648 Treaty of Westphalia. And the notion of justified preemption runs counter to modern international law, which sanctions use of force in self-defense only against actual, not potential, threats." "The idea that Americas would be transgressing the laws of man and God by invading Iraq, unless and until Saddam - or others - nuked Manhattan was absurd." Did no one remember Clinton's misadventure in the Balkans? Liberals loved that war because Slobodan Milosevic posed no conceivable threat to the United States. To the contrary, President Clinton put it, "This is Americas at its best. We seek no territorial gain; we seek no political advantage." "Deposing Milosevic, Clinton explained, vindicated no national interest, but was urgent because it was akin to stopping "hate crime." One searched in vain for a description of some American interest in the Balkans. For the record, there was no U.N. Security Council approval for Clinton's unprovoked attack on the Serbs."(77).

Special Task Force Chief Charles "La Bella called for a comprehensive investigation into "the entire landscape" of campaign finance allegations. He said that there were schemes "conjured up by sophisticated political operatives to circumvent" election laws during the 1996 presidential campaign." La Bella contended that there was sufficient evidence to warrant the appointment of an independent counsel for the following persons: President Clinton, Vice President Al Gore, first lady Hillary Rodham Clinton, and former White House aide Harold Ickes. With respect to Hillary Clinton, La Bella said that an independent counsel should investigate the extent of her knowledge about the illegal foreign fund-raising activities of Johnny Chung and Charlie Trie. La Bella and Freeh argued that the Justice Department didn't sufficiently investigate whether Bill Clinton had, in essence, controlled the Democratic Party's advertising campaign and used it and its funds in support of his reelection effort." "It is hoped that this report will place in context the abuses uncovered in our investigation: a system designed to raise money by whatever means, and from whomever would give it, without meaningful attention to the lawfulness of the contributions or the manner in which the money was spent," La Bella wrote. "The intentional conduct and the 'willful ignorance' uncovered by our investigation, when combined with the line blurring, resulted in a situation where abuse was rampant, and indeed the norm. At some point the (Clinton-Gore reelection campaign) was so corrupted by bloated fund raising and questionable contribution that the system became a caricature of itself." "He maintained that his Justice Department superiors were "intellectually dishonest" and practiced "gamesmanship" to avoid an independent counsel investigation. "The contortions that the Department has gone through to avoid investigation of these allegations are apparent." At one point, La Bella suggested that the department was so intent on protecting the president that it had made its decision and then tried to apply its reasoning retrospectively. "In Loral avoidance of an ICA* was accomplished by constructing an investigation which ignored the President of the United States – the only real target of these allegations. It is time to approach these issues head on, rather than beginning with a desired result and reasoning backwards." As reports about the Freeh and La Bella memos surfaced, Senate and House investigators requested, and then subpoenaed, copies of the memos, but Janet Reno adamantly refused to turn them over. Finally, the House Government Reform and Oversight Committee voted to hold Reno in contempt of Congress for refusing to comply with the subpoenas. When he concluded his duties with Special Task Force, La Bella was planning to return to San Diego as interim United States attorney. He was in line for a permanent appointment and was earnestly seeking it. But just fifteen days after he recommended an independent counsel, Democratic Senator Barbara Boxer, related by marriage to Hillary Clinton, rejected his appointment. Even the *New York Times* editorial board registered outrage. "Democrats", wrote the *Times*, "have added a new tactic to the campaign to protect President Clinton from an independent investigation of White House fund-raising during 1996. The tactic is to punish the truth-tellers. (This) will send a chilling message to assistant United States attorneys throughout the country. Democratic fund-raising is a no-go area for Federal law enforcement, and if you push too hard, the President's Congressional allies will make you pay."(78)

In an April 5, 1998 article in the *Washington Times* (National Weekly Edition) by Paul Bedard, "The White House has twice as many lawyers on its staff as previously announced, meaning nearly one in every five employees in the executive office of the president is a lawyer. New documents show the White House has 84 taxpayer funded lawyers on staff or borrowed from other agencies – 33 lawyers in the White House Counsel's Office and 51 assigned to other offices. The counsel's office also has 14 paralegals and assistants. Previously, the White House said it (had) 34 paid legal staff, as well as 15 lawyers borrowed from other agencies."

There are many more examples that could be cited, but those provided above should give the reader an understanding of the wide range and types of abuses that the Clinton-Gore administration – and Janet Reno's complicity as Attorney General – foisted on the American public. That the Liberal (left) Media – and the Liberal (left) members of Congress – were party to these travesties should have brought outrage, but the spin-masters and the willingness of the Mainstream Media to suppress such stories has made this nation far weaker than at any other time since the early 1930s. Our "Ship-of-State" was close to foundering with the Clinton crew on board!

Many have criticized William Jefferson Blythe Clinton, and raised some interesting points in doing so. For instance:

• "Air Force Major General Harold N. Campbell called the president a "dope smoking," "skirt chasing," "draft-dodging," commander-in-chief during a banquet speech." Campbell was subsequently fined and forced to resign." (pg. 95, *Dereliction of Duty*) But that's accurate, unbiased reporting!!!

• *Wall Street Journal* contributing editor Mark Helprin described Clinton as "the most corrupt, fraudulent and dishonest president we have ever known."

• "Relaxing on Air Force One after the 1996 elections, Bill Clinton told a pool of reporters that he owed his political revival to the Oklahoma Bombing; "It broke a spell in the country as people began searching for our common ground he said." He castigated talk radio for broadcasting "a relentless clamor of hatred and division." The 'Right', he said, "was sowing distrust of government institutions and creating a climate that fostered recourse to violence." He did not name the Republicans as co-conspirators; he didn't have to. The media clerisy made the connection for him." (page 4, *The Secret Life of Bill Clinton)*®

• *"Something about Bill Clinton - his ineffable caddishness, perhaps – is changing the political discourse of the country. Every year that he continues in power, he eats a little deeper into the eroded legitimacy of the political order. The importance of this cannot be exaggerated." (*Emphasis added) pg XIII, *The Secret Life of Bill Clinton.*

• "Clinton is a constant liar who has, on occasion, devised lies that are truly extraordinary. He tells huge lies when a little white lie would be sufficient and perhaps even pardonable. He applies a modest lie when no lie is needed. ...He lies like a man totally unencumbered by conception of right and wrong. He lies like a sociopath." Pg 155. *Boy Clinton.* And same source, Pg 154: "after historians have had sufficient time to assess Clinton's career, I believe they will rate him the greatest presidential liar in the history of the country."

• "It is implausible that the administration of a blameless president would use the FBI for a political prosecution, be caught with nine hundred FBI files in its possession, or use the IRS to audit the president's enemies. It defies logic that all three of these abuses of power could occur in one administration without the concurrence of the president. That much corruption can be the responsibility only of the man at the top – even under a less exacting standard for president that the one Hilary Rodham delineated for a different president in 1974." Pg 163, *High Crimes and Misdemeanors.*

• "How could Clinton-Gore have tolerated such a dramatic increase in the number of illegal aliens living in the United States? Did Bill Clinton or Al Gore ever bother to distinguish between legal and illegal immigration – or were their only concerns in this area to discredit Republicans and pander to the various immigrant constituencies? How could the Clinton-Gore administration have ignored the issue of border security even after Gazi Ibrahim Abu Mezer (also spelled "Maizar") – a Palestinian terrorist convicted of trying to blow up the New York City subway system to kill scores of Americans, particularly Jews, in 1997 – was found to be an illegal alien who had been repeatedly apprehended by U.S. Border Patrol agents as he tried to sneak in from Canada?" pg.56. *Let*

Freedom Ring. Pg. 18. Same source. "The silence of the Clinton Administration in responding to these offers (for bin Laden) was deafening," wrote (Mansoor) Ijaz in an op-ed piece in the *Los Angeles Times.* "As an American Muslim and a political supporter of Clinton, I feel now, as I argued with Clinton and Berger then, that their counter-terrorism policies fueled the rise of bin Laden from an ordinary man to a Hydra-like monster." Pg. 19, *Let Freedom Ring.*

• "Clinton's an unusually good liar. Unusually good. Do you realize that?" Sen. Bob Kerry (D.-Neb.), as quoted in Esquire magazine. pg. 21, *Human Even*ts, August. 16, 1996.

• "The Clinton (tax) plan raised taxes on small businesses, family farms and Social Security beneficiaries. It hiked the gas tax and the inheritance tax. For the first time, it made income tax increases retroactive to a date prior to its own administration's taking office. By any realistic definition, it hit the middle class hard. The plan included a $30 billion hike in the gasoline tax. The gas tax was increased a total of 6.8 cents per gallon." Pg. 25. *Human Events,* August 16, 1996.

• "Clinton's sterling "legacy" is built on spin, lies, and misunderstanding. His defenders invest so much in puffing it up not just to have a club with which to beat George W. Bush, but to provide a basis for the eventual "Hillary for President" campaign. Bill Clinton's legacy is essential to Bill and Hillary's ongoing political project, and correctly understanding that legacy is essential to resisting it. The common thread to Clinton's time in office is smallness, of him personally and of his politics. Clinton's malleability and his fear of risk kept him – after the crash of Hillary's health-care plan – from proposing or thinking anything grand. Clinton shrank liberalism and the presidency. It was liberalism largely shorn of its idealism and of its ambition." *National Review,* October 27, 2003. article by Richard Lowry.

• "Rather than face Judicial Watch lawyers in court, the FBI and the Department of Justice offered an $85,000 settlement to JW client and former FBI Special Agent M. Dennis Sculimbrene. Mr. Sculimbrene testified against the Clinton White House in Judicial Watch's lawsuit, and as a direct result of his politically explosive testimony, he was subjected to a number of intimidating tactics including an illegal FBI background check, had his White

House pass removed, and was ultimately forced into retirement. The Clinton-Gore Administration offered to pay for all Judicial Watch's legal cost (totaling over $2,000,000) if Judicial Watch agreed to drop its first Chinagate investigation and lawsuit. Judicial Watch refused the "bribery" attempt and instead dramatically "ramped up" its Chinagate investigations and lawsuits, which continue to this day! In court testimony secured by Judicial Watch, Nolanda Hill, the business partner and confidante of the late Commerce Secretary Brown, swore that Hillary Clinton conceived of the fundraising scheme involving the sale of seats on official trade missions, with the knowledge and participation of Bill Clinton and former Vice President Al Gore." (Excerpts from Judicial Watch newsletter, November, 2003.)

- "Clinton hasn't just broken the law. He's a disturbed and disturbing personality. Two years ago, Michael Kellett, author of a book charging that Vincent Foster was murdered, wrote a pamphlet describing Clinton as a "Psychopath." At the time it sounded like another overheated charge. But recent events make it seem a coldly clinical description. The essence of the psychopath is his sheer absence of conscience. He lies and manipulates without remorse, often with charm and suavity. He isn't insane or delusional, but highly rational. He knows right from wrong, but doesn't care. He's a proficient actor who can simulate emotions appropriate to the immediate situation, but he doesn't feel them deeply (if he feels them at all) and he abandons them when their utility has passed. He's egocentric, seeing others purely as instrumental to himself. He craves power. He is likely to be sexually promiscuous, with no stable attachment to any individual. He loves to take risks; he isn't embarrassed when caught in outright lies. He'll take amazing gambles, committing serious crimes to cover up lesser ones: For him, raising the stakes is irresistible. But Clinton, whatever else he is, is not a normal man. You have to marvel at the legions who are still supporting him, even though he has made it clear that he is as willing to destroy friends as enemies, if it serves his interest to do so. The psychopath may be sane, but his flunkies must be nuts." (Excerpts from July, 1998 article by Joseph Sobran, *Middle American News.*) Wouldn't this make a great eulogy when the time comes?

The "Axis of Evil" was coined by President Bush wherein he

identified those nations that presently pose the most serious threats to the safety of the US and the Free World. Most cheered, some jeered, but it was an accurate shot over the bunkers of the leaders of those countries named, as well as a wake-up call for others. This was not just another Clinton-Gore "spin" item, or passive remark; thankfully, those days are gone.

In turn, allow me to offer the equivalent internal triangle, if not an axis of evil, that we had in the U.S. prior to the 2000 election. The first leg of the triangle is labeled "Clintons and Gore"; the second is Janet Reno and those many entities within the Justice Department that did her bidding to protect the Clintons-Gore activities; the third leg is co-shared by Ron Brown (posthumously), Hazel O'Leary, and Sandy Berger. The supporting White House staffs, including those sheep who followed the "Pied Piper" from Arkansas, and those dozens of add-on lawyers that played supportive roles for Clinton, plus those enablers in Commerce and State and Energy and Defense that quietly allowed the transfer of vital technology abroad, are readily recognized as deserving "Most Shameful Performance" Oscars for their consistently collusive efforts. It is doubtful any of them will come forward for their well earned awards. The Liberal Media? With few exceptions, they too deserve their own fame/shame 'award' for being the increasingly dangerous, irresponsible, and anti-Republic evil they have willingly become. Would that we could reward all of them with what they have so richly earned.

The Hillary Follies:

Hillary Rodham Clinton is basically a mirror image of her husband Bill. She has raw ambition, has serious problems with the truth, is power hungry, is very shallow beyond her own ambitions, is a control freak, venal, wants her own way, uses people, is opportunistic, talks a good game, is short on meaningful accomplishments, disdainful of others, conniving, and is the consummate liberal leftist, if not a full-fledged neo-communist.

In her high school days she apparently fell under the influences of a young "socialist-wanna-be", Rev. Don Jones, which led to a relatively quick change in her political leanings from Republican to Democrat. Her Wellesley senior thesis, which is still under lock and key, is based, at least in part, on an interview she did of Saul

Alinsky, a well known leftist who advanced "no-holds barred" tactics, encapsulated in his motto "whatever works to get power to the people, use it." "One of his organizational tactics was the purposeful incitement of political opposition, which he called "mass jujitsu". "Others were disturbed by Alinsky's antinomian emphasis on attaining power through any and all means."(79) "Hillary's attraction to the Marxist theoretician Carl Oglesby, who became the head of Students for a Democratic Society, underscored the political tone of her Methodism. In a November 1994 article in *Newsweek,* Hillary cited a 1966 piece in *Motive,* "World Revolution and American containment" by Oglesby, as having made an indelible mark on her. "I still have every issue they sent me," she said."(80)

"A careful evaluation of Hillary's legal career published by the *American Lawyer* in 1992 was a rare exception to the usual coverage in the media. The investigation found that her reputation as a litigator was quite inflated; in fifteen years, she had tried only five cases in family law, patent infringement, and intellectual property. The publication could identify only two regular clients of hers." "Contrary to some press reports, she's simply spent too little time practicing to be considered much of a lawyer," the report concluded." "Hillary handled no noteworthy cases during her legal career, she was rarely seen in court."(81)

"In December 1977, President Jimmy Carter, in whose campaign Hillary had worked, appointed her to the independent Legal Services Board (LSC), a group of politically connected lawyers that distribute government legal aid money to local programs nationwide. Six months later the Carter White House put out word that Hillary Rodham would be its candidate to chair the board" "Equally intoxicating must have been the chance to grasp the levers of government power and to exercise that power in the service of her strongly held ideas and beliefs. The question of whether the federal government should provide free legal services to the poor has been bitterly contested since the government first began spending $1million a year on the program during the heyday of LBJ's Great Society in 1965. The LSC became an independent corporation by an act of Congress in 1974 and was given an annual budget of $90 million. By the end of Hillary's tenure, that budget had ballooned to over $300 million a year. Previously, legal aid had been thought of as helping low-income people apply for government benefits, deal with landlords, or fight for child support. The LBJ program was run by professional activists whose idea of social justice meant litigating and lobbying to change laws, and extending the welfare state, as well as Alinsky-style organizing of the poor into political pressure groups."(82

"Ample evidence of the goals of the LSC is contained in training manuals dating from Hillary's tenure, one of which described the LSC-funded attorney as "an arm of community organizations., that is, the lawyer was to function as part of a political effort, at times as a lawyer, at times as an organizer, an educator, teacher and PR man." "An end to be attained through courts and legislatures was "power over the distribution of wealth, power over the means of production and distribution."(83) Examples of frivolous and outrageous law suits filed against the federal government by LSC grantees is impressive (see pages 100 and 101®).For example, a case in southern Louisiana filed against the state's Wildlife Commission, which had promulgated rules on oyster collecting that were said to be discriminatory against black oyster collectors, the LSC gave a special grant for a three day workshop for the oyster collectors in which Marxist speakers and folk singers described a pattern of oppression. And several affirmative action lawsuits sought to classify drug addicts, alcoholics, and gays as "handicapped." (84)

"While Hillary chaired the LSC, the organization forged close links with the National Lawyers Guild (NLG) which had been founded in 1937 with the assistance of the International Labor Defense, the American section of the International Class War Prisoner's Aids Society, an agency of the Comintern. It remained an active affiliate of the International Association of Democratic Lawyers (IADL), an international Communist front controlled by the Communist Party of the Soviet Union. The IADL was described in a CIA report on Soviet propaganda operations published by the House Intelligence Committee in 1978 as "one of the most useful Communist front organizations at the service of the Soviet Communist Party." "The report noted that at its 1975 conference in Algiers, "the real and ideological interests of the IADL were covered by the agenda which considered law to be a function in the struggle against imperialism, colonialism, (and) neocolonialism. The IADL's thrust was to do battle with the large international companies as a way to gain adherents and backing the developing world. In 1975, the NLG executive board voted to "provide legal support and resources to aid the Palestine Liberation Organization."(85)

The paragraphs above relate a significant amount of data on the true intent of the organizations mentioned, and Hillary's active role with each. That the Media did little to expose the 'games' involved is puzzling, for the Comintern and the interests of the Soviet Union were not exactly parallel tracked to those of the U.S. But the names and associations, including the wife of the Governor of Arkansas, and the programs she supported, were known, as were the questionable nominations by Jimmy Carter These all but begged for investigative reporting. Until the Reagan administration challenged and questioned the organizations there was nary a ripple. Amazing! But there is more, much more:

"The New World Foundation is not a traditional philanthropy; it funds political movements rather than research, or educational, or charitable efforts. The foundation's literature says that it "seeks to join others in the struggle for social justice and peace," and holds that "genuine world security will be accomplished through equitable and just relationships instead of through military strength." "During Hillary's tenure, which lasted from 1982 through 1988 and included stints as vice chair and chair of the board of directors, the organization sharpened its focus on political activism. Hillary wrote in one annual report, "The New World Foundation has sought to use its resources to strengthen the bridge – between the system and the neediest and least represented – and to encourage other foundations to focus on the fragility of activists' efforts on behalf of the poor in general and people of color in particular. In this effort we have made mostly general support grants rather than special project grants, so as to provide core support to organizers and advocates." "The New World Foundation has given many grants to mainstream liberal groups like the Children's Defense Fund." But much of its money went to the hard left. The foundation supported the Christic Institute, which spent hundreds of thousands of dollars bringing charges against supporters of the Nicaraguan Contras that were later held in court to be frivolous. Many of the foundation's grantees, such as the National Lawyers Guild and the Institute for Policy Studies, have ties to the Communist movement. Another grant recipient, the Africa Fund, lobbied on behalf of the most extreme elements within the African National Congress. Under Hillary's chairmanship, the foundation supported the Committee in Solidarity with the People of El Salvador, better known as CISPES, which was founded in New York City in 1980 by Farid Handal, brother of El Salvador's Communist Party chief Shafik Handal. Farid had come to the United States to mobilize American support for the FMLN, the major Communist guerrilla organization in El Salvador."(86)

"The New World Foundation was also a benefactor of Grassroots International, which described itself as a "people-to-people partnership for social change." While a New World grantee, this organization was giving money to two Palestinian groups with ties to the Communist faction of the PLO – the Union of Palestinian Working Women's Committees and the Union of Palestinian Medical Relief Committees. Hillary apparently became involved with the 'New World Foundation through Marian Wright Edelman. When Hillary joined the organization, Edelman's husband, Peter, was a member of New World's board, as was the Democratic lawyer Vernon Jordan. When she signed onto the board in 1982, Hillary had already been serving since 1976 on the board of the Children's Defense Fund. She became chairman of the CDF board in 1986 and remained until she stepped down during the 1992 presidential campaign. Hillary has said that through her work with Edelman, "the world opened up to me and gave me a vision of what it ought to be." Her experiences with Edelman appeared to confirm her own belief that reform could best be achieved coercively, through top-down government edicts. Edelman would provide her with a vision of big government as an instrument of virtue and moral progress, a crucial element of Hillary's intellectual seduction."(87)

"Edelman's success as a lobbyist stemmed from her ability to turn all policy debates into debates about children. The strategy "requires that CDF reduce every issue of anti-poverty to a question of 'protecting children who can't speak for themselves'. Not only are Head Start, WIC (the Special Supplemental Food Program for Women, Infants, and Children) and immunization cast as children's issues, so are welfare, day care, housing and employment."(88) "While Hillary chaired the board, CDF was a member of the Alliance for Justice, the group that fought to politicize the courts in the activist spirit of the Yale Law School. The 'Alliance' orchestrated campaigns against conservative jurists like Robert Bork and Clarence Thomas, and many other Reagan and Bush judicial nominees in an effort to preserve the power of the courts to enact liberal policies. (Bork was publicly opposed by then Governor Clinton."(89)

To reduce the foregoing to basics, the Hillary profile that emerges is: she is an activist with an agenda, including that she views the tenets of socialism as requisite to her ambitions. This of course requires that she be the 'High Priestess', both before and after the dust settles. She wants to work the system from the inside, top down to at least mid-level, thereby insuring that the final form and functioning support structure is to her liking, which means filling the key positions with like-minded friends of hers - much like the key positions she filled in her husband's administration. For example:

• Donna Shalala, Health and Human Services; Marian Wright Edelman's husband was named Shalala's deputy; Olivia Golden, another Children's Defense Fund colleague, was named to head the department's section of Children, Youth and Families, including Head Start.

• Sheldon Hackney, National Endowment for the Humanities; Hackney's wife was on the CDF board with Hillary.

• Madeline Albright, named Ambassador to the U.N., then Secretary of State, was an assistant to the national security adviser in the Carter administration.

• Maggie Williams, chief of Hillary's staff of 18; had been a press aide at the Children's Defense Fund.

• Melanne Verveer of "People for the American Way" became Deputy Chief of Hillary's staff.

• Bypassed by both Clintons were Vernon Jordan, Warren Christopher, and Walter Mondale. All three had 'expected' key posts, but apparently were the wrong gender.

• Johnetta Cole, a friend of Hillary's and president of Spelman College in Atlanta, was destined to be named Secretary of Education by Hillary until A.M. Rosenthal of the *New York Times* described Cole a "propagandist for dictatorships." "Cole's ties to Marxist groups, including membership in a committee connected to Cuba's intelligence forces and to the World Peace Council, a Soviet front organization that was also vehemently anti-Israel. Cole was also a member of the Grenada Friendship Society, a support group for the Communist regime in that country." "(Hillary) did not seem to understand that people she admired for their commitment to "progressive" causes would be viewed by mainstream America as anti-American left-wing ideologues." (For more, see page 301, 302, *The Seduction of Hillary Rodham.)*

• Mickey Kantor was named United States Trade Representative. He had served with Hillary on the Legal Services Corporation board in the late 1970s.

• Hillary in effect served as the White House counsel, while four others served in that capacity during the Clinton administration. As such, "Hillary had a chokehold on the entire government." And she was also in charge of damage control.

• The official Chief Counsel job went to Bernard Nussbaum, with whom Hillary had worked on the Watergate Committee and on the effort to impeach President Nixon.

• William Kennedy, a Rose law firm associate of Hillary's and Foster, joined the staff as an assistant to Vince Foster.

• Hillary appointed 'close friend' Vincent Foster as deputy White House counsel. He also served as the Clinton's personal lawyer.

• Under Kennedy was Craig Livingstone, who had worked for Susan Thomases in the Clinton campaign as part of the 'advance operation'.

• Hillary also named Margaret Richardson as director of the Internal Revenue service. Richardson had advised Hillary on women's issues and raised funds for the Clintons. Her husband was a friend of Vince Foster's.

• Renaissance Weekender, and friend of Hillary, Ricki Tigert was named to head the Federal Deposit Insurance Corporation (FDIC), thereby providing insulation for the investigation of the Rose Law Firm's work for Madison Guaranty.

• Hillary named Jan Piercy, former Wellesley roommate, to be Bruce Lindsey's deputy in the White House personnel office.

• Janet Reno was approved by Hillary after two earlier choices for Attorney General fell by the wayside. Reno's record in Florida as a prosecutor in Miami and Dade counties was less than impressive, but she fit the other "liberal left oriented" requirements of the Clintons, and was faithfully pliable.

• Web Hubbell, the fourth member of the Rose Law Firm to join the Clinton administration, was named Associate Attorney General.

• Vince Foster's sister, Sheila Anthony, the wife of a former Arkansas congressman, was named the department's top legislative strategist.

• "With the exception of the head of the civil division, who was Al Gore's brother-in-law, Justice's key divisions under Hubbell and Heymann were headed by women or minorities who shared Hillary's expansive rights-based view of the law." (90) Obviously, Liberal Activists were sought and welcomed in the nest.

• "Hillary wanted even tighter control over the Supreme Court and top federal judgeships. Rather than relying on Justice to review and recommend potential Supreme Court nominations, as was the procedure in the Reagan and Bush years, the White House put together a "secret" team of seventy-five lawyers who scrutinized potential nominees' writings and judicial decisions. The team was headed by Washington lawyer James Hamilton, Hillary's friend from the House Watergate Committee. With Whitewater damage controller Jim Lyons, Hamilton had been the legal adviser to the Clinton transition. The highly unusual process – foregoing the seasoning and the institutional memory of the Justice Department – raised eyebrows in Washington legal circles and the legal press, but drew little notice otherwise. "Who are the Clinton Vetters and Why the Big Secret?", the *Legal Times* asked."(91)

• A Supreme Court vacancy occurred in early 1993. "The first nominee to the high court was Ruth Bader Ginsberg. As an advocate with the ACLU's Women's Rights Project, Ginsberg won several landmark gender discrimination cases and had become the first woman law professor at Columbia shortly before Susan Thomases attended the school in the mid-1970s. Though she has said she had nothing to do with the appointment, Thomases was once Ginsberg's research assistant." (also 91)

• "Ideological uniformity was also highly valued in the selection of appellate judges. H. Lee Sarokin of the Third Circuit, who opposed all pretrial detentions of criminal defendants; and Rosemary Barker of the Eleventh Circuit, who struck down parental consent laws on abortion and used the justification of "emotional deprivation during childhood" as grounds for sparing convicted murderers from the death penalty." "Janet Napolitano, named as U.S. attorney for Arizona, was a former legal services lawyer who represented Anita Hill during the Clarence Thomas Supreme Court confirmation. Napolitano was accused by ABC's *20/20* of having thwarted an undercover law enforcement operation to catch child molesters and child pornographers because she believed the sting unfairly targeted gay men."(92)

• "The issue of Hillary's influence arose over the question of whether Clinton would sign an international convention on Children's political, economic, and civil rights that had been unanimously adopted by the General Assembly of the United Nations in 1989 and ratified by more than 170 countries. True to Hillary's agenda on child welfare, the treaty created "international standard(s)" governing adoption, sexual abuse, child labor, health care, and education and guaranteed an "adequate" standard of living for all children, encompassing "physical, mental, spiritual, moral and social development." Conservative Christian activists attacked the treaty and tied its ideas, implicitly, to Hillary's views on children's rights. *Arizona Republic* columnist Kathleen Parker wrote that a "careful reading of the 21-page international treaty suggests a dangerous erosion of parental rights. Children have the right to decide who constitutes acceptable company regardless of what Mom and Dad say? Children have the right to enjoy any media, including, say, electronic pornography, despite parental protestations? Children have the right to join a religious cult that contradicts their parents' beliefs? In the hierarchy of the treaty, parents are reduced to instruments of propagation. The state does the parenting" "Hillary announced Clinton's decision to sign the treaty in February, 1995." (It has yet to be ratified by the Senate.)"(93)

• "Most of the general counsels were liberal women – Jamie Gorelick at the Pentagon, Jean Hanson at Treasury, and Judith Winston, of the Women's Legal Defense Fund, at the Department of Education."(94)

• Hillary's brazen attempt to ram through a "Health Plan" in the early days of the administration was typical Hillary. "To Hillary and her allies, central planning meant eliminating inequalities in the system, providing for the disadvantaged, and sharing the costs of health care equally among all Americans. But to critics, the plan's state-run "alliances", mandatory membership in health maintenance organizations, caps on insurance premiums, and annual limits on overall health care spending meant more government, fewer services, and less choice. The plan was loaded with politically appointed government boards and oversight agencies, designed to enforce corporate and individual compliance with federal standards, which specified in intricate detail the kind of health care services that could be provided, to whom, under what circumstances, and at what prices. Employers would pay for the bulk of the coverage, establishing a new middle-class entitlement, while government covered the uninsured. Most everyone would be required to join state or regional health "alliances, government controlled middlemen who would contract with various health plans, collect and negotiate insurance premiums, and oversee the allocation of benefits."(114)

• "While heading the (Health Care) task force, Hillary owned an interest, through a limited partnership, in several health care stocks, exposing her to potential liability under the 1978 Ethics in Government Act, (established by President Carter) which makes it a crime for government employees to knowingly influence any policy that might have a "direct and predictable" effect of their financial interests." "The Clinton's 1992 financial disclosure report, released in May 1993, showed a net worth of about $863,000. The largest investments, nearly $100,000, was a stake Hillary held in a special growth fund which held more that $1 million in health care stocks and which, at least during 1992, sold pharmaceutical stocks short. The fund, called ValuePartners (sic), was run by Smith Capital Managements, Inc. of Little Rock."(115)

The attack-dog assemblage in the White House was a strong defense team, assuring Hillary that she would not likely be maligned or challenged. Her socialistic pursuits and leanings of the past would be 'covered' by her select team, like house-broken cats dutifully taking care of their respective litter boxes. The past would remain hidden by multiple band-aids to the extent possible and there would be little left to chance for any serious political fallout. Bill and Hillary had the trained seals act in place and, given the unusually large legal team they had mustered to protect them, they thought their fiefdom was nearly as well protected as that they had organized in Arkansas. The FBI files they had illegally acquired plus the instant "War Room" response team/data, plus the White House political base information on hand was akin to bullet-proof vests for key staff and well armored tanks for Bill and Hillary. 'Field Marshal' Reno and a very compliant Media completed the picture. But bit by bit there was erosion.

• ""I have never experienced a cooler or more hardened operator than Hillary Rodham Clinton. The investigators working for Independent Counsel Kenneth Starr found, as we did, that in one White House scandal after another all roads led to Hillary. We came to see that, essentially, Hillary is a woman animated by a lifelong ambition. That ambition is to make the world accept the ideas she embraced in the sanctuaries of liberation theology, radical feminism, and the hard left. We came to see her as a politician who invented her own strategies of protective coloration, who learned to mask her true feelings and intentions. She has become a master manipulator of the press, the public, her staff, and - likely – even the president."(116)

• "Over the years Hillary Clinton has assembled and skillfully used an arsenal of opposition researchers and private detectives that her one-time mentor, Dick Morris, now identifies as "secret police" that has been used in a "systematic campaign to intimidate, frighten, threaten, discredit, and punish innocent Americans whose only misdeed were the desire to tell the truth." "Hillary is not merely an aider and abettor to this secret police operation. She has been its prime instigator and organizer. In the political life of the Clintons, it was she who pioneered the use of private detectives. It was she who brought in and cultivated the professional dirt-diggers and smear-artists. It was she whose obsession with secrecy was so intense that when White House Counsel and former judge Abner Mikva finally bowed to the law and delivered subpoenaed documents, she and her White House scandal team lashed at him with such a vicious streak of humiliating profanity that he resigned. And then there is the Hillary of *It Takes a Village* –gentle, mother-earth like and caring, sweet–talking the American people into socialism for their children."(117) In most of this, James Carville played a major role, specializing in the muck and bad mouth aspects.

• Much of what noted radical Saul Alinsky has written has

become gospel to Hillary, particularly in following his *Rules for Radicals*. For example: "Power is not only what you have but what the enemy thinks you have." "Whenever possible go outside of the experience of the enemy." "Make the enemy live up to their own rule book. "You can kill them with this, for they can no more live up to their own rules than the Christian church can live up to Christianity." " Ridicule is a man's most potent weapon." "Alinsky was an advocate of "mass jujitsu". In many of his forms of attack, he advocated letting the enemy move first, and then use his own momentum against him. But Alinsky also had a rule for pure attack. "Pick the target, freeze it, personalize it, and polarize it."(118) In Alinsky, we have, in essence, the modern political equivalent of the famed Sun Tzu.

• It was at Yale Law School "that she established a friendship with future Labor Secretary Robert Reich; future U.S. Trade Representative Mickey Kantor; future Deputy Secretary of State Strobe Talbott; and future Justice Department nominee Lani Guinier; and of course, the ultimate connection, her future husband and future president of the United States."(119) I wonder how Guinier ever got to say anything in that ego-driven gathering of 'orators'?

• One of Hillary's professors at Yale was Thomas Emerson, known as "Tommy the Commie." It was through Professor Emerson that Hillary had been introduced to defense attorney Charles Garry, who guided her involvement in the support and defense of the Black Panther Party. Members of the Black Panther Party, including the infamous Bobby Seale were being tried in New Haven for murdering one of their own. The victim, Alex Rackley, had been suspected by the Panthers as a police informant. Two Panthers confessed to shooting Rackley as part of a plea bargain. Possibly intimidated by the still fresh images of Kent State, and consistent with Yale's supine reaction to virtually every radical student initiative of the era, Yale President Kingman Brewster issued a statement sympathizing with the student and the plight of the Black Panthers. Then he opened the college dorms to demonstrators, whether they were students or not, and allowed Black Panther lawyer Charles Garry to make his residence on campus. Hillary formed a close association with Garry, and manifested no misgivings by the violent rhetoric of his clients, who called for police assassinations and said, "If Bobby dies, Yale dies." As part of her coursework with Professor Emerson, Hillary attended the Black Panther trials and put her considerable organizational skills to work in organizing shifts of fellow student to monitor the trial and report alleged civil rights abuses."(120)

• "Hillary's friendship with radicals like Garry and Robert Borosage (student writer, radical leader, and close acquaintance of Hillary's) led Hillary to all the Fashionable "Chic", as writer Tom Wolfe put it, including Robert Treuhaft and Jessica Mitford. Treuhaft was a former lawyer for the Communist Party. They were both committed Communists, Stalinists, in fact. Brock, in *The Seduction of Hillary Rodham*, quotes historian Stephen Schwartz as saying that this was not a pair of cuddly modern-day Fabians." "This was a group of hard Communists who had been running the Communist Party of Northern California. It was a political organization whose loyalty to the Soviet Union was explicit, whose discipline was Stalinist, and whose intellectual attitudes were mainly Stalinist. Treuhart is not like the Black Panthers. Treuhart is a man who dedicated his entire legal career to advancing the agenda of the Soviet Communist Party and the KGB" In 1972, Treuhart offered Hillary a summer internship working on behalf of indigent criminal defendants in Berkeley. Hillary accepted. She has never repudiated her connection with the communist movement in America or explained her relationship with two its leading adherents."(121)

• "Hillary was a budding Leninist, Menshevik, Bolshevik, Trotskyite – they were all debating societies. What really mattered to Lenin and what Saul Alinsky taught Hillary to value – *power.*"(122)

• Hillary served "a stint (in Boston) as a lawyer for the CDF. She joined the CDF board in 1978, and eventually served as its chairman for six years. At the Carnegie Council, Hillary worked as a research assistant to a panel chaired by Yale psychology professor Kenneth Keniston. Hillary assisted Keniston to write several chapters of the Carnegie panel's report of the Children's rights, *All Our Children.* This report became a kind of compendium of left-wing pie-in-the-sky wish list, one in which Keniston advocated a national guaranteed income, a universal entitlement state, and greatly expanded procedural right for children." "What would matter in the future of children," the Carnegie Council stated, "was not the family structure, but the larger village of teachers, pediatricians, and social workers who would socialize the task of raising, supporting, and nurturing children." "The time had come for society, we were told, to see the rearing of children as less of a parental task than as a social one." "It was here that the full panoply of Hillary's beliefs can be seen in microcosm, from generous family leave to universal health care. Most utopian and ultimately insidious of all were the council's proposals to develop "public advocates" who could intervene between parents and children on the latter's behalf, reducing parents to subunits of the state. There would even be "child ombudsmen" who would represent the rights of children in public institutions, helped by a new call of public interest lawyers to advocate the rights of children. Thus the parents would be subordinate to judges, social workers and bureaucrats – the real experts in the raising of children."(123)

• "The 1992 campaign criticism of Hillary as a dangerous ideologue is now long forgotten. But in fact, a careful reading of her work reveals a not-ready-for-primetime Hillary as radical as the Republicans said she was. Hillary's writings reveal a leftist ideologue, dedicated to centrally directed social engineering, dismissive of the traditional roles of the family, and interested in children primarily as levers, with which to extract political power."(124)

• "The extent of Hillary's involvement in a spectrum of investments, corporate relationships, and high-risk deals is nothing short of breathtaking. The woman who had so recently and publicly "indicted" Corporate America was busy consorting with the enemy, drawing down a small fortune in director's fees. In 190, Hillary joined the board of the French chemical giant, Lafarge, controversial for its toxic waste disposal contracts. She joined the board of TCBY Enterprises, Inc., the yogurt franchise housed in the tallest building in Little Rock. A client of the Rose Law Firm, TCBY paid the firm up to $750,000 when Hillary served on TCBY's board. The chairman of TCBY was one of Bill Clinton's most frequent and enthusiastic donors. Owning $80,000 of Wal-Mart stock, Hillary served on its environmental committee." (125)

• "Hillary would also do everything she could to avoid being called to account when Whitewater became the subject of a Congressional investigation. Shortly after the Whitewater investigation had begun, two college students were hired to shred piles of materials from the Rose Law office of Vince Foster. Foster – then six months in his grave – had been in charge of Clinton's personal legal affairs once they moved into the White House. His files could not conceivably have been shredded without the Clintons' consent. Investigators cannot prove, but do not have any doubts, about who ultimately ordered the shredding of those files....I* deposed (Carolyn) Huber during congressional investigations into the White House Travel Office firings. Huber testified that in the course of transferring correspondence between her office and the first lady's, Huber took a stack of papers from the third-floor living quarters of the White House, just off Hillary's office, and returned them to her East Wing office. Huber didn't look at them again until several weeks later. What she saw forced her to call Hillary's lawyer, then her own lawyer. They were Hillary's itemized legal work for Madison Guaranty, under subpoena for several years, the existence of which had been repeatedly denied by the White House. Hillary had told the investigators that she had ordered all Whitewater records and Madison records shredded in the mid-eighties. Moreover, Huber said she was certain the papers had been placed there recently, which suggested that Hillary had been reviewing the papers all the while she had been denying their existence. The White House spin strategy quickly began to dismiss as "old, recycled, rehashed news" on a "failed land deal" all the exposes of Hillary's denials. The records, even with Hillary's fingerprints on them, ultimately proved nothing, the spin doctors said."(126) The Mainstream Media? Golly, gee, etc.

- "As president, when facing a trial in the Senate on impeachment charges, Clinton's accusers in Congress saw the full force of the Clinton counter-assault. Sexual secrets (on Republicans) were revealed by a troika of private detectives, disseminated by professional smear artists (like Sydney Blumenthal), and paid for by pornographers (like Larry Flynt). The strategy has been replayed over and over."(127)

- "The Travel Office fiasco showed one more thing about Hillary Clinton and her husband. They had every right to replace the Travel Office employees. They did not have to smear them or lie about why they were being replaced. But it has become second nature to do both. The abuse of power, the destruction of people who are in the way, the lying, even when the truth might be entirely acceptable – those are the Clintons' *modi operanadi.*"(128)

• "We already know that in her relentless quest for power Hillary has been as financially acquisitive and ethically agile as any politician. But she also has pursued a politics of vendetta, deceit, and extremist ideology to a remarkable degree." "She has also proven that for those in her way or who fail to help her, there will indeed be "hell to pay". Never have the stakes been so high – both for her and he opportunity to change this nation."(129)

• After attacking Giuliani incessantly, assisted by the *Times* "sleaze-fest" under the headline, "The Rudy Chronicles", Hillary attacked Rick Lazio, including causing an investigation by the Bureau of Labor Statistics of him, following her remarks of possible improprieties by him. Lazio was fully cleared. .But Hillary's ploy to smear him had had its desired effect, as had her pitch to Rangel and Sharpton for the Black vote in New York City. Despite the obvious NYC and White House pressures involved, the investigation substantiated that Lazio had done no wrong. And certainly his very modest stock market gain was far more ethical that what Hillary had done in Arkansas in gaining her $100,000 windfall. That Hillary made the charges against Lazio the same day the Independent Counsel announced that she had testified falsely in the Travelgate scandal is certainly highly suspect, but the media seized only on the Lazio story, led by the *Times.*(130)

• "With the news that on his last day in office Clinton had pardoned Marc Rich, the worst tax fugitive in U.S. history – this within days of his ex-wife Denise's pledge to donate nearly a half million dollars to Clinton's presidential library – investigators swung into action. Indictments looked almost certain after Mrs. Rich invoked her Fifth Amendment rights to remain silent when she was called to testify before the House Government Reform and Oversight Committee in February 2001. That same month witnesses emerged with canceled checks claiming that first brother Roger Clinton had shaken down a Texas family for $235,000 in exchange for a presidential pardon. But despite compelling evidence in the Clinton pardon mess, more than two years of investigations have yielded not a single indictment, with the Bush Justice Department abandoning the case against Mrs. Clinton altogether. Some, like Fox News Channel's Bill O'Reilly, suspected a deal. In fact, the popular cable host told his *Radio Factor* audience in June 2002 that he had evidence Bush had made a backroom deal to keep prosecutions muzzled."(131)

• "The Clinton camp, however, did not universally appreciate Bush's indulgence. In fact, several veterans of the Clinton-Gore administration began actively pushing for Bush's impeachment in mid-2002 on the Web site 'democrats.com' which planned a voter registration drive to win control of Congress and make Bush's impeachment possible. 'democrats.com' headlines like: "Bushgate"- Which Scandal Will Bring About Bush's Impeachment?" complained, "Bush has run the most scandal-ridden administration since Richard Nixon produced Watergate." The "Bushgate" feature, democrats.com promised, would "track the progress of ALL of Bush's scandals as they make their way from Internet exposes to columns, books and films to mainstream stories to Congressional investigations and finally, to IMPEACHMENT." The anti-Bush Web site might not have attracted any attention except for the identities of the people who put it together, mostly hardened political veterans of the Clinton-Gore administration. 'democrats.com' co-founder David Lytel was President Clinton's personal Webmaster, who helped to develop and edit the award-winning White House Web site. Former Clinton pollster Stanley Greenberg sits on the "democrats.com" advisory board, along with Dick Bell, formerly the head of the Interactive Media Department at the Democratic National Committee. Jack Gill, another advisory board member, was director of special projects in the Office of Media Affairs in the Clinton White House. He also helped develop the White House's Web site. And democrats.com advisory board member Greg Simon was the chief domestic policy advisor to Vice President Al Gore. Shortly after NewsMax.com broke the news of the "democrats.com" drive to impeach Bush, the Democratic National Committee quickly moved to distance itself from any formal role in the project."(132) That this clique would have James Carville and Paul Begala, and a host of others, with their extensive attack and counterattack resources at hand, to assist Hillary at every turn as well as to attack Bush-Cheney, is not surprising. Hillary has already trod heavily on the Bush record, and tried mightily to obfuscate her own role and that of her husband in the still pulsing scandals of the Clinton years. Her distaste of Bush is probably only exceeded by her disdain for President Reagan.

• "Hillary's reference to Reagan is particularly revealing and may explain why, as a once and perhaps future White House occupant herself, she has put presidential etiquette aside to deliver broadside after broadside at the Bush tax cut program. One-time Clinton bodyguard, Arkansas state trooper L.D. Brown recalls a number of episodes in the 1980s where Mrs. Clinton expressed a visceral disdain for the fortieth president. While then Governor Clinton admired Reagan and got on well with then Vice-President Bush, Brown recalls Mrs. Clinton erupting in profanity at the news they'd been invited to attend a clambake at Bush's Kennebunkport estate. "Hillary flatly refused to go," reported Brown, in his own 1997 memoir *Crossfire: A Witness in the Clinton Investigation; "Fuck him, Bill. He's Reagan's Goddam Vice President!"* (emphasis added) (133) Another example of her anger and pettiness: "She blew up at a Secret Service agent for declining to carry her bags. When the agent explained that he needed to keep his hands free in order to protect her, she replied, "If you want to remain on this detail, get your fucking ass over here and grab those bags."(134)

• "Surprisingly, little has been written about what appears to be one of the Clinton administration's most outrageous abuses of power, one that mirrors Mrs. Clinton's penchant, during her days in Arkansas, for hiring private detectives to silence her husband's sex accusers. However, the relentless use of the Justice Department, the Internal Revenue Service and other federal investigative agencies by the Clinton White House to go after critics, witnesses, and even election rivals probably exceeds by far any other administration in U.S. history. Even more startling: the media's virtual silence on this phenomenon through two Clinton terms in office, a trend that continues even as the former first lady has moved to the front of the field of 2004 presidential hopefuls in survey after survey." (135)

• Hillary's miracle of the cattle futures was widely covered when the story surfaced in 1994 (though she hasn't been questioned about it lately). Less well known is the fact that she perpetrated an amazingly similar scam on the taxpayers of the state of New York, the very same folks she now represents in the U.S. Senate. The story of Hillary Clinton's $100,000 New York State rip-off surfaced briefly in early 1996, only to be lost in the news of Whitewater and Travelgate scandal developments at the time. Just two years before she moved to 1600 Pennsylvania Ave., Mrs. Clinton was paid $101, 639, for services rendered, from an agency funded with federal and state tax dollars. The first lady's windfall came as a result of a sweetheart deal between her Rose Law Firm and New York's National Center on Education and the Economy, (NCEF). What did Hillary Clinton do for that tidy bundle of cash? Answers weren't immediately forthcoming, so then-New York State Attorney General Dennis Vacco launched an investigation. Current New York Governor Pataki's reaction at that time (1990) was typical "To pay $100,000 to an Arkansas law firm out of scarce education dollars where it seems no vital services were performed is an outrage." "The story first surfaced in New York *Newsday* in April 1994. Back then reporter Lou Dolimar uncovered some very curious details about the NCEE. "The center, an educational think tank, was heavily salved with Democrat operatives and (the) president's political supporters, including Ira Magaziner. The center's chairman was John Sculley, then head of Apple Computer and a principal Clinton backer. *Newsday's* 1994 report failed to raise investigators' eyebrows at the time. Why? Probably because those empowered to investigate answered to the Clintons' No.1 New York Democratic Party ally, then Governor Mario Cuomo. In fact, it was Cuomo himself who created the NCEE with a $5 million state grant. The contract negotiated by NCEE Director Marc Tucker simply pledged to pay Hillary $12,500 at the beginning of every month, with absolutely no requirements that she account for her time. But expense reports show that she attended about a dozen Workforce Commission meetings, for which she was reimbursed an additional $10,797 for travel and other miscellaneous cost. Vacco, now in private practice in upstate New York, told NewsMax.com that the physical work product of Hillary's efforts amounted to a single 12-page report.

• The concern that Hillary has so publicly – and politically - embraced concerning children has a way of disappearing when it does not serve her political agenda. She was quiet when the Waco fiasco occurred, and again when the Oklahoma City Federal Building was mysteriously bombed, both instants resulting in multiple deaths of children, and again when Elian Gonzalez was forced by Janet Reno's "Storm Troopers" to return to Cuba. Hillary, 'Selective Silence' is a poor hiding place.

There is more, but there will always be more when the subject is the Clintons. The foregoing section on Hillary is only a summary of the more outlandish activities in which she has involved herself since leaving the Rodham nest. She has grown from the nerd category to become a thinly veiled, avid Socialist; from the less than popular high schooler into a dangerously ambitious, power hungry force. In her mind, Hillary has no equals. That she has reveled in her position as a contending senator is evident. Early on she eagerly sought out Sen. Robert Byrd to learn the Senatorial ropes, especially Senate parliamentarian procedures. Normally a function not sought by other Senators, she has chaired the Senate more than 200 times, which is easily understood, for on those occasions she is addressed by her peers on both sides of the aisle as "Madam President", truly a cherished and needed aphrodisiac, with no batteries included. Without doubt, she is the most self-absorbed, power struck, venal and ambitious senator in modern history. Influenced and guided by her ultra Liberal/Socialist/Leninist minded associates over the years, she is also driven by the humiliation she has endured by her husband's many years of philandering. She may have earlier deprecated Tammy Wynette for "standing by her man", but Hillary's reaction is now to become the "High Priestess" of the Democrats, enthroned in the White House. That could lead to a new TV series, perhaps with Hillary flitting about with broom, terrorizing all within the beltway, and Bill Clinton boiling away in a large pot.

How did this all come about? She is not the first 'carpetbagger', nor will she be the last, but in Hillary's case New York was a simple choice. There were only three states where she might be elected, California, Illinois and New York, with all three having the requisite plurality of Democrats. California was obviously out. Illinois was risky. With the retirement of Sen. Moynihan, New York became the *only* choice. Mayor Giuliani's marital and health problems made it easy for Hillary and her promises. Rep. Lazio's late entry into the race following Giuliani's withdrawal was not much of a challenge for the "attack dogs" at Hillary's disposal. The Media was overwhelmingly supportive, led by the *NY Times,* and other outlets toeing the liberal line. So what is next? Unless stopped by one or more of the pending law suits she faces (e.g., by Judicial Watch), Hillary is ready to roll: she has very well funded political action "committees"; billionaire George Soros is behind her; she has available the "attack dogs" and their extensive data bases; she has pocketed Terry McAuliffe and the Democratic Party; Harold Ickes already has a 'Hillary Clinton' office in NYC; she is attempting to soften her image, including complimenting U.S. military entities she previously so despised; she is less shrill than the Bella Abzug clone she was; she has been and is one of the leading Bush attackers; she is a capable fund raiser - and has shared some of the moneys she has raised with fellow Democrats thereby gaining their support. She may wait until 2008, but should the 2004 Democratic convention ask, she will blushingly, gushingly accept the party's draft nomination. If Kerry is the 2004 nominee, Hillary will accept the VP offer, if it's made. In her own bid, her VP choice will likely be from California, or another large electoral count state. I shudder to think of the cabinet choices should Kerry and she succeed. And what would happen to our now magnificent military forces? Our country needs the best of choices, not the worst. Such thoughts lead to outrage… and nightmares.

"The Clinton years might seem like a long national nightmare of scandal, sleaze, and ruthless pursuit of power. Hillary herself is a link from the demands of the Watergate staff, to the Whitewater, etc., fiascos, to the abuses of executive power, to the defense of her husband's perjury and obstruction of justice. But now it is Hillary's turn. The Clinton era is far from over; Hillary's ambitions are far from satisfied."(136) Barbara Olson could not have written a more poignant epitaph for the unfinished Clinton era. Certainly our future security has been seriously imperiled by Clinton's having prostituted this country for Chinese financial support, and for pretending that Osama bin Laden and like-minded Muslims posed little threat. As said before, this twin legacy will have impact for at least a generation to come, perhaps longer. Hillary's impact, and legacy, is less well identified at this point in time, but she has cast sufficient numbers of long shadows to give all of us concern until her political demise. Should she succeed in achieving the power and position she seeks, our Republic will clearly be in peril. Especially disturbing is the Media's being all but mummified on the subject of Hillary and her worrisome associations. Hopefully the readers will remember.

The Electoral College that our forefathers so wisely established as part of the process of electing a president saved us from Al Gore, thereby again proving its vital role. The 32 states that voted for Bush as opposed to the 18 states that Gore carried would have been muted by those that want only the popular vote to count. Remember, Gore wanted votes re-counted by hand in only three heavily Democratic counties in Florida. All the numerous recounts in Florida by the Media and others confirmed Gore still lost![3] Ringling Brothers was not the only circus act in Florida.

Gore had legions of lawyers who strived mightily to get the Florida votes Gore so desperately sought to become president, but by every measure he clearly lost. Further, the population of all the states' counties carried by President Bush: 143 million; the population of all the states' counties carried by Gore: 127 million.® Our electoral system needs to be preserved, not eliminated. According to "American Presidents", by David Whitney, some 16 times presidents have been elected with less than 50% of the popular vote, including Bill Clinton in both of his elections, and President George W. Bush. The freedom of our Republic has endured these many years with the electoral process as one of its needed pillars. There is simply no justification in pursuing the 'popular vote' argument. For example, both New York and California have very large pluralities of Democrats which raises the specter of them possibly dictating their choice(s) over the majority of the other 48 states, should those favoring the 'popular' vote system have their way. The Electoral College safeguard is the one entity in the process that remains tamper-proof. Again, the Media should be out front leading the band on this, for it is not a subject to be raised just every four years. Lest we forget, some of Gore's people hinted that they might attempt to 'tamper' with the Electoral delegates in 2000. The mainstream Media did not leap to decry the threat until after others raised a very strong fuss across the country. Chalk up another shame, Media.

But remember, the Liberals, especially the "Left", still scream that Gore won the election. After all, without fanning those arsonist set fires, how could the Left be kept energized by Terry McAuliffe, Jesse Jackson, Al Sharpton, Tom Daschle, Ted Kennedy, John Kerry, Howard Dean, James Carville, and the other Liberal/Radical Left sheep that readily bleat whenever given the cue to do so?

The name of George Soros, mentioned above, may not be familiar to all, for he did not surface until well into the Clinton years. A multi-billionaire financial entrepreneur, fervent "Open Society" advocate, anti-Jewish, and colleague of the Russian oil billionaire Khodorkovsky (recently jailed by Russian President Putin), Soros advised V.P. Al Gore, State's Strobe Talbott, and Lawrence Summers of Treasury, urging a strongly supportive U.S. policy towards Russia. In turn, he added to his wealth by engaging in speculative efforts involving the ruble, etc., which the U.S. at the time was reportedly quietly supporting. "But with the passing of the Clinton Administration, the golden era of Russiagate has passed."

• Openly defiant of the McCain-Feingold campaign reform bill, Soros now aggressively exploits "Section 527" regarding political donations and reportedly has already pumped well over $200 million into the Democrat's campaign to defeat President Bush. He has also supported efforts to establish a 'Liberal" media outlet, and contributed heavily to Hillary Clinton's money raising efforts (and office), and to MoveOn.org. Unless or until nailed for illegal activities, Soros will likely be part of our landscape until his death, or until his money has gone in supporting the political left. But the Mainstream Media remains largely quiet on Soros, and few of our politicians have the guts to challenge him.

And now for "Congress".

* from Cmdr. Hamilton McWhorter, USN (ret.), Sept. 29, 2003 E-Mail, "from FOIA requests, public laws and various published books, and not refuted by Clinton."

1 Unfortunately, this was long before Ann Coulter's book "*Treason*", Crown Forum Publishing, 2003, exonerated Sen. Joseph McCarthy. McCarthy cited and pursued actual government investigations of suspected Communists working against the government, including those in Hollywood, See pages 21-22, "*Useful Idiots*" by Mona Charen, Regnery Publishing, 2003, which also exonerates McCarthy. Both books debunk the McCarthy mythmaking created by the Liberal Left which for decades has used *McCarthyism* as a highly pejorative term and used it extensively to fend off or decry embarrassing challenges.

* paraphrase of data, pg.185, "*Useful Idiots*"

* Seal had learned that the Sandinista government was allow-

ing the Medellin cartel to use Nicaragua as a trans-shipment point for cocaine. To catch the Sandinistas in the act, a camera was concealed in Seal's C-123 military cargo plane to record the drug loading in Nicaragua. When Seal returned from his successful surveillance mission, his DEA handlers had received word that the conservative *Washington Times* newspaper knew all about his mission and was about to break the story. It was clear that the White House (or the NSC) had leaked the story after seeing an opportunity to tell President Reagan what he wanted to hear: to embarrass the Sandinistas and, at the same time, win some Congressional support for the Contras." Pg 100, "*Compromised*, Clinton, Bush and the CIA." A video of the Sandinista drug involvement was also shown on U.S. national TV.

® a close friend and financial supporter of Clinton, Lasater became involved in ADFA, and later was also convicted on drug charges, but pardoned by Clinton.

* "Gregg did not deny their friendship". Stephen Engelberg, NY Times, June 16, 1989, A3.

* On pages 342and 343 of *The Secret Life of Bill Clinton*, additional details regarding Rodriguez, L.D. Brown., and Clinton are given. On pages 12, 13, 14 and 15 of *Boy Clinton* are similar but expanded details.

* Linda Ives joined up with a California film producer named Pat Matrisciana to make a documentary on the deaths. It was called *Obstruction of Justice.* The video, tightly documented, was a heart-wrenching expose of "The Machine". (page 273, The Secret Life of Bill Clinton.)

* The Progressive Review and other sources such as the Washington Times, Washington Post, Arkansas Democrat Gazette, Insight, London Telegraph, London Times, Boy Clinton, and The Secret Life of Bill Clinton...

* The Progressive Review, etc., per the above sourcing. Available on http://emporium.turnpike.net/P/ProRev/connex.htm

* "*We're Right, They're Wrong*", by James Carville has to be one of the most eccentric books ever printed by Random House. For those interested in pure liberal rhetoric and "sick" leftist humor, it is worth reading. The names he cites in the 'Acknowledgements' section are more than sufficient to tell the reader that Carville is

not alone in his far left political rabidity. He might better have pursued a career as a carnival barker. But he has played the perfect deceiving political foil for the Clintons.

* "The survivor, Air Force Tech. Sgt. Shelley Kelly, a stewardess, survived the crash for some four hours. Kelly and another stewardess had been seated in a jumpseat at the very rear of the 737. That area was found basically intact after the crash. According to the Air Force, she received first aid from Croatian rescuers but died on the way to a nearby hospital. Her autopsy report states that Kelly died of a broken neck." Pittsburgh Tribune-Review, November 24, 1997. The investigative probe of Brown was closed by Independent Counsel Daniel Pearson six hours after Brown's death was confirmed.

* "The economist Herbert Stein on the op-ed page of the *Wall Street Journal* skewered Magaziner's numbers, calculating that the Clintons' health care package would be adding a $120 billion to the deficit annually by the middle of the 1990s." pg. 94, *Boy Clinton.* Pg 228 has additional details.

2 Sen. Glenn reportedly had sizeable unpaid debts from his earlier unsuccessful run for president, and agreed to impede the senate investigation of misdeeds by the Clintons in Arkansas and the investigation of charges of campaign irregularities involving illegal PRC contributions in return for payment of his debts. For Glenn's strong cooperation, Clinton subsequently authorized Glenn's space ride. The Media ignored both Glenn's committee "performance" and the connection of it to the space ride. Glenn retired in 2001.

• prorev.com/connex2.htm, page 39

* Clinton Investigative Commission, pg. 33. (*Report to Congress 1997)* The Commission also raised many questions regarding the death of Vincent Foster, including that he may have been about to go public with his knowledge about the Clintons' previous illegal activities in Arkansas. The frantic cover-up by Bernard Nussbaum, Maggie Williams, and Patsy Thomasson in removing evidence from Foster's office and preventing proper police procedures following Foster's death is also covered in summary form on

pg. 36 of the Investigative Commission's.report. Countless questions remain unanswered.

® Published in 1997, *The Coming Conflict with China,* Bernstein and Munro, covers a wide range of concerns the United States should have regarding China, extending well beyond such previous basics as Most Favored Nation, World Trade Organization membership, and human rights. China's goal of being the supreme power in Asia is well covered, as are the balances the U.S. needs to keep in focus as China seeks to achieve its desired status as the single most dominant power in all Asia.

* The brave but failed efforts by the Tibetans to defend their country from the swift and brutal invasion and occupation by the People's Liberation Army, resulting in the death of some one million Tibetans, is detailed in *Tears of the Lotus*, by Roger McCarthy, (McFarland Publishing C0., 1997)as is the story of the amazing escape of the Dalai Lama from Lhasa to India in March 1959 The pursuit and practice of genocide by the Chinese, both by extermination and assimilation of Tibetans, who are now all but a minority in their own country, is also told.

• Notes 67, 68, and 69 are paraphrased summaries from Bob Woodward's "Shadow" on the Riady family and Lippo Group connections to Clinton and Gore.

® Two weeks before he was to retire, the SAIC in NYC, James Fox, was suspended by FBI Director Freeh for 'appearing on television too much without (Freeh's) permission'. Fox had led the successful investigation into the 1993 World Trade Center bombing and had appeared on TV during the investigative process, assuring the citizens of NYC that the FBI investigators had the situation well in hand. I had worked with Fox in San Francisco where he was ASAIC. We formed an excellent working relationship as well as a personal friendship. I called Jim when I heard that Freeh had suspended him expressing my shock, but he simply said that "not all horses' asses are necessarily on horses". A great line. Fox is now deceased, but his sense of humor and impressive professional abilities are not forgotten by those who knew him well.

* "Investors Business Daily reported March 30 and April 9 on Energy Secretary Hazel O'Leary's efforts to trim the number of security guards at the nation's nuclear labs and to loosen background check. Of 2,686 federal requests for wiretaps from 1993

to 1997, only one was turned down – that for suspected Chinese nuclear spy Wen Ho Lee. TV network coverage of these revelations? Zero." Source: The Media Research Center, Alexandria, Va... Article carried in the *Las Vegas Review Journal,* April 2, 1999

® After retiring in July 1982 from the CIA, I became active in the Association of Former Intelligence Officers (AFIO) and served as the San Francisco Bay Area Chapter president in 1984 and 1985. I invited the Soviet and PRC Consul Generals, separately, as our dinner guest to address our AFIO chapter. The Soviet Consul General accepted twice and the Chinese Consul General once. The common emphasis that each of them strongly stressed was the threat that the Self Defense Initiative (SDI) represented, and that SDI was clearly in contravention of the ABM treaty…and that the Soviet Union would be forced to develop a similar program if the U.S. persisted. (The Chinese Consul also included his view that PRC-U.S. economic interests would flourish in time.) The basic absurdity of the Mutually Assured Destruction concept versus the promising benefits to all of SDI was not accepted reasoning by either of them. Obviously, this confirmed that the U.S. was on the right course in pursuing the SDI concept!

* Independent Counsel Act

® "It did not happen. In violation of its "Brady" responsibilities, the prosecution withheld material from the defense that was exculpatory or impeached the credibility of government witnesses. It delayed a year in handling over FD-302 witness statements that were critical to the defense. .It also told a series of demonstrable lies. If this is how the Justice Department behaves in a high profile case after the President and the Attorney General had both made explicit promises of transparency, I dread to think how it conducts itself when nobody is paying attention." Pg. 6, *The Secret Life of Bill Clinton.*

® "*The Seduction of Hillary Rodham*"

* Barbara Olson, author of *Hell to Pay,* was the Chief Investigative Counsel for the congressional committee that investigated "Travelgate": and unearthed "Filegate".

* Bill Sammon's "*At Any Cost*", (Regnery Press), provides excellent reading on how Gore "tried to steal the election", including the Democrat plan to disallow/disqualify military ballots.

® Professor Joseph of Hamline University School of Law, Minnesota tallied 19 states for Gore, 29 for Bush. Apparently he did not wait for the final results of all 50 states.

- From the May 2004 issue of NewsMax.

CHAPTER 6

CONGRESS

(Sublime and Ridiculous)

Almost ninety feet above the Potomac River in Washington D.C., on what once was known as Jenkins Hill, sits the U.S, Capitol with its some 540 rooms. Around the Capitol are the House and Senate buildings, and in these buildings "the most shocking, outrageous, and often corrupt activities take place. The activities highlight not only the hypocrisy of members of Congress and their staffs, but also what is wrong with the political system Congress itself has fashioned." (1)

"Each year, 3 to 5 million Americans traverse the corridors of this mosaic lined treasure box, wearing out the Minton tiles underfoot. Yet the tourists never see large portions, which are roped off or closed to the public. These areas include nearly one hundred hideaway offices the congressional leadership and more senior members maintain for themselves in the Capitol. Marked only by door numbers, the hideaway offices are often fabulously appointed rooms with gilded crystal chandeliers, ornate floor-length mirrors, working fireplaces, and frescoed walls and ceilings. Ostensibly, the offices allow members to escape the demands of their regular offices so they can think, or to save time getting to a vote on the floor of the House or Senate. Daniel Webster used his hideaway – S-326-A – to store his wine. More recently, hideaways have been used for "hideaway honeys."…"Capitol Police officer Gregory M. Lacoss will never forget turning the doorknob to one of Lyndon B. Johnson's seven Capitol hideaways when LBJ was majority leader of the Senate. It was 3 a.m., and Lacoss was making his rounds, checking to make sure all offices were locked. Lacoss opened the massive door and gasped. On the sofa, Johnson was having sex with Carole Tyler, a blond, curvaceous, secretary."(2) At least LBJ wasn't sleeping on the job!

Unfortunately, the United States Congress has 535 members, most of whom fancy themselves as fully qualified cabinet members, or exceptional policy makers, or brilliant military tacticians, or highly qualified on a number of subjects which need and deserve his or her own review and modification. And many have presidential ambitions. It is interesting that reportedly within this same august group there are: 29 who have been accused of spousal abuse; 7 have been arrested for fraud; 19 have been accused of writing bad checks; 117 have bankrupted at least two businesses; 3 have been arrested for assault; 71 cannot get a credit card due to bad credit; 14 have been arrested for shoplifting/drug-related charges; 21 are currently defendants in lawsuits; 84 have been arrested for drunk driving. Add to these figures, at least one was caught plagiarizing in college (Sen. Biden); one was caught cheating on a Spanish exam (Sen. Ted Kennedy) and forced to leave school for a semester; one was involved in a negligent driving accident that resulted in the female passenger drowning (Sen. Ted Kennedy); two were rebuked separately on the House floor for their highly questionable involvement in homosexual related activities (Reps. Studds and Frank); one served prison time for his involvement in a House stamp fraud activity (Rep. Rostenkowski); and one was forced to resign from an intelligence oversight committee for leaking classified information relative to the Iran-Contra hearings (Sen. Frank Leahy). And that's just what has surfaced publicly.

"For more than one hundred years, the House Bank had quietly cashed the members' pay checks – tinted a distinctive pale orange – out of a little first floor Capitol office. The bank neither paid nor charged interest. It made no loans. Other than member's salaries, it took no deposits. Yet to House members, it was another way to chisel money from the federal government. As in the case of the House Post Office, Congress had been covering up the practice for decades. According to John Cronin, Jr. the GAO auditor who did some of the reports, the House bank was covering checks that members had written on their overdrawn accounts. Meanwhile, the members got to use the money interest-free. In effect, the members had sanctioned their own misconduct. It seemed to be the perfect scam."(3)

"In February 1990, the GAO made public its report on the House Bank. This time, instead of referring in footnotes to sums "due from members", the GAO made it clear that members were kiting checks. A spate of news stories followed. In September 1991, another GAO audit said members had written 8,331 bad checks in the fiscal year ended June 30, 1990."(4)

The subsequent investigation by the ethics committee found that 269 current and 56 former members of the house had been allowed to routinely overdraw their accounts without paying interest or penalties. Among the then current members, the biggest abusers were Stephen J. Solarz, a New York Democrat, who had written $594,646 in bad or overdraft checks and was overdrawn for thirty months; Carl C. Perkins, a Kentucky Democrat, who had written $565,651 in bad checks and overdrawn for fourteen months; Harold E. Ford, a Tennessee Democrat, who had overdrawn his account by a total of $552, 447 and was overdrawn for thirty-one months, and Robert J. Mrazek, a New York Democrat, who had written $351,609 in bad checks and was overdrawn for twenty-three months. Charles Hatcher, a Georgia Democrat, had written 819 bad checks and had been overdrawn a total of thirty-five months. Ronald D. Coleman, a Texas Democrat, had written 673 bad checks and had been overdrawn a total of twenty-three months. Among the abusers from the Republican side was Newt Gingrich, the minority whip, who had written twenty-two bad checks with a face value totaling $26,891." "The scandal and its initial cover-up was but another example of how members of Congress expect everyone but themselves to play by the rules."(5)

The advent of television in Congress, first in the House in 1979 and then in the Senate in 1986, has had an impact on both houses. Some damn it, others praise it, but most see it as an opportunity to preen for the viewer. "Even television doesn't show what really goes on in floor debates. The basic C-SPAN channel presents debates to 63 million homes, while C-Span 2 covers the Senate, reaching 39 million homes. "If the people who elect members of Congress sat in the gallery for one day and watched their members in action or in inaction, they never would elect them," said Richard Xander, the Metropolitan Police officer who was detailed to the Capitol

Police. "Time is wasted, some members never come to the floor. A good portion of them don't do anything. They talk to empty chambers and a couple of doorkeepers and Capitol policemen."(6)

"Members also include in the *Record* statements that they never gave on the floor. House members include these statements – often sappy tributes to constituents – under "extension of remarks." Senators call the bogus material "additional statements." "Taxpayers wind up paying for the extra verbiage. Each day, 13,000 to 14,000 copies of the *Record* are printed, requiring 700,000 to 800,000 pounds of newsprint a month. The publication goes to members and staff, libraries, lobbyists, and newspapers."(7)

"For most Americans, Congress is a distant abomination, out of sight and out of mind. We don't think about it much, we barely know what it does, and we can hardly even name our elected representatives. The jarring reality, however, is that Congress is directly involved with your life every single day. You literally cannot go to your supermarket, the local drugstore, the hospital emergency room, or the nursing home without being directly affected by decisions made in Congress. You simply cannot tune out or escape the fact that Congress is a significant, relevant force in your life. For example, we found that:

• For years, Congress has protected the food industry from tougher safety standards. At the same time, from 1987 through 1996, the food industry has fed Members of Congress more than $41 million in campaign contributions. Meanwhile, thousands of Americans die each year from poisoned food, and many millions more become ill.

• Even though 400,000 Americans die each year from tobacco-related illnesses, and despite new evidence that company executives lied about smoking for decades, Congress has protected the tobacco industry, and it continues to allow the U.S. government itself to operate programs to help grow tobacco, insure it, export it, and protect it against foreign imports. From 1987 through 1996, the tobacco companies have blown smoke and more than $30 million in contributions to Members of Congress and the two major political parties.

• Roughly 10 to 20 percent of all drug treatments cause adverse reactions, which may kill as many as 160,000 Americans a year and cause more than 1 million people to be hospitalized.

Congress has done nothing about this problem, except to protect the pharmaceutical industry by proposing weaker regulations. From 1987 through 1996, the pharmaceutical and medical device companies injected nearly $28 million into the re-election coffers of Capitol Hill lawmakers.

• In 1980, 54 percent of workers at large and medium-size U.S. firms had the full cost of their families' health insurance paid by the their employers; by 1993, the number had dropped to 20 percent. Meanwhile, we pay the full health-insurance costs for Members of Congress. It has been "visiting hours only" for Congress on this and many other worsening health-care issues, and the health-care industry has inoculated Congress against any reform or regulatory tendencies by giving Members more that $72 million from 1987 through 1996.

• The food you buy at the grocery stores costs more because Congress has done sweetheart deals for some of its major patrons. For example, the price of milk went up by twenty cents a gallon in New England because of a special subsidy for the region's dairy farmers passed by Congress. Dairy interests in general poured $10 million into Congress from 1987 through 1996.

• Have you noticed that your monthly cable TV bills are higher? You can thank Congress for that. When the historic Telecommunications Act became law in 1996, the industry and Congress said that cable rates would fall because of the new law and market competition. Well, so far, cable prices are 8.5 percent higher than before the law was enacted. And millions of dollars in new revenue have accrued to the cable companies. The cable television industry invested $15.9 million in political contributions to Members of Congress from 1987 through 1996.

• As ranked on the Fortune 500, the nation's largest corporations, from number one, General Motors Corporation through number 500, Vencor, Inc., contributed at least $182 million to Members of Congress from 1987 through 1996 through their political action committees (PACs). "Soft money" contributions to both parties during the same period totaled more than $73 million. It is hardly surprising that over the years Congress has repeatedly cut taxes for corporations and wealthy individuals. In 1956, corporate income taxes accounted for 28 percent of all federal tax revenue. Today that number is down to 10 percent."(8)

• "In the 1996 congressional elections nationwide, candidates raised $791 million and spent $766 million, more than ever before, all from private sources. According to the Center for Responsive Politics, in 1996 the average cost of winning a seat in the U.S, Senate was $4.7 million, and the average cost of winning a seat in the U.S. House of Representatives was $674,000.(9)

So what does all this mean? Simple: those elected want above all to stay in office. And why not? Where else would they get the easy income for the little work they have to perform, or have the many and substantial perks they enjoy, and a lifestyle that only Washington, D.C., would applaud or condone? This means the incumbent must accumulate enough money to run an effective campaign against a challenger. It then follows that he or she must look first to those who contributed the most to their campaigns, and add to that base to the extent possible. It also means the incumbent wants the current system of financing campaigns (PACs) to continue, thereby putting challengers at a financial disadvantage. "Soft money" translates into funds raised and spent outside the restrictions of the federal election laws. There are few restrictions on "soft money" and the advantage to the incumbent is obvious. The abuses of "soft money", the illusory accountability of PACs, and the extensive financial rewards by lobbyist to politicians, have increased substantially these past years, leading to the marked decrease in confidence American voters have in Congress. We may soon see "Only Millionaires Need Apply" signs for Congress. There are more than 50 millionaires in the House; more than 40 Senators are millionaires. And the lobbyists represent open cash registers to all members.

"It is not particularly mysterious why people have soured on their elected officials. In June, 1989, House Majority Whip Tony Coelho resigned from Congress rather than face formal ethics hearing and charges over the favors he had received from a savings-and-loan operator who had helped him buy a $100,000 junk bond. Within days, House Speaker Jim Wright also resigned after a year and a half of constant revelations of ethical improprieties. In November, 1990, in hearings led by Democrat Henry Gonzalez of Texas, the chairman of the Committee on Banking, Finance, and Urban Affairs, five Senators – Democrat Alan Cranston of California, Democrat Dennis DeConcini of Arizona, Republican John

McCain of Arizona, Democrat John Glenn of Ohio, and Democrat Donald Riegle, Jr., of Michigan – were accused of taking a total of $1.4 million from Lincoln Savings and Loan Association operator Charles Keating and of pressuring federal regulators on his behalf."(10) "Americans are still paying for the $480.9 billion savings and loan debacle, which occurred largely because a Democratic-controlled Congress, inundated with $11.6 million in contributions from the industry, freed savings and loan associations from traditional constraints in the 1980s."(11) What happened? Virtually nothing to the "Keating five"; Cranston was reprimanded and the other four Senators were judged as having "exercised poor judgment or to have given the appearance of impropriety". "As we look back today, it is interesting to note that more than 80 percent of the money Keating raised for or on behalf of the five Senators was unregulated, almost impossible to discern, "soft money."(12)

Yet Senator McCain chose to again venture into muddy waters in 1999, according to a May, 2002 article in NewsMax.com. "Before accepting $30,000 in presidential campaign donations from collapsed telecommunications giant Global Crossing, campaign finance crusader Sen. John McCain used his powerful position as chairman of a key Senate committee to weigh in on the company's behalf with the Federal Communications Commission. In March 1999, Senate Commerce Committee Chairman John McCain wrote a letter to the FCC raising questions about plans by Global Crossing's competitors – MCI Worldcom, Sprint, and AT&T Corp. – to lay broadband cable under the Pacific Ocean, the *Wall Street Journal* reported recently. Within days, "20 employees of Global Crossing contributed $30,000 to Mr. McCain's presidential fund", the paper said."

As John Kennedy wrote in 1956 in his *Profiles in Courage*, "a Member of Congress realizes that once he begins to weigh each issue in terms of his chances for reelection, once he begins to compromise his principles on one issue after another for fear that to do otherwise would halt his career and prevent future fights for principle, then he has lost the very freedom of conscience which justifies his continuance in office. But to decide at which point and which issue he will risk his career is a difficult and soul-searching decision."(13)

It is difficult to understand how the voters of this country arrive at some of the choices they do. Clinton's election was thanks to Ross Perot. His re-election was more because of a sorry Republican campaign and less than an inspiring candidate. But who could believe either "Slick" Willie Clinton or 'carnival-barker' Perot? Carter's election was also a fluke, thanks largely to Ford's incredible debate gaff regarding Poland's status which literally saved Jimmy's faltering campaign.

But the re-election year-after-year of mediocre incumbents to Congress is another matter. Many in Congress stay well beyond their productive years, becoming so entrenched in their partisan ways they no longer benefit the country. The Congressional system bestows chairmanships on those who keep being re-elected, and party positions are similarly accorded. But is America truly well served by this process? For example, why do West Virginia voters repeatedly return former KKK member Robert Byrd to the Senate? There can't be much pork left that he can deliver, and his often extended venting and long lost oratorical skills impress few, if any, on either side of the aisle - or the viewing audience. Isn't he simply an over-qualified has been? Speaking of has beens, what of bulging Ted Kennedy? Is Massachusetts bereft of a quality replacement? No wonder the Red Sox haven't won a World Series since 1918! Far better that Ted Williams had been elected to represent Massachusetts! He was a four year War Hero jet fighter pilot, a magnificent ball player, and a straight shooter whose words and deeds brought fame, not shame, to Massachusetts. Can Kennedy and Barney Frank make similar claims, or pretend that they truly represent their constituents more than their personal agendas? By most accounts, Kennedy has brought multiple shames, embarrassments, and dishonor to both the Kennedy name and to the state he represents.* And what about Sen. Leahy? He has wrought enough damage to cause most Vermont voters to wonder and even non-liberals to blush. His studied posturing and deceitful practices deserve rebuke, not imitation. New York traded Sen. Al D'Amato for the even more repugnant liberal Sen. Charles Schumer, basically a classless exchange. Kennedy, Leahy, and Schumer now sit in judgment of well qualified judicial nominees and have the temerity to reject those that do not share their liberal left, stilted, sophomoric views? Outrageous! Then upon the retirement of statesman Sen. Moynihan, the large plurality of Democrats in New York City blindly chose rabidly ambitious Hillary Clinton. This effectively put a large "No Vacancy" sign in that state's political window and acknowledged victory for the liberal Democratic attack-dog kennel, assisted, of course, by Reps. Nadler, Rangel, Rev. Sharpton and other like-minded from the Liberal Left. Dick Gephardt? The "Union's favorite Congressman" is on his last run for the presidency, and hopefully will realize it is time to retire. And acerbic John Dingell? What would the automobile industry do without him? Apparently his only other constituents are members of the Black Caucus, although he was quick to damn Judge Starr and defend Bill Clinton. Sen. Tom Daschle? His senate 'floor' and television speeches have been consistently obdurate, just as his studied, often whining, attempts at civility are not convincing. And what would he do if he no longer had the hefty income from his wife's lucrative lobbying efforts? David Bonior? He and Maxine Waters make a wonderful duo, although neither impresses as having much talent beyond their shared acrimony, but perhaps Henry Waxman could join them in forming a uniquely "aggravating trio", playing to the politically deaf. And with all due respect, the late Strom Thurmond should have retired twenty-five years before he did. It was painful to watch him on the floor of the Senate, and certainly his long tenure kept younger, equally able and deserving candidates from serving. Most antiquities gain in true value and worth, but not those in Congress. There are many in Congress not yet in the antique category, of course, but these examples should underline for them the points at hand.

"According to investigative reporter Sheila Kaplan, in the summer of 1995, Republican Party "Eagles" – donors giving $20,000 or more – were led on private "midnight tours" inside the Capitol, which is closed to the public at that hour. Robert Dole spoke to the "Eagles" and even gave them a special tour of his State Majority Leader Office. We all know that many of the Democratic Party's wealthiest donors were 'invited' to sleep overnight in the Lincoln Bedroom of the Clinton White House. Less well known is Senate Majority Leader Tom Daschle's 1997 fund-raising event at Mount Rushmore in South Dakota. According to Ruth Marcus of *The Washington Post,* for $5,000 each, a group of well-heeled Democratic party donors, along with Daschle, were led on a hike up to the top of George Washington's head by the national park superintendent himself. If ordinary folks try that, they're fined $500. Just as these magnificent memorials are symbols to the rest of the world of the United States' remarkable success, our politicians' exploitation of them to raise campaign cash and to reward their patrons symbolizes what's wrong with our political process."(14)

An ever lengthening shadow is also being cast by the gaming interests. The gambling industry is hard at work, expanding rapidly across the United States. Estimates have Americans spending nearly $500 billion in 1994, "an increase of 2,800 percent since 1974. More than 200 American tribes have set up gambling operations. Not surprisingly, campaign contributions from gambling interests have shot up, with at least $6 million given to the two major political parties, presidential candidates, and congressional candidates in the 1995-1996 election cycle alone." ..."Not only is there no policy about gambling today at the federal level, but there is also no serious legislative proposal to tax gambling profits as a potential new source of revenues at the federal level. According to 1996 lobbying registration records, the gambling industry employs more than 180 lobbyists."(15) By the 2004 cycle, these figures could easily double if not triple. The "greed" factor in the gaming industry has no bounds and will not diminish, nor will the many other negatives that need to be considered as part of the gambling package. Who will be responsible for these negatives? Not the gaming industry, you can rest assured of that! Gaming interests exert tremendous political influences, if not control, of most entities in Nevada. We can expect that as gaming continues to spread, the model of Nevada will be largely replicated across the country.

Does Washington have room for more lobbyists than the 15,000 or so that are already in place working hard to peddle their influences on the 535 members of Congress and their staffs?

The ratio of lobbyists to Congress persons is obviously skewed, but lobbyists also provide "services" to their "clients", such as "raising" money, providing timely updates on issues, junkets (free trips), luxury vacations, free airplane trips, gifts, etc. Even the Motion Picture Association of America has more than 30 lobbyists in Washington, and maintains a plush seventy-seat theater for Members of Congress. Reminds one of the old cliché: "I can resist anything but temptation".

In addition to lobbyists as a source of revenue, each party has members who have "leadership" political action committees, by which campaign money passes from Member to Member. For example, "According to the Center for responsive politics, at least (read more) than fifty-one lawmakers have such PACs today. In the 1995-96 election cycle, House speaker Newt Gingrich dispensed at least $751,500 to Republican candidates from his Monday Morning Pac. House Majority Leader Dick Armey gave at least $733,558 to Republican candidates from his Majority Leader's Fund PAC. House Minority Leader Richard Gephardt gave at least $505,645 to Democratic candidates from his Effective Government Committee PAC. Senate Majority Leader Trent Lott, through his New Republican Majority Fund, gave at least $655,142 to GOP candidates."(16) The idea behind this largesse by party leaders is not just to help a colleague, but to influence (buy) his or her allegiance, thereby keeping his/her leadership role safe, or to climb the party ladder towards more power. For example, this is simply an integral part of the true "blood sport" played on Capitol Hill, not dissimilar to the mayhem of Australian Rules football, but with few if any rules, no referees, and no public spectators or investigations.

"Since 1929, the number of House members has been fixed at 435. The number of senators has remained at 100 since Hawaii became a state in 1959. Yet congressional staff has grown briskly. Since 1960, the number of employees in the legislative branch has grown by 50 percent; while total legislative branch employment has increased sharply, members' personal and committee staffs have increased the most: 165 percent for the Senate and 170 percent for the House" "Personal and committee staffs number 17,566 employees – 32 per member – of the total legislative branch complement of 33,396."(17) "I think the biggest problem we have is too much staff," Rep. Collin C. Peterson, a Minnesota Democrat said. "There are too many subcommittees. Nobody really knows what is going on. Every member has a staffer involved in every legislative issue. You can't get in the committee rooms because there are so many staffers." "We are more dependent on staff than we used to be. That is not good," said Rep. Lee H. Hamilton, an Indiana Democrat. "It makes consensus-building more difficult. You have to keep in mind that the role of the Congress is to build a consensus behind an approach or a solution to a problem."(18) In Canada, Parliament has over 400 members and functions with a staff of 3,500.

"House members are allowed to spend an average of $900,000 a year on salaries, postage, equipment, travel, and district-office rent. This does not include allowances for franked mail. Each senator is allowed to spend an average of $2 million on franked nail, depending on the population of his or here state and other factors."(19)..."Committee staffs obtain additional help from House and Senate offices of legislative counsel, which provide advice on the technicalities of drafting legislation, and from the Congressional Research Service (which has a staff of seven hundred). The House legislative counsel has thirty-five attorneys, while the Senate side has twenty-five. Both offices have a strictly nonpartisan tradition."(20)

"As part of its legislative function, Congress conducts investigations and hearings before proposing solutions. It also conducts oversight hearings to determine if the executive branch is carrying out lawfully, effectively and efficiently the laws Congress has passed" "Hearings are part show business. Usually, key findings are leaked to the press before hearings to attract more coverage. If members can grab headlines to illustrate findings, they have a greater chance of passing legislation to correct abuses"(21) "You have hearings that are an absolute joke," a former Senate Judiciary Committee staffer said. "It will be two or three members. A couple of senators will make an appearance, ask a few questions, and then leave. The idea that they can devote any real attention to more than a handful of issues is complete folly. It's alarming to see them voting on issues they know nothing about" "Congress runs on an issue cycle that lasts about a week. One week you are working on proliferation, the next week you are getting an appropriation for some company in your member's state. It's totally political. It is so mind-numbingly difficult to get real legislation passed."(22) And anyone that has watched either the House or Senate "in action" on C-Span has to wonder at the miracle that our country has managed to tolerate or survive under this strained system, or can afford such poor service by under-employed, over-rated elected officials and their large, similarly biased staffs." "Of course, "Congress has made sure few (campaign) violators will ever be called to task. The Federal Election Commission, which is supposed to enforce election laws, is so understaffed that it can only audit a dozen campaigns per election. The commission could not possibly sort out which expenses are legitimately related to campaigns. Moreover, Congress gave the agency no power to levy fines. The FEC must negotiate penalties with those who violate election laws or hope the Justice Department prosecutes the offenders. Because the FEC has six commissioners split evenly between Republicans and Democrats, controversial cases before the commission go unresolved. It would be difficult to imagine a more useless agency, but that's because Congress wanted it that way."(23)

"Members can also set up their own state PACs that do not have to abide by the federal limitations and do not have to report to the Federal Election Commission. States generally have less stringent reporting requirements and contribution limits. While foreign governments and corporations are barred from contributing, they can easily hide the source of the funds by funneling them through foreigners living in the United States or legally funnel the money to campaigns through the American subsidiaries. In this way, Democratic operative John Huang raised more than $1 million from Lippo Group, an Indonesian conglomerate, and other Asian sources. Finally, candidates can orchestrate so-called "independent expenditures" to buy campaign advertising, totally evading the legal limits on contributions."(24) Whereas bundles of money used to be delivered directly to senators and representatives, "By all accounts, the amount of corporate and special interest money now supporting candidates far exceeds the sums parceled out in the old days. An entire industry with its own jargon – maxing out, salting, sewer money – has been founded on guiding the money flow from special interests to candidates. While some of the contributions have to be disclosed, they are so pervasive and widely accepted, it is impossible for a voter to make any decisions based on the data. (Some) Newspapers run dizzying lists of PAC contributions, and public interest groups list contributions to members on the Internet. But the sad fact is that most citizens do not know who represents them in Congress. Asked in a survey taken by the *Washington Post*, the Kaiser Family Foundation, and Harvard University, only 24 percent knew who their congressman and senator were. Only 53 percent knew who the Speaker of the House was."(25)

"The net effect of the campaign finance system is that it "shifts enormous power to the moneyed interests," Rep. Lee Hamilton, (former) Indiana Democrat, said. "It decreases the power of the ordinary person. There are so many decisions made here by all of us, none of us is free from sin." "So all kinds of decisions are distorted here because of our dependence on money." "All of us would say we are not affected by it, but as a practical matter we are affected by it. The agenda is affected by it. What bills come forward, what bills don't come forward. Amendments are affected by it. What bills are permitted, what bills are not permitted. The pervasive influence of money is much more than people think."(26) "You ran for office with a concept of making necessary changes. Forget that concept, you will never get around to it," said Cec Heftel, a former Democratic House member from Hawaii; "to get by, your have to belong to the system. In Congress, the system is based on spending money. From my first days in office, I was astounded at how easily Congress could spend money that didn't exist. And it was being spent on behalf of the special interests which were the source of campaign contributions."(27)

According to the listings of the "Top Ten Career Patrons of Congressional Leaders (in *The Buying of the Congress*) the more flagrant offenders in the Senate benefiting from "special (corporate) interests from 1987 to 1996 in "donations exceeding $100,000" are: Trent Lott, who received $367,498 from the National association of Realtors, Washington, and the Auto Dealers & Drivers for Free Trade, Jamaica, NY, which provided $332,126; Connie Mack, who received $339,550 from the Auto Dealers & Drivers for Free Trade, Jamaica, NY; Larry Craig, who received $139,363 from the National Association of Realtors, Washington; William Roth, who received $143,839 from MBNA Corporation, Wilmington, Del; Alfonse D'Amato, who received $250,000 from Andrews Group, Inc. New York; $180,400 from Goldman, Sachs & Company, New York; $151,300 from Fisher Brothers Management Company, New York; $118,500 from MBNA Corporation, Wilmington, Del.; $118,250 from D.H. Blair & Company, New

York; $117,001 from Bear. Stearns & Company, New York; $111,754 from CS First Boston Corporation, New York; $102,000 from Sanford C. Bernstein & Co., Inc., New York; and $100,000 from David H. Feinberg, New York; John Kerry received $106,996 from Mintz, Levin, Cohn, Ferris, Glovsky & Popco, Boston; Newt Gingrich received $816,107 from Terry and Mary Kohler/Windway Capital Corporation, Sheboygan, Wis.; $383,513 from Capital Formation Counselors, Inc. Belleair Bluffs, Fla.; $370,300 from Milliken family/Milliken & Company, Spartanburg, S.C.; $300,755 from Schwan family/Schwan's Sales Enterprises, Inc., Gary, S.D.; $239,300 from Philip and Elizabeth Gelatt/Northern Engraving Corporation, La Crosse, Wis.; $230,000 from Jesse and Sylvia Thompson/Clermont Development Company, Charlotte, N.C.; $227,000 from K. Tucker and Karen Andersen/ Cumberland Associates, New York; $222,000 from Alfred Lee and Virginia Loomis, Bluffton, S.C.; and $213,110 from Golden Rule Financial Corporation, Lawrenceville, Ill.; Richard Gephardt received $215,300 from Anheuser-Busch Companies, St. Louis; $177,930 from Thompson Coburn, St. Louis;$128,850 from Bryan, Cave, St. Louis; $107,325 from McDonnell Douglas Corporation, St. Louis; $101,944 from International Brotherhood of Electrical Workers, Washington; and $100,000 from United Automobile Workers, Detroit, Mich.; Vic Fazio received $302,085 from American Medical Association, Chicago.

Democrat Sen. Miller, of Ga. has publicly stated this year that Sen. Kerry "leads the pack" in accepting money from 'special interest groups', despite Kerry's denials. A Feb. 5, 2004 *Associated Press* article cites some of Kerry's windfalls, including dollars from Boston's "Big Dig" and it multi-billion dollar cost overruns. Kerry reportedly received very large payments for arranging for his cousin to be awarded the multi million dollar contract by Vietnam to refurbish the deep water ports at Vung Tao and Cam Ranh Bay, and for other lobbying efforts he has made on behalf of Vietnam.

"Capital Hill lawmakers huff and puff endlessly about the unfairness of the tax system, and they promise that corrective action is just around the corner. In fact, they spend a great deal of time conniving to lower taxes that fall on a tiny minority – the wealthy – and dreaming up new loopholes for corporations. Restoring progressivity to the tax system and helping out the middle class are tasks that Congress never manages to get around to. It is not hard to see why. Corporations and the rich pay the bills for Members of Congress, and lawmakers with oversight of the tax system always receive special attention from the deep-pocket crowd. Contributions to members of the tax-writing House Ways and Means Committee grew from a total of $4.4 million during the 1977-78 election cycle to nearly $16.7 million in 1995-96."(28) That the system needs fixing is an understatement, but very few of the ills cited in this chapter have been the subject of Media coverage. It certainly appears that the Mainstream Media has been more a part of the hiding than of exposing the problems. For example, "The tax system encourages (such) outrages by allowing corporation to write off all executive compensation, no matter how high. In 1993, President Clinton proposed capping the write-off at $1 million a year. Corporations, he said, had a choice: reduce salaries or pay more taxes. Then Washington's army of corporate lobbyists went to work. First, they persuaded President Clinton to limit his proposal to publicly held companies. Clinton agreed to exempt executives at subsidiaries, so that the $1 million cap applies only to executives of the parent corporations. To skirt the proposal's application to the latter group, many companies began deferring pay above $1 million until after retirement. Further undermining the compensation cap is the increasing use of stock options to supplement executive pay. Stock options give executives the right, but not the obligation, to buy shares in their company at a fixed price."(29)

"Just about every lawmaker on Capitol Hill claims to want to make the nation's tax system fairer and easier to understand. Each call for reform, though, ends with a tax system that is ever more convoluted and confused. The original income-tax code of 1913 was twenty-six pages long. Today it has ballooned to 4,760 pages. It is no accident that the tax code means more billable hours for the Washington tax lawyers and lobbyists who are among the richest sources of reelection revenue for congressional incumbents. Every new corporate tax loophole that is proposed provides employment for a herd of $500-an-hour K Street lobbyists. Every new tax break that lawmakers maneuver through Congress translates into more campaign contributions from grateful deep pockets."(30)

Does our Congress speak with a forked tongue? You can bet your last tax return that with very few exceptions it does! And what has the Media done to help? Close your eyes and what do you see? NOTHING! So the gullible voter gets the same litany of promises from every candidate, the latter knowing that few voters, much less political rally attendees, really know their assets from a hole-in-the-ground about what is being spun. The candidate creed: "Look good, rehearse your pitch until you feel comfortable, avoid or gloss over traceable facts; speak briefly and convincingly, tell 'em what they want to hear, be adept at dodging direct questions, and toss in a little humor for good measure"...then pack for Washington to get your share of that special pie that Congress bakes there .

We should not leave the role and impact of the lobbyists without mentioning NAFTA, the North American Free Trade Agreement of 1993. Except for Ross Perot and a few others, NAFTA was billed as bringing immense benefits to all Americans. "As it turned out, NAFTA did create plenty of jobs. Unfortunately for American workers, most of (the jobs) were in Mexico. From 1993 to 1996, employment at (Mexican) assembly plants, most of them owned by U.S. companies, soared by 48 percent, while the average hourly wage of worker in these factories fell from $1.80 to $1.21; following Congress's ratification of NAFTA, "hundreds of thousands have been fired as a result of government trade policy – from the Caribbean Basin Initiative to expanding commercial ties with dictatorships such as China and Indonesia – that favors free trade above protecting jobs at home or promoting human rights abroad. The business community has lobbied long and hard to ensure that Congress and the White House support its free-trade agenda. The price has not been cheap. USA NAFTA, the chief business coalition lobbying for the treaty, spent $5 million to $8 million on advertising alone. The Mexican government and Mexican business interests spent lavishly as well, unbuckling some $30 million to advance the treaty. That money paid for the services of sixty three Washington Lobbying and public-relations firms and thirty-three former U.S. government officials."(31)

"Economic globalization has been a gold mine for American corporate executives, but it has not been nearly so good for American workers. Because it is CEOs and their companies that fill campaign coffers, though, Congress and the White House have enthusiastically promoted free trade. In fact, one government agency (U.S. Agency for International Development) not only encourages U.S. companies to relocate abroad but also helps pay their moving expenses."(32)

"Congress has been miserly about helping employees who have lost their jobs because of NAFTA, allocating just $120 million for worker retraining from 1994 to 1996. Washington was far more generous when Wall Street – which holds billions of dollars worth of Mexican government stocks and bonds – took a hit in the wake of Mexico's 1994 economic collapse. At that time, Congress and the Clinton Administration hastened to put together a $50 billion international bailout. By enabling Mexico to make good on its commitments to foreign bondholders, the bailout amounted to a government financed rescue of Wall Street investment houses. Indeed, at least $4 billion of the initial $20 billion in bailout money went to Wall Street investment houses and the richest of Mexico's rich."(33)

"In addition to killing off jobs in the United State, NAFTA has been a disappointment on many fronts. NAFTA's proponents promised that the U.S. trade surplus with Mexico would hit $9 million by 1995. Instead, the $1 billion trade surplus of 1993 turned into an $18.3 billion trade deficit three years later. The U.S. trade deficit with Canada, the other NAFTA partner, grew from $12.2 billion in 1993 to $24 billion in 1996. Congress and President Clinton vowed to protect communities that had lost jobs because of NAFTA by creating the North American Development Bank, which would offer cheap loans to hard-hit areas. The bank was endowed with $3 billion, but it did not make a single loan during its first three and a half years in business."(34) Ross Perot, where are you, and why so silent? You were right about that great sucking noise comment you made on Larry King's program about NAFTA!

"Big business has devoted more time, money, and effort to its all-out lobbying campaign to maintain China's most-favored-nation trade status than it has to any other trade issue. Imports from China have ballooned the U.S. trade deficit. In 1995, the United States exported more to Taiwan's 21 million people ($19 billion) than it did to China's (nearly) 2 billion people (less than $12 billion), but the United States runs an annual trade deficit of $45 billion with China."(35) "So who has benefited from freer trade with China? The most obvious winners are the elite group of huge corporations that head up the China lobby, especially AlliedSignal; AT&T; Boeing; Cargill Inc.; Caterpillar, Inc.; Motorola; and the Big Three automakers."(36) And don't forget the many U.S. companies in China that Kissinger, Scowcroft, Haig, and Eagleburger have enabled. Even the Walt Disney Company hired Kissinger to smooth China's ruffled feathers when it produced *Kundun*, a movie reflecting Tibet's plight under its severe occupation by the People's Republic of China. Add also, TRW, Nike, General Electric, Dresser Industries, IBM, and a host of more modest sized companies. The trade deficit continues to spiral upward for U.S. exports to China, and are estimated at a measly 2 percent. Remember, imports from China between 1986-1996 cost the U.S. 680,000 jobs and one billion dollars a year in tax revenues.* Add NAFTA costs to that and even the Mainstream Media might recognize that we're talking real money and job losses!

"If operated with the efficiency of a profit-making business, the federal government could probably be cut in half. Businesses cut costs by weeding out unnecessary staff, mandating across-the-board cuts, or contracting out jobs that other can do more cheaply. But that requires effort and thought. Instead, Congress took swipes at entire programs and departments. Meanwhile, Congress left largely intact the Government Printing Office, a part of the legislative branch that spends $30.3 million a year. Besides printing the *Congressional Record*, the GPO prints government publications, congressional bills (at least 1,400 copies of each) and stationery and calling cards for members of Congress and their staffs. The House and Senate don't want to give it up."(37) "While the GAO (General Accounting Office) is also part of the legislative branch, it regularly comes up with findings that make lawmakers who want to allocate more money to the Defense Department or to their pet project uncomfortable. Through its recommendations and its audits of procurement contracts, the GAO actually returns money to taxpayers. In 1995, the GAO saved the government $16 billion, a return of more than $35 for every $1 spent for the GAO' budget. If the executive branch acted on more of the GAO's recommendations, the savings would be far higher. But Congress slashed the GAO budget anyway, reducing its staff to 3,700 from 5,325 in

1992. GAO had to disband its program evaluation division."(38) One would think the GAO staff would have been increased, not decreased, but that probably was too much for the dilettantes in Congress to comprehend. After all, if it does not involve re-election 'interests' by lobbyists, other special interest groups, heavy-duty donors, PACs, politically important constituents, and televised committee hearings, can there be much else of importance in the life of a Congressperson? Remember, there are nearly 4,000 PACs; and only so many hours in a day. How about the nation's business? How did that slip in here?

"The Constitution gives Congress legislative and impeachment powers. It says Congress may determine its own rules and punish or expel its members. Congress may collect taxes, borrow money, regulate commerce with other nations, coin money, establish post offices and roads, declare war, provide for an army and navy, establish courts, and protect authors and inventors by giving them exclusive right to their works. But the Constitution makes no mention of solving constituent problems. Nevertheless, Congress devotes a large share of its resources to that function. Since 1972, the proportion of members' staffs assigned to their district offices – where most constituent service is done – has more than doubled. While helping constituents is laudable, it is also another way members win reelection. It detracts from the time members need to use to serve the country as a whole. Moreover, it usually means that one citizen's problem is solved at the expense of another's, whose application is moved lower in a pile."(39) "Congress is increasingly in danger of becoming an auction house where public policies are sold to the highest bidder. That is why America desperately needs courageous, independent leadership and integrity in the political process – elected officials who are not cowed by the interests, and who have the backbone and conscience to the right thing for the country."(40) But in the interim, clearly the uppermost issue for most members of Congress is money.

"Aside from its corrupting effect, most members acknowledge that the campaign finance system prevents Congress from operating effectively. Even the proliferation of committees and subcommittees goes back to the influence of money. The more committees a member sits on, and the more committee chairmanships they win, the more likely they are to receive 'donations' from the special interests affected by those committees. Yet, just as members seem incapable of rationally scheduling their time, they are unwilling to step off the treadmill and reform the campaign finance system."(41)

That the former Jenkins Hill complex is "protected" by a Capitol Police force of well over 1,000 officers, with an annual budget estimated at more than $75 million is the tip of the iceberg. "It is only a small part of Congress's operations. With 33,396 employees and an annual budget of $2.5 billion, the legislative branch is a self-contained mini-city that sprawls over 274 acres. On this turf Congress has established it own police department, barbers, restaurants, health spas, public works department, office buildings, subway, post offices, parking lots, printer, travel agencies, florist and power plant – all subsidized by taxpayers."(42) "The legislative staff includes 10,372 employed by the House; 7,194 employed by the Senate; and 2,127 employed by the Architect of the Capitol, who runs and maintains the buildings. Within the legislative branch, Congress runs the Library of Congress, which employs 3,902, and has a copy of every book copyrighted in the United States; the General Accounting Office, which employs 4,408 and is the audit arm of Congress; the Government Printing Office, which employs 4,158; the Congressional Research Service, which employs 775; the Congressional Budget Office, which employs 218; and the National Botanic Garden, which employs 51 and provides flowers and plants to Congress."(43) *

Perhaps the most damning statement is: "Yet members of Congress are not about to risk being replaced by clean candidates. Congressional interest in investigating foreign money in campaigns is but another diversion to deflect attention from the real problem."(44) And what about the problem of all the other money sources? "Among twenty Western countries, the United States is the only one that places no restrictions on the total amounts that may be donated to legislative races. Canada, France, New Zealand, and Great Britain limit candidates' campaign spending. Most of the countries have some form of public financing for legislative elections."(45)

"The mainstream press needs to realize that campaign financing and candidates' character are intertwined issues that require more aggressive coverage. The *Washington Post* considered it front-page news that Indonesian businessman Mochtar Riady, whose relatives and associates had provided hundreds of thousands of dollars to the Democrats, had written a 1993 letter to Bill Clinton outlining policy initiatives he wanted. Yet lobbyists and PACs responsible for bringing in far more money make such requests of members of Congress every day. Such requests, considered routine, rarely make their way even to the inside pages of newspapers."(46) The first sentence in this quotation bears repeating daily, and should be on the bulletin board of every Media newsroom.

"By exempting itself from the laws and standards that apply to every other facet of American life, members of congress have created for themselves a haven of power and privilege without precedent in America. By refusing to enact for themselves the public funding plan they created as an option for presidential candidates, members have deprived citizens of honest representation. From turning the Capitol Police into a corrupt force to legalizing bribery for themselves, members have institutionalized dishonesty and deceit. By providing themselves with $20,000 custom-made silk-covered chairs while shutting down the government to cut the budget, members have engaged in obscene hypocrisy, pampering themselves like kings and queens in a monarchy."(47)

Members of Congress may wonder why they are held in contempt by a large majority of the voting public in this country. Perhaps an explanation will help. Most of you have not earned our trust; your blatantly partisan actions deserve our contempt; you have been found to be woefully dishonest; your concerns are not for the good of our country but based more on your own self-interests and the interests of those who "fund" you; your deeds and voting record do no begin to match your publicized promises and words; your posturing and whining are disgusting; you carelessly disregard your oath of office; and, finally, you are selfish, over-paid, under worked, greedy, and little more than partisan puppets endlessly reciting your party's mantras. There are, of course, some Congresspersons who do not deserve this sweeping condemnation, but indeed they are a unique few. Most in Congress traveled to Washington already financially secure but not, as yet, millionaires. Certainly that status changes within a short period of time. Does anyone know of a Congressperson who did not soon become financially far better off than before he or she first took office? Under the cover of 'cost-of-living adjustments, Congress votes itself pay increases annually, adding to their other varied and substantial perquisites and benefits. However, these do not meet the tempting smorgasbord appetites of most for long. And that "smorgasbord" includes almost an endless list of special interest money and perquisites available from all major and medium sized "interest" categories, ranging from trial lawyers to doctors, to insurance companies, to all unions, to gas and oil industry interests, to corporate businesses, to environmentalists, etc., etc. So the many hog-trough options become increasingly difficult to resist, not unlike the temptations that most first experienced in earlier public pursuits "back home". Certainly none who were involved in "sleaze" in any form in their local politics are going to get religion when they get to Congress! And the longer they serve, the easier it is for them to rationalize their actions. Term limits? Meaningful campaign finance reforms? Tough regulations, e.g., full disclosures on lobbyists and special interest moneys? Hallelujah! Desperately needed, but don't hold your breath. And in the event there is a tightening of sorts on campaign finances, we can be certain there will be loopholes left through which money can flow favoring the incumbents. Only a drastic overhaul of the campaign system will result in any meaningful reforms or any leveling of the contesting fields. There is far too much money at play now, and even greater amounts at stake. In politics, there are few, if any, miracles beyond the dollar.

The American Spectator's September 2003 issue carried some interesting, but inconclusive, data on the male make-up of Congress. "Of the 150 men in Congress with military service to their credit, 87 are Republicans, and 63 are Democrat. There are 39 Republican and 18 Democrat hunters; there are 28 Republican and 10 Democrat divers or pilots; there were 22 Republican and 9 Democrat "Heroes" (not further defined); 22 Republican and 8 Democrat "Outdoorsmen" "Since Bill Clinton won in 1992, over 400 elected Democrats have become Republicans, including two senators (Richard Shelby of Alabama and Ben Nighthorse Campbell of Colorado), and five members of the House of Representatives. The number of party switchers spiked at 82 in 1995, but 39 switched in 2001, 31 in 2002, and 15 in the first third of 2003."

What does this mean? Probably not much other than those who switched did so because they thought it would bring them more votes. Or perhaps because the principles of the Democratic Party of old were being defaced if not abandoned, or because they did not agree with Bill's and Hillary's political persuasions, especially the definitive 'left', if not socialist, tack of the Democratic Party during the Clinton years. There might very well have been more switches to the Republican side following the "impeachment" process, but the ferocity of the "defense" mounted by the Clinton lawyers and the unbelievably weak-kneed Senators posing as 'jurors' on both sides of the aisle likely served to dissuade such political decisions. Many simply followed John Conyers' lead in blaming Starr for Clinton's 'problem'.

"For all the titillation about thongs and cigars, the story of the impeachment and trial of William Jefferson Clinton was not so much about sex as it was about power. It tested the boundaries of America's constitutional framework and challenged the precepts of nearly every major institution of public life. And the consequences went far beyond whether Clinton would stay in office or not – for good or ill, the decision made by the House and Senate would redefine the limits of Presidential and congressional power for decades to come. The impeachment and trial had their origins not merely in the lustful pursuits of a chief executive or the animosity he generated in his foes, but in the fast-changing, scandal-soaked Washington culture that had grown coarse and corrosive over the years. The impeachment and trial represented the triumph of partisanship on both sides of the aisle. Led by the White House, the Democrats made a calculated decision to promote partisanship and then use is as a shield for the president – demonizing his critics, resisting every GOP move, and even manufacturing disagreements where none existed, all to prove the political perfidy of the other side."(48)

"Clinton's problems on Capitol Hill were exacerbated by his long history of rocky relations with congressional Democrats. Democrats in Congress did not trust their president much more than the Republicans did. From the start, they had learned the hard way to watch what Clinton did, not what he said. They watched with exasperation as Clinton cut deals with Republicans over welfare reform, trade agreements, and the budget, often leaving Hill Democrats out of the picture. And then in the ultimate insult, Clinton, on the advice of consultant Dick Morris, cemented his reelection in 1996 through a "triangulation" strategy designed to set him apart from both Republican and congressional Democrats. Now Clinton could no longer afford to keep his distance from his party. He needed them to come to his rescue."(49)

"The 453-page (Starr) report outlined eleven counts against Clinton – four alleging that he committed perjury in his January 17 deposition in the Paula Jones case, one alleging perjury in his August 17 grand jury appearance, five alleging obstruction of justice, and one asserting abuse of office. According to the report, the president had lied when he said he had no specific recollection of being alone with Lewinsky, he had lied when he said he could not remember any specific gifts he gave her just three weeks after presenting her with a half dozen presents, and had lied when he said he did not engage in "sexual relations" as defined by Jones's lawyers. Starr charged that Clinton orchestrated a broad campaign to illegally impede the Jones lawsuit, and the subsequent criminal investigation, by coaching his secretary, Betty Currie, to lie; by encouraging Lewinsky to lie in an affidavit in the Jones suit and simultaneously arranging for a job for her at the Revlon cosmetics company through his friend Vernon Jordan; by helping Lewinsky to hide gifts that had been subpoenaed; and by lying to aides, knowing they would repeat his false statement to the grand jury. In view of the enormous trust and responsibility attendant to his high office, the President has a manifest duty to ensure that his conduct at all times complies with the law of the land," Starr wrote. "In sum, perjury and acts that obstruct justice by any citizen – whether in a criminal case, a grand jury investigation, a congressional hearing, a civil trial, or civil discovery – are profoundly serious matters. When such acts are committed by the President of the United States, we believe those acts 'may constitute grounds for an impeachment.'"(50) That Clinton chose to mount air strikes against Iraq on the eve of the impeachment vote in the House after earlier standing down when similar strikes had been urged by his key advisors gave rise to the "*Wag the Dog*" debate that followed.

"In the wake of the Starr report, Podesta and other top Clinton aides decided to abandon their usual practice of keeping Clinton's attorney off the talk shows and opted to mount a full-scale television offense. Chuck Ruff, David Kendall, Cheryl Mills, and Special Counsel Larry A. Breuer – all inside players who eschewed the limelight – were booked on the major Sunday network talk shows, along with Podesta himself. The defense they offered might have been effective in a court of law but immediately fell flat in the court of public opinion. Tom Daschle and other senior Democrats around town watched with increasing exasperation as the lawyers did what lawyers do and fell back on legalistic answers. The president did not lie under oath to Jones's attorney, they said, he simply evaded and gave technically accurate half-answers. He was not trying to tamper with a witness when he fed false stories to Betty Currie because she had not yet been called to testify. He did not commit perjury because he did not consider oral sex to be "sexual relations."(51) "He (Abbe Lowell, one of Clinton's defense team) was convinced the one person who had clearly lied before the grand jury was Vernon Jordan. None of the committee's Democratic lawyers who had reviewed his testimony considered his story plausible. Still, Lowell had concluded that most of Jordan's statements that seemed to be obvious lies were not about central events."(52)

And true to form, the strategists for the Democrats, noting Clinton's still high approval ratings and general support of the Mainstream Media, instructed: "Democratic candidates should want their election campaigns to engage the impeachment issue. Do not run from it. The impeachment inquiry is an opportunity."(53). The Clintons also attempted to arouse anger at the impeachment effort by taping some 250 radio spot and telephone messages aimed at African-Americans and Latino communities.

"The arguments were predictable – the GOP experts insisted that the allegations involving Clinton were at the core of what high crimes were thought to be, while the Democratic academics maintained "the very heavy artillery of impeachment" would be a disproportionate response to the president's misdeeds. With the assistance of White House aide Sydney Blumenthal, the Democrats had tried to stack the deck in the days leading up to the hearing by recruiting more than four hundred historians and more than four hundred law school professors to sign statement opposing impeachment because the charges lodged against Clinton did not add up to high crimes."(54) Given the Liberal Left bent of most faculties in this country, as well as the predominantly pro-Democrat lawyers, this gambit was no challenge to the White House gurus. They could as easily have had a similar or larger number of the Liberal Media from within the Beltway turning cartwheels for them! An impressive defensive move? No. Obvious? Very! The Media should have had a field day with it, but chose instead to highlight the defense "offered" by the historians and lawyers, and their own.

But the White House pit bulls had basically lost their concerted effort to substitute "censure" for "impeachment". December 18, 1998 began two days of debate which ended with the House voting along party lines, 228 - 206 in favor of impeachment on grand jury perjury charges (Article 1). Article II, accusing Clinton of perjury in his Paula Jones deposition, was rejected 220 – 205. Article III, the obstruction of justice charge, passed 221 – 212. Article IV, the abuse of power charge, was defeated 285 – 148. President Clinton did not think that the House had been "fair"; "This was not about what *he* had done; this was about what *they* had done. The president defiantly refused to feel shamed by being impeached. He was the victim, not the culprit."(55)

Shortly before the Senate trial was to begin, three Democrat and three Republican Senators met (Daschle, Biden and Levin, and Stevens, Domenici, and Thompson) to resolve questions regarding witnesses, procedures, etc. They became known as the Gang of Six, and, in a meeting with the House Managers, made it clear that Senators would not be supporting the impeachment. Biden referred to a lengthy trial as "a self indulgent process" and Stevens declared "The President isn't going to be removed."(56)

On "Thursday, January 14, 'Manager' Hyde, opened the trial. "Members of the Senate, what you do over the next few weeks will forever affect the meaning of those two words "I do". You are now stewards of the oath. Its significance in public service and our cherished system of justice will never be the same after this. Depending on what you decide, it will either be strengthened in its power to achieve justice or it will go the way of so much of our moral infrastructure and become mere convention, full of sound and fury, signifying nothing."(57)

To defend Clinton, Charles Ruff had decided he had to undercut the Managers. "What you have before you," Ruff told the senators, "is the product of nothing more than a rush to judgment." "The managers had concocted a "Witches brew of charges" and constructed their case "out of sealing wax and string and spiders webs on top of "shifting sand castles of speculation."(58) "We are not here to defend William Clinton the man," he said. "He, like all of us, will find his judges elsewhere. We are here to defend William Clinton the President of the United States, for whom you are the only judges. You are free to criticize him, to find his personal conduct distasteful. But as whether this is the moment when, for the first time in our history, the actions of a president have so put at risk the government the framers created that there is only one solution. You must find not merely that removal is an acceptable option, that we will be okay the day after you vote", he added in an implicit rebuttal to Lindsey Graham's powerful argument from the week before. "You must find that it's the only solution that our democracy should not be made to sustain two more years of this president's service."(59) And Ruff had more than a card or two up his sleeve.

"Chuck Ruff arranged (a) secret signal with the Senate Democratic leadership – and in characteristic fashion kept it so secret that he did no even tell his White House colleagues. When the Senate opened its two-day question-and-answer period on Friday, January 22, Ruff wanted the option of rebutting anything the managers might say, but the rules did not provide for any direct back-and-forth. So he and Tom Daschle's aides agreed that if he needed a chance to reply, the Senate Democrats would submit a question to the chief justice asking the White House, "Would you please comment on any of the legal or factual assertions made by the managers in their response to the previous question?" The aides would simply fill in the name of one of several "default senators," who had agreed in advance to let themselves be used in this way, such as liberals Chris Dodd and Patrick Leahy. And Ruff would indicate when he needed such help simply by laying his pen down on the table in front of him on the Senate floor. The signal was so subtle that when it came to put it into use, Senate aides sometimes missed. Left in the dark, Ruff's fellow White House lawyers correctly assumed he had made a deal with Senate staff and played guessing games among themselves trying to figure out what his cue was."(60) The Republicans attempted to plant telling questions, but the rules required that all questions from Senators be in writing, with none to be asked directly, reflecting that no time had been allowed for rebuttal. But the chicanery of Ruff, Daschle, Dodd, Leahy, and certain of their staffs was trumped by Senator Byrd who had issued a news release statement "barely two hours into the question-and-answer session" calling for 'Dismissal of the Charge' and 'End of the Trial', because "I am convinced that lengthening this trial will only prolong and deepen the divisive, bitter, and polarizing effect that this sorry affair has visited upon this nation." "Rankled by the media-saturated atmosphere surrounding what he saw as a sacred duty, Byrd scoffed that calling witnesses would "only foster more of the hallway press conferences and battle of press releases that are contributing to the divisions of our parties and our nation."(61)

In simple terms it was all over but the celebration at the White House. The Senators knew it, the House Managers knew it, and so did Henry Hyde, who could only say "We tried to be fair because we understand you need a two-thirds vote to remove the president. We needed Democratic support. So far we had none. That is okay. Let the process play itself out. But we were fair. By dismissing the articles of impeachment before you have a complete trial, you are sending a terrible message to the people of country. You are saying, I guess, perjury is okay, if it is about sex. Obstruction is okay, even though it is an effort to deny a citizen her right to a fair trial. I know, oh do I know, what an annoyance we are in the bosom of this great body. But we are a *constitutional* annoyance and I remind you of that fact."(62)

There was to be one more gasp by the Managers: it involved calling Lewinsky as a witness, and calling on Kenneth Starr to testify. When the 'trial' resumed the next day, "Democrats immediately launched into their strategy of making the day about Starr. The first question was directed in Senator Harry Reid's name: Did they notify the White House or the Senate that they planned to go to court with Starr's office to force Monica Lewinsky to talk with them? The second Democratic question came from Senator Pat Leahy: Did they talk with any member of the Senate about this or consider that it would violate Senate rules? The third was from Senator Chris Dodd and went to the White House: "Wasn't there a fundamental due-process question raised by the unilateral actions of the Managers? The fourth was from Daschle, again to the White House: Shouldn't the president's lawyers be allowed to participate in any debriefing of Lewinsky? The Democrats, of course, knew the answers to all of their questions: the point was to make the managers squirm."(63) While the Republicans were trying to formulate a renewed effort, the Democrats were also pursuing Byrd's ploy to end the "trial". And they finally convinced Byrd to soften the statements he had made condemning Clinton, reducing his long draft motion simply to: "The Senator from West Virginia, Mr. Byrd, moves that the impeachment proceeding against William Jefferson Clinton, President of the United States, be, and the same are, duly dismissed."(64) Not a bad try by the 'Wiley' Bird, but his motion was defeated 56-44. Concerned about public reaction, the Senators much preferred to allow the charade to continue, with the House Managers finally yielding to Senate pressure to call only three witnesses: Lewinsky, Jordan, and Blumenthal. Dick Morris and Betty Currie were dropped from consideration after considerable discussion.

Meanwhile, Clinton had a plan of his own. In a conversation with Dick Gephardt at the White House, "Clinton made clear he had his sights set on the 'Managers' and wanted Gephardt to target the president's enemies. "You have a chance against Barr," Clinton said of the Georgia conservative who had crusaded for impeachment before anyone else. "He only got fifty-five percent against a nobody." In fact, Clinton rattled off numbers and names of opponents in several of the managers' districts, indicating how much he had thought about it. He named four Republicans in particular he wanted to defeat – Barr, Jim Rogan, Steve Chabot of Ohio and Jay Dickey of Arkansas. Dickey was not one of the managers, but he represented the president's hometowns of Hope and Hot Springs and Clinton felt intensely about the congressman's betrayal in voting for impeachment. Beating them, Clinton believed, would show for history's sake that their prosecution of him was illegitimate from the beginning."(65) Could we be talking about being in denial, and/or perhaps psychotic? The Media did not pursue either topic.

During her "testimony", the House Managers were thwarted by Lewinsky at nearly every turn, aided by the presence of her lawyers and those from the White House. She did admit that Betty

Currie had assisted in the cover-up of the gifts that she had received from Clinton but little else. From Jordan they finally got a condescending admission that he had met with Lewinsky in New York City, which he had previously denied, and that he had tried to help find employment for her as a favor to the White House. From Blumenthal they got nothing but denials of his having leaked stories to the press. In short, the result was much-ado-about-nothing; long on talk but very little substance, thanks to the vigorous defense efforts and the disgustingly cowardly senators.

On February 12, 1999, all forty-five Democrats and ten Republicans voted "not guilty", thereby acquitting Clinton on Article I (perjury), 55 to 45. On article II ("obstruction"), the vote was 50 – 50, with Senators Specter, Chafee, Collins, Jeffords and Snowe breaking ranks with the Republicans. The "trial" was over. Shortly after 1239 P.M., Rehnquist said "The chair directs judgment to be entered in accordance with the judgment of the Senate as follows: "The Senate, having tried William Jefferson Clinton, President of the United States, upon two articles of impeachment exhibited against him by the house of Representatives, and two-third of the Senators present not having found him guilty of the charges contained therein: It is therefore, ordered and adjudged that the said William Jefferson Clinton be, and he hereby, acquitted of the charges in this said article."(66)

Two months to the day after his acquittal in the Senate, a federal judge handed down a different verdict. Judge Susan Webber Wright, who oversaw the Paula Jones case from start to finish, held Clinton in contempt for giving "intentionally false testimony" in his January 17, 1998 deposition when asked about his relationship with Lewinsky. Never before had a president been held in contempt of court. Weber said: "It is simply not acceptable to employ deceptions and falsehoods in an attempt to obstruct the judicial process, understandable as his aggravation with plaintiff's lawsuit may have been." She ordered the president to pay the Jones team \$90,000[1] to compensate for his false testimony and referred him to the Arkansas Supreme Court disciplinary committee, which in May 2000 would recommend that Clinton be disbarred."(67)

"But the president remained unrepentant. In interviews after the trial, Clinton made clear he saw the impeachment process as nothing more than an illegitimate vendetta. "I'm not ashamed of the fact that they impeached me. That was their decision, not mine. And it was wrong."(68)

David Schippers, the lawyer who had worked very hard on behalf of Henry Hyde and the House Managers in putting together their case against Clinton, has often noted that the facts were all there in the evidence room that had been established when Starr's 'library of information' was delivered to the House, but not all of the Representatives took the time to read the material. Even more disturbing was that not one Senator chose to read any of it. Schippers had also wanted to expand the investigation into other areas, such as the obvious 'Chinagate' misdeeds of Clinton and his cohorts, but time restrictions imposed by the Democrats did not permit. Thus, in the Washington, D.C., version, Caesar escaped unscathed from the Forum, his toga only slightly soiled, while the bemused Senators congratulated themselves for having saved Rome, at least temporarily. The Mainstream Media was simply relieved that the process was over, and still wondering why the House had impeached Clinton. Duh!

What did the impeachment cost the tax payer? I have no idea and it is not likely that historians will begin to find an accurate figures until all records are released many years from now. The legal costs alone have to be monumental, for none came cheap dollar wise, just politically so. And don't forget the 'private investigators', staff time, travel, etc. that both Democrats and Republicans pursued. (The Clintons tried to get approval to charge their legal bills to the tax payer as well, but we are told that this was not approved.) The Clintons have repeatedly escaped justice as most of us know it, for more than 30 years, and are still wheeling and dealing, as are many from that administration. The tax payers are being stung for the Clinton's New York home, their house in Washington, D.C., Bill's Harlem "office", part of his 'library' costs, and of course the high cost of 24 hour Secret Service protection. Chew on that a bit, and remember the advanced technology benefits rendered to China by the Clintons and Gore that are gifts that will keep on giving for more than a decade.

In the week of April 8, 2002 issue of *Human Events* and in the July 2002, the *Middle American News* The National Taxpayers Union listed the "Top Ten" Taxpayer's friends in both houses of Congress as well as the "Biggest Spenders". Leading the Senate 'spenders' were Democrats Hillary Clinton, Paul Sarbanes, Barbara Mikulski, Kent Conrad, Mark Dayton, Byron Dorgan, Carl Levin, Harry Reid, Dan Inouye, and Chris Dodd. In the House the top spenders were Democrats Rangel, Lucille Roybal-Allard, Matsui, Nita Lowey, Robert Wexler, Patsy Monk, Elijah Cummings, Juanita Millender-McDonald, John Olive, and Tom Sawyer. The Top Ten NTU "Friends" (most frugal) in both Houses were all Republicans. Sen. Hillary Clinton's rating of 3 percent was the lowest NTU rating ever earned. The 'mainstream' Media apparently did not think any of this was worth noting.

Washington is a unique setting for former household names. The *Las Vegas Review Journal* of April 19, 1998 noted that "After losing the presidential election, Bob Dole joined a Washington law firm brimming with big names and high-priced lobbying business. He's there with another former Senate leader, George Mitchell, former Treasury Secretary Lloyd Bentsen, and former Texas Gov. Ann Richards." But the swinging doors are kept well oiled within the beltway enabling the favored to come and go with little publicity, especially when toiling as lobbyists, as noted earlier.

Before closing the chapter on Congress, the following are raised as points of information which may be of interest:

• According to an *Associated Press* story in the August 3, 2002, *Las Vegas Review Journal*, "Some members of a congressional inquiry into the Sept. 11 attacks are rejecting FBI suggestions that they undergo lie detector tests as part of an investigation into leaks from the panel. Among those objecting are two leaders of the inquiry, House Democratic Whip Nancy Pelosi of California and Sen. Richard Shelby, R-Ala., who have said the tests intrude on the separation of powers between the legislative and executive branches. Pelosi, the top Democrat on the House committee, said "an investigation under oath is one thing, but a polygraph is quite another." Interesting. A good polygraph examiner can detect deceits and falsehoods; perjury is indeed "another matter". It has to be definitive, and semantic fig-leafs can become an art form, just as Clinton rhetorically and ridiculously questioned the definition of "is".

• Sen. Shelby said Senate leaders have told him and other committee members that it was not a good idea to submit to polygraphs. Sen. Lott said he expects most members of Congress wouldn't take the test, "and I don't know if that's a necessary or appropriate thing." It is interesting that CIA officers are subject to polygraph tests, as are FBI, Treasury, and DEA agents, and most 'investigative' personnel in the military. Why should members of Congress or Congressional staffs or Foreign Service Officers or White House Staff personnel not be willing or required to take polygraph tests as a prerequisite to, and periodically during, their service? History has provided horrendous examples of how our government has been penetrated at various levels over the past decades, including of the CIA, the FBI, the State Department, Cabinet positions, and our military services. Has everyone forgotten John Huang and the Department of Commerce? Or the Livermore Laboratory scandals? Is Congress trying to say that there could be no penetration(s) of the Senate, or the House, or of their respective staffs? Or of the White House? Nonsense, to politely understate it. The Mainstream Media has not yet touched this subject in any meaningful way, and, given its liberal bent, it is unlikely that it will…but it should.

• The excusing argument of crossing judicial, legislative, and executive lines is both transparent and holds no water. The Chinese – and others - certainly saw and seized the opportunities and vulnerabilities of the Clinton years, and there can be no assurances that Congress or State or other cabinet offices are free from similar "viruses". Homeland Security? Yes, it is vital, including at all levels if we are to be serious. Politics per se should not be involved in such an important issue except to embrace the concept. It is our nation and our freedom and the welfare of future generations that are involved. The simple questions of: "Have you in the past or are you now working for a foreign intelligence service or foreign government?" would be the logical place to start. From that basic point, the questions could, where appropriate, then branch out to "leaks", which Washington ostensibly decries, either intended or accidental, and thence to illegal financial gains, but then we might be without a Congress!

• Remember, Sen. Joe McCarthy simply pursued and publicized information that the FBI already had. Alger Hiss and Lawrence Duggan were Soviet spies; White House aide Lauchlin Currie was a spy; Harry Dexter White was a spy; Judith Coplin headed a spy ring at Justice; as did William Remington at Commerce; the military services have discovered spies in their midst. There have been spies in our nuclear facilities in New Mexico and California. How can anyone assume there are none on Capitol Hill or in other U.S. government offices? Has the Media made any of these points? Why not? Doesn't America need and deserve protection both on the ground and by "Star Wars" technology? Why should Congress and its staffs not be polygraphed?

Over recent years, certain politicians and individuals have maintained positions that have proven to be grossly misleading and inaccurate, or detrimental to our country. For instance, from *Human Events*, week of October 27, 2003:

• "When President Reagan refused to agree to Soviet Premier Gorbachev's proposal that the U.S. abandon plans to pursue its Self-Defense initiative in return for indefinite arms-control concessions, Sen. Ted Kennedy said, "That grand and historic opportunity was there in Iceland, but it has been sacrificed – at least for the moment – on the uncertain altar of SDI". "Gorbachev, years later, identified this act by Reagan as the beginning of the end of Soviet Communism." From June to September 1987, "Democrats worked to undercut him (Reagan) by de-funding SDI. Senate Republicans are forced to filibuster the Defense Appropriation bill to remove from it restriction on SDI funding. Sen. Robert Byrd (D-W.Va.) leads the charge to de-fund the program." And the Media snoozed, or joined in the chorus of doubters, and did not note Kennedy's and Byrd's sustained record of serious mistakes.

• August 3, 1988. "Once again, Democrats try to keep SDI funding down. This time they succeed in passing a bill that cuts funding by about 20% below the President's request. Reagan vetoes, saying "The bill would signal a basic change in the direction of our national defense, a change away from strength and proven success and back toward weakness and accommodation of the 1970s."

• September 23, 1991. "President George H.W. Bush and a Democratic Congress approve a $3.8 billion in spending on development of a missile defense. Bush's plan is to expand missile defense beyond U.S. borders, to protect U.S. allies and the armed forces deployed abroad from missile attacks by rogue nations. Sen. Tom Harkin (D.-Iowa) leads the unsuccessful effort to cut the funds."

• December 5, 1991. "President Bush signs the Missile Defense Act of 1991, which mandates the Defense Department to develop and deploy a ground-based missile defense system by 1996. The measure easily passes a Democrat-controlled Congress whose members recognize the popular appeal of the Patriot missiles resulting from the Gulf War.

• November 11, 1993. "President Clinton signs a defense bill dramatically cutting missile defense funding to $2.6 billion. The new spending also guts research into a strategic national missile defense, dedicating funding mostly for theater missile defense, referred to as LDS or a "Limited Defense System. This narrow focus leads to the squandering of technological advances that had been made toward a national or global system." The Mainstream Media lauds Clinton.

• May 14, 1997. "Sen. Jesse Helms (R.-N.C.) blocks Clinton Administration plans to re-negotiate and re-legitimize the ABM treaty with the Soviet Union's successor states. Eight years after the demise of the Soviet Union with which it had had been made, the ABM Treaty is championed by Liberal Democrats."

• July 23, 1999. President Clinton signs the National Missile Defense Act, which states simply that "it is the policy of the United States to deploy a national missile defense." It passes the House easily, over the objection of Dennis Kucinich (D.-Ohio). It passes the Senate by unanimous consent, without a peep from Democratic Senators Kennedy and Kerry (Mass.), Joe Lieberman (Conn.), or John Edwards (N.C.)" Talk about stand-up guys firm in their convictions!

• September 1, 2000. "As the election approaches, President Clinton declines to authorize deployment of a national missile defense. "The technology for a missile shield," he argues, "has not been proved." The question the Media should have asked: "What was China's influence in this decision?"

• April 10 - May 1, 2001. "Russia, China, and North Korea tell the U.N. Disarmament Commission that a U.S. missile defense system would cause a new arms race and undermine the ABM Treaty. President Bush, undaunted, makes a public statement three weeks later-"We need a new framework allowing us to build missile defenses to counter the different threats of today's world."

• June 27, 2001. "Bush submits his 2002 Defense budget to

Congress, with $7 billion – later amended to $8.3 billion – for missile defense."

• December 13, 2001. "Over Democrat's objections, President Bush announces that the United States will unilaterally withdraw from the ABM treaty, effective June 2002. The move will allow the U.S. to develop and deploy a missile defense system. Liberals complain: "I think that it presents some very serious questions with regard to the future." Days of Whine & Roses -

• May 24, 2002. "Despite the "arms race" predicted by Daschle, President Bush and Russian President Vladimir Putin sign an historic arms control treaty, calling on each country to reduce its nuclear warheads from roughly 6,000 to between 1,700 and 2,200."

• June 25, 2002. "National Missile Defense funding survives a covert, futile attempt by Senate Democrats to strip-out key elements of its funding (led by Sen. Levin). Congress ultimately provides $7.6 billion for research and development for fiscal 2003."

• December 17, 2002. "President Bush orders the Pentagon to start deploying a national missile defense system by 2004. The system is expected to be the first layer of a system with several redundancies for safety in stopping enemy missiles."

• March through May, 2003. "During the Iraq War, Saddam Hussein's forces launch a series of missiles toward Kuwait City and U.S. military targets. Patriot 3 missiles prove even more effective than their predecessors in the Persian Gulf War."

• September 30, 2000. President Bush signs a Defense appropriation bill with $9 billion set aside for missile defense. This was a fitting closing response to the various liberal Democrat nay sayers – and their negative Liberal Media supporters. Their motto should be "Often, wrong, but never in doubt."

Equally, if not more, unsettling activities by certain Liberal Democrats well deserve mentioning such as:

• In the Week of December 8, 2003 edition of *Human Events*, Herbert Romerstein writes, "a KGB report ... reveals that "In 1978, American Sen. Edward Kennedy requested the assistance of the KGB to establish a relationship "between the Soviet apparatus and a firm owned by former Sen. John Tunney (D.-Calif.). KGB recommended that they be permitted to do this because Tunney's firm was already connected with a KGB agent in France named David Karr. This document was found by the knowledgeable Russian journalist Yevgenia Albats and published in Moscow's *Izvestia* in June 1992.

• Continuing, Romerstein wrote "Another KGB report revealed that on March 5, 1980, John Tunney met with the KGB in Moscow on behalf of Sen. Kennedy. Tunney expressed Kennedy's opinion that "nonsense about the Soviet military threat and Soviet ambitions for military expansion in the Persian Gulf" was being fueled by Carter's (National Security Advisor) Brezinski, the Pentagon, and the military industrial complex." "Kennedy offered to speak out against President Carter on Afghanistan. Shortly thereafter he made public speeches opposing President Carter on this issue. This document was found in KGB archives by Vasily Mitrokhin, a courageous KGB officer, who copied document from the files and then defected to the West. He wrote about this document in a February 2002 paper on Afghanistan that he released through the Cold War International History Project of the Woodrow Wilson Center." It would appear that Kennedy was still upset at Carter's winning the nomination.

• Romerstein wrote that "In May 1983, the KGB again reported on a discussion in Moscow with former Sen. John Tunney. Kennedy had instructed Tunney, according to the KGB, to carry a message to Yuri Andropov, the General Secretary of the Soviet Communist Party, expressing Kennedy's concern about the anti-Soviet activities of President Reagan. Kennedy offered to "undertake some additional steps to counter the militaristic policy of Reagan and his campaign of psychological pressure on the American population." "Kennedy asked for a meeting with Andropov for the purpose of "arming himself with the Soviet Leader's explanations of arms control policy so he can use them later for more convincing speeches in the U.S. He also offered to help get Soviet views on the major U.S. networks and suggested inviting "Elton Rule, ABC chairman of the board, or observer Walter Cronkite or Barbara Walters to Moscow." "This document was first discovered in the Soviet archives by London *Times* reporter Tim Sebastian and a report on it was published in that newspaper in February 1992." Our Media? Asleep at the switch.

• Romerstein's article notes that "Sen. Kennedy played a major role during the 1970s in crafting the restrictions that made it so difficult for the FBI and CIA to do the job of protecting the

American people. One of the most pernicious restrictions was the Foreign Intelligence Surveillance Act (FISA) passed in 1978. President Franklin Roosevelt, in 1940, had ordered the FBI to wiretap Nazi and Communists because they were operating in the United States on behalf of hostile foreign powers. Every president after him used the inherent power of the President to order wiretapping for national security purposes. Kennedy told the Senate Intelligence Committee in 1976 that "For at least the last five years, I and others in the Senate have labored unsuccessfully to place some meaningful restrictions on the so-called inherent power of the Executive to engage in surveillance." "When Congress discussed legislation to require a court warrant to wiretap enemy agents and terrorists, Kennedy and the ACLU began a campaign to raise the barriers as high as possible. Kennedy introduced the concept in the Foreign Intelligence Surveillance Bill that required evidence that someone was providing classified information to a foreign intelligence service. Someone who "only" had a clandestine relationship with a foreign intelligence officer and carried out covert influence operations for a foreign power could not be wiretapped. When we see the KGB reports we can understand why Kennedy would want this provision in the law. Kennedy was not a KGB agent. He also was not "a useful idiot" who was used by the KGB without understanding what he was doing. Kennedy was a collaborationist. He aided the KGB for his own political purposes."

- Romerstein added that "The restrictions that Kennedy successfully put in the Foreign Intelligence Surveillance Act were so tight that when the FBI arrested Zacarias Moussaoui (the so-called 20th high-jacker) in August 2001, they could not get permission to download his computer since FBI headquarters understood that they did not have enough evidence to get a warrant from the Foreign Intelligence Surveillance Court. After 9/11 when they did download his computer they found, among other interesting things, information on the air currents over New York." "After 9/11 Kennedy and other demagogues in the Congress blamed the FBI and CIA for the intelligence failure. The slogan was "they didn't connect the dots." There was no way to connect the dots when they weren't allowed to collect the dots." Ahhh Teddy, have you no shame?

I would suggest that Mr. Romerstein is being far kinder about the tawdry sustained performance(s) than Sen. Kennedy deserves. And was the FBI or CIA aware of his and Tunney's contacts or of the French Connection? Hopefully others will further question Kennedy and Tunney regarding their "KGB channel(s)", authorized or otherwise. The whiff test will not be complete until that happens. But don't count on the Mainstream Media to get these answers.

There are additional 'whiff" or "smell" tests that need to be made and publicized, For example:

• In an article in The Week of July 1, 2002, issue of *Human Events* written by David Freddoso, "Senate Armed Services Chairman Carl Levin (D-Mich.), an outspoken opponent of a national missile defense system, had removed $815 million from vital elements of the program on a party-line vote in his committee May 9. Last week, he almost fended off a GOP attempt to restore the funding in the 2003 defense authorization bill (S.-2514) – despite a threat from President Bush that he would veto the measure if the missile defense money was left out. "Levin, who has repeatedly asserted that a missile shield will precipitate a new arms race with Russia or China, denied that he was trying to kill Missile defense. "The $815 million," he told Human Events, "is money which, after a very thorough analysis, was determined to be redundant and not needed this year." "(On) June 25, Levin lined up his Democratic colleagues to vote against a GOP attempt to restore full funding for missile defense on the Senate floor." "Levin, along with most his Democratic colleagues and some two dozen Republicans, had voted in several roll-calls two weeks earlier to keep more than $1 billion dollars in non-emergency pork-barrel earmarks in a 2002 supplemental appropriations bill intended to fund the war on terror. Among the items allowed in that bill were: $175 million for a laboratory in Iowa that will take eight years to build – or more than $85 million in excess of what was requested by President Bush; $2.5 million in "emergency" funds to map Hawaiian coral reefs; $1 million to study "self-administered high frequency chest oscillation therapy" at Johns Hopkins Medical School in Maryland (the state of Democratic Senator Barbara Mikulski); $11 million to bail out struggling New England fishermen; $5 million for a Mathematics and Science Education Clearinghouse in Ohio (home of Republican Senate appropriator Mike DeWine); $26.8 million for the U.S. Geological Survey to map U.S. urban areas; $3 million to drill for water in New Mexico, (home of Sen. Pete Domenici (R.), and a member of the appropriations committee); $1 billion for Pell grants. The Senate eventually passed this pork-laden bill by a vote of 72 to 24 on June 6, but three weeks later could barely find a majority to save the proposed national missile defense system." Obviously, our Congress people have their priorities straight. Re-election above all else! Did our Media jump on this one? Not exactly.

• According to an article by Mr. Lambro in *Human Events,* The Week of May 26, 2003, Sen. Frank Lautenberg of New Jersey was outraged by Bush's budget for fiscal 2004, calling it "grossly irresponsible, especially as costs mount for postwar Iraq and our national deficit increases", according to a story in the Ocean County (N.J.) *Observer* on May 4. However, Lautenberg wasn't worried about the deficit when he offered two amendments to the budget bill that would have spent $118.5 billion. All of these amendments sought more money for schools, general aid to the states, transportation, public works projects, agricultural assistance, health care, law enforcement and environmental projects – programs with budgets that already have risen substantially over the past decade." Perhaps Sen. Lautenberg should have checked with former and ethically disgraced Sen. Torricelli to see what had been happening in those same sectors while Torricelli was a senator from N.J. Some habits are very hard to break. Did the Mainstream Media bother to note this hypocrisy?

• The "leaks" just keep coming. In January 2004, we learned of the "suggested" planning memo written by a Democrat, not yet publicly identified, outlining suggestions of how to demonize the Republicans in the Presidential campaign next year. The 'surprise' was that it surfaced from the Intelligence Oversight Committee, touted to be the one bed-rock non-partisan committee in the Senate. Well, historically speaking, that's not exactly a correct description. The senatorial leak of the CIA mining operation in Corinto Harbor in Nicaragua was a damaging, treacherous leak, especially in that the CIA had briefed Congress 11 times on the mining, starting in January, 1984. The Editorial in the *Wall Street Journal* quoted in the spring issue of Periscope states "Our view of this affair is that congressmen could not forgive the CIA for notifying them about the mining, forcing them to explain their acquiescence or deny that they knew. The Corinto leaks and the diversionary debate about who knew what (and) when are good arguments against any restrictive notification requirement. Nor was Corinto the only leak from the Senate committee. The successors to Messrs. Goldwater and Moynihan on the Intelligence Committee were Dave Durenburger, who is under investigation for leaking that the US recruits Israelis to spy on Israel, and Patrick Leahy, who resigned from the committee after admitting that he'd leaked a preliminary draft on the Iran – Contra report. It is of course good to build confidence, and the President may very well want to experiment with the 48-hour notification of a different intelligence committee headed by David Boren. But it is quite another thing to promise the same thing for all future Presidents and all future Committees." Did the Media review the checkered history of members of this committee? And doesn't the "who knew what, and when, sound vaguely familiar? Some politically inspired Democrats, including Sen. Clinton, chose to infer, or charge, that President Bush knew of the attacks in advance of the terrorist attacks of 9/11/2001. The Media did not ridicule Clinton.

One of the sorriest spectacles by the Liberal Democrats was the so-called "memorial" service for Sen. Paul Wellstone at the University of Minnesota basketball arena. First, the Wellstone family relayed to the White House that "they did not want Vice President Cheney to attend because Cheney's Secret Service detail would disrupt the event." But "They did not express similar ('security') reasons about the attendance of former President Clinton and former Vice President Gore (and their Secret Service protection)." The *Human Events* article (week of November 4, 2002 issue) also noted "The next indicator of the event's crass nature was when the crowd booed (Gov.) Ventura, an independent, and Senate Majority Leader Trent Lott (R.-Miss.), but cheered Clinton and newly minted Democratic Senate candidate Walter Mondale." Ventura walked out, saying "Certainly to turn a memorial into a political rally, I found quite disturbing." "The memorial service seemed more like a political rally than a funeral, with the most impassioned political appeal coming from Wellstone's youngest son," Jonathan Karl reported on CNN." "The same reason every Democratic senator voted that Bill Clinton was "not guilty" in the 1999 impeachment trial. The same reason the party backed Al Gore in his disgraceful demands after the 2000 election. There is only one thing that matters to the Democrat elite today: political power. We have seen before, and we will see again, that they will pay any cost to keep it."

And there are additional examples of what has happened to the Democratic party that once was strong and far more moderate than the very Liberal or Left bent it now has. Neither Harry Truman nor John Kennedy would recognize the present Democratic Party, and even social-programs-minded Franklin Roosevelt would be perplexed at the drastic changes that occurred under Johnson, Carter, and Clinton. The party that once gently or otherwise urged all new arrivals seeking to settle in this country to vote for the Democrats because "we will take care of you", and had the unions to rely on to further persuade the new "citizens", now depends on raw power and money at all levels to gain their goals. From the ports of entry to the employment offices to the welfare offices to the unions, etc., the Democrats seek to add to their rolls, primarily with persuasions and promises of money and services and benefits, all tax payer money, of course.

To counter some of what the Democrats spew, perhaps all should heed what Thomas Sowell wrote in one of his recent commentaries titled "Lott, race and hypocrisy" carried in the *Las Vegas Review Journal* in late October, 2003. "Republicans could start by setting the facts of history straight. A higher percentage of Republicans than Democrats voted for the landmark Civil rights Act of 1964 and Voting Rights Act of 1965. It was Republicans who, for better or worse, created racial quotas for blacks in the 1969 "Philadelphia Plan". "It was Democrats who put confederate flags on Southern state capitols as a sign of defiance against the advancement of civil rights. Today, when going after minority voters, Republicans can offer things that Democrats cannot possibly offer – the most important being vouchers to provide a decent education for their children. Democrats have to oppose vouchers because the teachers' unions oppose vouchers – and Democrats are beholden to the unions, both for money and for their ability to get out the votes on election nights. It is Democrats who are disarming law-abiding citizens in the face of heavily armed criminals. Hard data show that crimes against minorities and women have declined the most where state laws have allowed citizens with clean records to carry guns." Sowell has also written "The liberal brand of compassion has almost invariably meant spending the taxpayer's money. When it comes to spending their own money, it is by no means clear that liberals do more of it than conservatives, though they make more noise about it."

"When Sen. John Kerry pontificated at the Vietnam Veterans Memorial on Veteran's Day, a group of veterans turned their backs on him and walked away. They remembered Mr. Kerry as the antiwar activist who testified before Congress during the war, accusing veterans of being war criminals. The dust jacket of Mr. Kerry's book, The New Soldier, features a photograph of his ragged ban of radicals mocking the U.S. Marine Corps Memorial, which depicts the flag-raising on Iwo Jima, with an upside-down American flag. Long after he changed sides in congressional hearings, Mr. Kerry lobbied for renewed trade relations with Hanoi. At the same time, his cousin C. Stewart Forbes, chief executive for Colliers International, assisted in brokering a $905 million deal to develop a deep-sea port at Vung Tau, Vietnam – an odd coincidence." (Letter in the March 2003 edition of NewsMax.com by Michael Benge, Foreign Service officer and former Vietnam POW (1968-1973), titled "The Real John Kerry"). Chameleon Kerry was also linked to the infamous Christic Institute. Ask the Media to check it out.

In the Week of January 14, 2002 issue of *Human Events*, Stephen Moors wrote in an article titled "The Daschle Deception", "In his tirade against tax-cuts, Daschle defined himself at least nine times as an advocate of "fiscal discipline." Allowing taxpayers to keep more of their own money is "fiscally irresponsible" Daschle charges, because it unbalances the budget. The "single largest reason for the decline in the surplus was the Bush tax cut," he argues. "Unfortunately, Daschle's credentials as a fiscal conservative are highly suspect. For many years now the National Taxpayers Union has rated Daschle as one of the biggest spending senators in the history of the Congress. If Daschle suddenly is concerned about deficit spending, that is a radical change. After the Republicans took over Congress in 1995, Daschle led the fight to torpedo the GOP plan to balance the budget. He voted against the balanced budget amendment – three times. He is one of the chief sponsors of this year's obese agriculture subsidy bill – the most expensive farm bill ever." In short, Tom Daschle is a fiscal conservative like Jesse Ventura is a ballet dancer.

• That Liberal Democrat Senators in both North and South Dakota talk "Conservative" when visiting their states but vote the Liberal Democratic line when in Washington. "When it comes down to the role of the federal government, Dakotans like the way things are." "Both states are net consumers of tax dollars, receiving more in federal spending than they send to Washington in taxes. There are Air Force bases, B-52s, missile silos, and other Pentagon spending. There is Bureau of Land Management spending. There is Bureau of Indian Affairs spending. There is heavy highway spending for the widely dispersed population. And there is lots and lots of agricultural spending." "It really is fascinating with these guys," says one frustrated White House aide. "They're in states where tax relief is popular, education reform is popular, ballistic missile defense is popular; it's "astounding" that Conrad and Dorgan (and Daschle) were able to vote against the tax cut." (National Review, June 25, 2001.)

• "Last June, when Sen. Tom Daschle (D.-S.D.) became leader of the new Democratic majority in the Senate, he pledged a "spirit of bipartisanship." "Polarized positions," he said, "are an indulgence the Senate cannot afford and our nation cannot tolerate." "Since then, however, the record shows that Daschle has been relentlessly partisan and polarizing. Meanwhile, the Republican controlled House, led by Speaker Dennis Hastert (R.-Ill.), has passed a long list of excellent bills." In comparing the performance of the House and the Senate on eight issues, the stark differences unclothe Daschle's promises of bipartisanship. (1.) Taxes: On April 18, 2002, by a vote of 229 to 198, the House passed a bill to make President Bush's $1.3 trillion tax cut permanent rather than have it expire in 2011. On June 6, by a vote of 256 to 171, it passed a bill permanently abolishing the death tax. On June 13, by a vote of 271 to 142, it passed a bill permanently establishing the reduction in the tax marriage penalty. In the Senate, on April 16, Daschle told a group that calls itself the Fiscal Responsibility Coalition that he would "never bring up the permanent tax cut the President has advocated. On June 12, by a vote of 54 "yes" to 44 "no" (60 "yes" votes needed for cloture), the Senate voted not to permanently repeal the death tax. (2.). Limited Spending: On March 20, by a vote of 221 to 209, the House passed a budget resolution to limit discretionary spending to $759 billion in fiscal year 2003. For the first time since 1974, the Senate failed to pass a budget resolution that would limit the amount of money that could be dumped into discretionary spending bills. In 1998, when the Congress as a whole did not pass a joint budget resolution, Daschle chastised the Republican House leadership for the failure, demanding that they "do what the law requires." (3.) Ensuring Energy Supply: On August 2, 2001, by a vote of 240 to 189, the House passed an energy bill that would allow drilling in the Arctic National Wildlife Refuge (ANWAR) and reduce U.S. dependence on Middle Eastern oil. On April 24, the Senate passed its own version of an energy bill, heavily influenced by Daschle, that prohibits any drilling in ANWR and is focused instead on providing incentives for cost-efficient, environmentally correct alternative energy sources. "We are going to stop ANWR," Daschle vowed at the time. (4.) Banning Human Cloning: On July 31, 2001, by a majority of more than 100 votes, the House passed a bill to ban all human cloning. Despite an earlier promise to bring a cloning ban up for a vote, Daschle prevented the Senate version of a complete human cloning ban from coming to the Senate floor by proposing a rule that would have ensured its defeat, and passed in its place a bill that would legalize the cloning and killing of human embryos. "I feel like I've fulfilled my obligation." Daschle told a press briefing June 13. (5.) Restoring Welfare Reform: The 1996 welfare reform act expires at the end of September. In June, the House passed a bill reauthorizing the reform and increasing welfare work requirements, requiring 70% of welfare recipients to work at least 24 hours a week and do 16 hours of other activities that can include school. The Senate Finance Committee has approved its own version of welfare reform that includes loopholes that completely gut the current welfare work requirements. For example, it would allow a person to go to college full time, do no work at all, and receive welfare. (6.) Securing Homeland: On July 26, the House passed a bill, 295 to 132, to create a new Homeland Security Department, allowing the department the flexibility to hire, fire and promote many employees according to their merits rather than according to scierotic civil service employment rules. Hours after President Bush said he would veto a Homeland Security bill that did not include his provision for creating a meritocracy in the new department, a Senate committee approved a bill that would maintain sclerotic civil service employment rules. (7.) Banning Partial-Birth Abortion: On July 24, by a vote of 244 to 151, the House passed a ban of partial-birth abortion. Two previous bans passed by both the House and Senate were vetoed by President Clinton. President Bush pledged to sign a ban on partial-birth abortion. No one expects Daschle to bring the ban up for vote soon. He voted against the ban when it was enacted by a Republican Congress under Clinton in 1995, then for it in 1997. Now he says the ban "may not be in keeping with court interpretation of the rights of woman." (8) Confirming Judges: Since May 9, 2001, President Bush has nominated 32 people to U.S. Circuit Courts of Appeal and 90 people to Federal District Courts. The Senate so far has confirmed only 13 of Bush's 32 appellate court nominees, and 59 of his 90 district court nominees. Among the remaining 19 nominees for the appellate courts, only four have had hearings in the Judiciary Committee. One highly qualified appellate nominee, Judge Charles Pickering, has already been "Borked", failing to get a majority vote in the committee." *Human Events*, Week of August 12, 2002.

• "The ultra-left-wing Slate magazine, which traditionally is aligned with the Daschles, wrote in last January's edition that "Linda Daschle shoehorn(ed) into the Transportation Department budget a requirement that the FAA purchase at least half of its baggage scanners from L-3 International which is one of her clients. The L-3 scanner turned out to be a lemon." (Judicial Watch)

• From Judicial Watch, September, 2002. "Linda Daschle is one of Washington's most powerful lobbyists on behalf of corporate interests, counting American Airlines, Northwest Airlines and Schering-Plough among her high profile corporate clients. Linda Daschle's client, American Airlines, has lobbied for years to water down safety and security regulations that might have helped foil the World Trade Center attacks. Further, with the support of both Tom and Linda Daschle, the company received a whopping $583 million taxpayer bailout in the wake of September 11." "As reported on CBS *60 Minutes*, Tom Daschle was charged with inappropriately intervening to reduce safety inspections of an air-charter company owned by a family friend after one of the planes crashed in 1994, killing four. While at the FAA, Linda Daschle acted to exempt the friend's airline from intensified safety inspections." "Linda Daschle's client, Boeing, is attempting to strike a sweetheart deal to lease 100 aircraft to the U.S. military. According to the Congressional Budget office, the lease plan will cost taxpayers $37 billion, while outright buying the planes would only cost $25 billion. Tom Daschle will be responsible for scheduling the vote" "The air transport industry donated more than $100,000 to Daschle's campaign in the last election cycle. Linda Daschle's client, Northwest Airlines, was the second largest contributor to the Senator's campaign in 1998." According to the *New York Post*, Daschle, who has called on President Bush to release documents related to the Harken matter, refuses to allow the public to see how much his wife makes from her big business deals by releasing his own tax documents." "Daschle's wife, Linda, also represents Loral Space and Communications, which paid a $14 million fine last January to settle charges of illegally sending sensitive missile technology to China."

• *Human Events* carried a list of other Linda Daschle lobbying interest in its week of May 20, 2002 issue, including the years served and the amounts paid to the firm. "American Airlines, 1997-2001, $1,260,000; Northwest Airlines, 1997-2001, $870,000; Loral Space and Communications, Ltd., 2000-2001, 740,000; United Technologies Corp., 1999-2002, $640,000; Schering-Plough, 1999-2001, $470,000; L-3 Communications, 1997-2001, $440,000; Boeing, 1998-2001, $440,000; American Trucking Association, 1999-2000, $240,000; Cleveland Hopkins Int'l. Airport, 2000-2001, $180,000; ICF-Kaiser, 1997-1998, $160,000; American Assn. of Airport Executives, 1998-2001, $120,000." Of course Daschle has assured anyone that would listen that he is not involved in any way with his wife's lobbying efforts. Bonnie and Clyde should have been so fortunate.

But Daschle has lots of company, both as a hypocrite and one who can hedge with the best of them when necessary. For instance:

• Judicial Watch cited three quotes well worth repeating. In chronological order they are: "Senator Tom Daschle on October 5, 1999 stated "I find it simply baffling that a Senator would vote against voting on a judicial nomination." "And then there is this Daschle quote from Oct. 28, 1999, "Hispanic or non-Hispanic, African American or non-African American, woman or man, it is wrong not to have a vote on the Senate floor." Senator Patrick Leahy on October 3, 2000 stated "Governor Bush and I, while we disagree on some issues, have one very significant issue on which we agree, that is what we are paid to do in this body. We are paid to vote either yes or no – not vote maybe. When we hold a nominee up by not allowing them a vote and not taking any action one way or the other, we are not only voting maybe, but we are doing a terrible disservice to the man or woman to whom we do this." "And Senator Kennedy: "Justice delayed is justice denied." Add Schumer and we have four of the biggest hypocrites ever assembled in the Senate. Want to add a female? Perhaps Barbara Boxer? No applause, please, although she readily qualifies. More? Sure!

• About Attorney General David Pryor, Sen. Schumer said "His beliefs are so well known, so deeply held, that it's very hard to believe –very hard to believe – that they're not going to deeply influence the way he comes about saying , "I will follow the law." "There are two grounds on which one's views on personal moral issues such as abortion should disqualify a candidate for judicial office: if they are indeed extreme or eccentric, of if these personal beliefs distort the candidate's judicial actions and compromise his fidelity to the Constitution and the rule of law. Pryor's views on abortion meet neither of these criteria." Is Schumer therefore anti-Christian or anti-Catholic? No. But the net effect of Schumer's "deeply held views" litmus test, now slavishly followed by his fellow Senate Democrats, is to disqualify from the bench anyone whose personal views of abortion coincide with those of traditional Christianity, Judaism and Islam. The test is not a religious test. It's an ideological test – that has the obvious effect of excluding from the bench tens of millions of believers who suffer from "very, very deeply held views." (from a *Washington Post* article by Charles Krauthammer, August 31, 2003.)

• In the case of attorney Miguel Estrada (to the federal appeals bench), the *Las Vegas Review Journal* editorial of February 13, 2003 noted: "During Mr. Estrada's five-hour confirmation hearing last year, Democrats complain the nominee refused to answer hypothetical questions concerning how he would rule on specific cases, including those on abortion rights. This gave "little to review," Democrats complain. Senate Judiciary Chairman Orrin Hatch, R-Utah, replied that in fact, Mr. Estrada answered every question the Democrats had. "The truth is he didn't answer questions in a way that could be used against him."

• "There's no reason to mince around this, we're not going to allow an up or down vote on Miguel Estrada," Sen. Harry Reid told Senate Majority Leader Bill Frist. "What do the Democrats have against him? What they really resent, of course, is that Mr. Estrada is a conservative, and a member of the law firm that represented Mr. Bush in his successful Supreme Court fight to certify his election to the presidency."

• In the *Human Events* week of April 8, 2002 issue, an article by Terence P. Jeffrey noted Senate Judiciary Chairman Patrick Leahy, (D-Vt.) "has revealed, again, the ideologically driven and amoral approach he takes to reviewing Republican appointees to the federal bench. On March 16, Reuters reported that Leahy was claiming that unnamed Republican were "concerned" about alleged connections between Enron and Texas Supreme Court Justice Priscilla Owen, whom Bush has nominated to the 5th Circuit Court of appeals in New Orleans. "I have heard from a lot of Republicans who are concerned about her Enron connections," Leahy told Reuters. Lest there be any doubt this was the beginning of a Leahy-led smear campaign against the judge, *Time* magazine confirmed two weeks later that Owen was one of the "conservative nominees" Senate Democrats had "targeted" to block." The bottom line of the article is that after running the story to ground, Mr. Jeffrey determined that Leahy's intended smear had no basis whatsoever. Not only had Owen not been a tool of Enron, individual Enron employees had formed a PAC, and Owens received three contributions in 1994 totaling $5,000 from it, plus miscellaneous donations from Enron employees that brought the grand total to $8,600." "The concerned "Republicans" that Leahy glibly referred to have yet to be found."

• In the week of September 16, 2002 issue of *Human Events*, Thomas L. Jipping wrote: "Sen. Patrick Leahy (D.-Vt.) said he was "glad" Justice Owen was "able to have this hearing. Before you think that sounds like a nice gesture, remember he is chairman of the committee. President Bush nominated Justice Owen on May 9, 2001. Both her home state senators strongly supported her. The ABA gave here its highest rating. Leahy could have granted a hearing at any time. Instead, he stalled for 440 days and had the gall to say he was glad she could have a hearing. The previous three Presidents saw their first batch of appeals court nominees confirmed in an average of 81 days. Today, after nearly 500 days, only five of Bush's first 11 nominees have had a hearing and only four have been confirmed." Other Democrats, e.g., Feinstein and Durbin, have been nearly as obnoxious as Kennedy, Leahy, and Schumer, but none of them have been challenged for their performances in or by the "Mainstream Media."

• "Since 1968, the Senate has taken 26 cloture votes on 17 different judicial nominees. In the case of 10 of those nominees, the lone cloture vote passed, and the Senate voted to confirm. In the case of three nominees, the first cloture vote failed, but the Senate eventually confirmed them – demonstrating that those filibuster were not intended to defeat the nominees. The four other nominees who faced cloture votes were Supreme Court Justice Abe Fortas, Miguel Estrada, Texas Supreme Court Justice Priscilla Owen and Alabama Atty. Gen. Bill Pryor. Fortas withdrew his 1968 nomination to be Supreme Court Chief Justice after a 45-43 cloture vote. Thus every filibuster in the past 45 years designed solely to defeat a judicial nominee who would otherwise be confirmed by a Senate majority is going on right now. Seven cloture votes have failed on appeals court nominee Estrada, three on Appeals court nominee Priscilla Owen and one on appeals court nominee Bill Pryor. In October 1999, Senate Majority Leader Tom Daschle (D.-S.D.) found it "simply baffling that a senator would vote against even voting on a judicial nomination." "Daschle has now baffled himself 12 times." "In June 1998, Sen. Patrick Leahy (D.-Vt.), the ranking member on the Senate Judiciary Committee, promised to "object and fight against any filibuster on a judge, whether it is somebody I opposed or supported." "He has broken that promise 15 times." *Human Events,* week of August 11, 2003, article by Thomas L. Jipping.

That three senators on the Judiciary Committee have records of shame seems not to matter to their fellow Democrats in their bigoted and hypocritical approach to "passing judgment" on nominees. As noted previously, Senators Kennedy, Leahy and Biden clearly fail the "clean" litmus test they so rigorously attempt to place on others. Now that 'James Carville' has been cloned in the person of Sen. Charles Schumer, the "Injustice Committee" might as well close down, for little, if anything, positive will be accomplished, thanks to the machinations of Kennedy, Leahy, and Schumer, the Liberal Left's 'three stooges'. It is merely a matter of time before that committee irreparably disgraces itself or simply implodes. Perhaps it would be more appropriate if it meet only on April 1 and October 31. That the so-called Mainstream Media has said very little about this repugnant political carnival and brazen affront to our Constitution is equally disgusting as the conduct of the Liberal Left Democrats perpetrating these shames!

Why doesn't Congress have to abide by the same standards, rules, and laws that all of us across the land are required to meet? That we allow Congress to dance around or ignore laws is ridiculous. Nepotism is rampant in Washington and should be disallowed, as should the extensive practice of wives, relatives, siblings, etc., being allowed to assist or serve as 'lobbyists' This is unethical, shoddy, and irresponsible, but without the Mainstream Media's help it will continue just as it has in the past.

Why does Congress have its own Medicare and Health program which bears little resemblance to the programs the rest of us have? If members of Congress had the same programs we have they would soon improve ours. Does Congress really deserve the inflated income its members enjoy? Look at their work schedule and weeks away. Who perpetuates their automatic system of raises? Congress, of course! Why can't we taxpayers have some sunshine on what it really costs this country to maintain Congress and their respective staffs? It's a wasteful, failing 'business' that can't be justified. Their endless bickering is only self-serving, not for the good of our country. For example, why so many sub-committees? The answer is more financial opportunities! Is Congress simply creatively making work? Certainly! How can you not favor "Term limits'? Congress doesn't, but that's a strong argument for it in and of itself! When was the last time you heard a Congress person mention term limits?

Why can't each state pay for half, or at least a third, of the cost of each of its elected members of Congress, including salary, personal staff costs (Washington and in-state), official travel, health care and retirement? Don't the Representatives and Senators from each state represent the interests of their respective states, and not those of others? Would most of the present 'crop' in Congress be outraged? Absolutely, but all the more reason to have it adopted! Think of the accrued benefits to the U.S.G., all the States, and especially the voters.

Speaking of term limits, why is it set in concrete that House terms are for two years? Why not three, or possibly four? A two year term consists of at least six months as a learning and adjustment period, followed by a year or more raising funds for the next 'campaign'. How can that impact positively on the country, the state, or the voters? Make the House term a three – or four – year term; Representatives can then spend time on legislation rather than just grubbing for political dollars. Sixteen years in the House or a total of eighteen in Congress are far more than enough for any elected Congressperson, and underscores the argument for term limits. The days of perpetual office holding should be proscribed. Enforced retirement at age 70 - at 50% pay - strikes me as 100% logical. The present and expected pace of events in this and future millenniums is well beyond the scope and command of most beyond that. As it is, far too much time is spent by members of Congress away from Washington raising money, or on "junkets" or so-called "Fact Finding" trips, 90 percent of which are largely a bad joke, a waste of time and tax payer money. When family and staff are included, the trips are even more costly and simply undercut the official U.S. representation already assigned there. (Having served in an official capacity in U.S. Embassies abroad, I can personally attest to the wastefulness and fallacious value of most such trips by members of Congress.) These practices need to be eliminated or seriously reduced in scope and numbers, but won't be unless demanded by taxpayers and regularly monitored/supported by the Mainstream Media.

Similarly, by cutting out the practice of appointing 'political Ambassadors' (often based on their dollar contributions or political connections), and reorganizing the State Department's Foreign Service into a well trained, multifaceted, politically balanced, and professional entity our official presence overseas would be vastly improved. This would hopefully lead to reducing the number of underemployed and the habitually petulant liberal careerists now employed there, and pay measurable dividends in the future. Our country has been poorly served by the antiquated practice of 'paying off' well meaning but poorly qualified, deep pocketed, politically connected hacks as Ambassadors. The world is no longer a suitable environment for such apprentices. Companion to this change would be the weeding out of literally hundreds if not thousands of union protected employees that wander the halls of the State Department – and other government agencies - playing at or putting in time in their designated job descriptions. This cleansing would hold true for all governmental agencies in Washington. This will take time and much effort, but think of the eventual dividends. My God, our government might become more efficient and productive, despite the politicians.

Voters need to keep in mind that the incumbent members in Congress and those aspiring to replace them aren't trained for these positions. They originally came from varied walks of life and most were products of on-the-job-training in the world of politics, or other opportunistic, selfish, power trip, pursuits. Almost all are glib and artful in dodging questions. Most in congress were lawyers. Some are not really qualified by education, temperament, inclination or previous experience. Most are simply greedy. A few may not even have the interests of our country in mind. The days of the self-anointed pompous holding sway, or the arrogant trying to usurp the will of the voters by their obvious and selfish zeal, should be well behind us, but are not as yet. If we are not careful and attentive, our Republic may not fully recover its health, thanks to those who seem not to care or understand, or wish us harm. Such things have happened in this world before, and certainly can again. Don't the staffs in effect run Congress? Do you believe Congresspersons actually prepare their own papers? Do you have any idea what your Congresspersons actually do or how many hours they actually work, or actually cost you as a tax payer? Think about it. Have you seen any of the tax returns of your Congresspersons? Why not? They work for you!

Why doesn't the Mainstream Media provide us with a detailed break-out of the costs the taxpayer is bearing for all our Congresspersons, including their perquisites and staffs? It could certainly make for a good start towards the reforms needed. Of course it would be very helpful if the Mainstream Media included the amounts that lobbyists and other "special interest" entities provide to all members as well. Or is that asking too much?

Why can't the Liberal Left bigots and hypocrites in Congress and in the Liberal Media understand they are a distinct danger and disgrace not only to themselves, but, more importantly, to the basic structure and meaning of our government? They disgrace our founding fathers! Unfortunately, they are not alone. As noted earlier, there are many organizations in America that seek to undermine our Constitution and that which has made our country the envied beacon of freedom throughout the world. If Congress persons choose to vaguely cite "interest" groups when addressing open sessions, why not identify them by name? Anonymity? Horse hockey! These are not terribly difficult questions to answer. But inference and innuendo and anonymity are the arts of politicians and need challenging. Perhaps someone will write about Congress in depth. A suggested title? Perhaps "In Contempt of Congress"? Or "Quorum Call"? My apologies, and sincere respect, to the fifteen or so percent of the 535 members of Congress who try to serve faithfully, admirably, honestly and in a bi-partisan manner. How sad that the percentage is so upside down.

Chapter 7 addresses some of the more worrisome activist organizations in our country that need to be investigated, audited and challenged on a regular basis, for they individually and collectively, like many of our elected politicians, are distinct threats to our Republic. Do you believe the Mainstream Media have the guts or gumption to do so? ...Neither do I.

* "Senatorial Privilege, The Chappaquiddick Cover-Up", by Leo Damore, and "The Senator", by Richard Burke, relate in detail the real Ted Kennedy, including his using the extensive Kennedy influences to cover his misdeeds. Included in "The Senator" are details of many years of extensive womanizing, alcohol and drug abuses, pettiness, and weaknesses. For the uninformed, both are recommended reading.

* Philadelphia Inquirer article: "America: Who Stole the Dream?"

* The reader needs to keep in mind that the data cited in *Inside Congress* is dated 1997 and that data cited in *The Buying of the Congress* is dated 1998. It is highly probable that the financial figures and numbers have increased substantially since then. It is

unlikely that there is a single source where one can find accurate data on the costs to the taxpayer related to maintaining members of Congress, their staffs, and the total infrastructure costs of Congress. Think "staggering"!

[1] The $90,000 was in addition to the $850,000 Clinton agreed to pay Jones

CHAPTER 7

"ORGANIZATIONS"

(Remember, Look Under the Rocks!)

Although the "Cold War" ended in the late '80s and early '90s with the collapse of the "Soviet Union", the Communist and Socialist countries of this world do not wish us well. Nor do most of the Muslim countries. The basic beliefs and goals of Lenin, Marx, Engels, Stalin, Andropov, Gorbachev, "Chairman" Mao, Fidel Castro, and the Communist Party of the United States focus on undermining our way of life. Fundamentalist Muslims preach hate and death to us in their mosques. 'Front', or 'cover', or offshoot organizations continue to work quietly to undermine our Republic. A complete listing of them could easily fill the telephone book of a fair sized city. Some began in the "Cold War", others are merely updated versions, and there are more recent ones, but none of them have the best interests of our country in mind. Indeed, quite the opposite. Some are more subtle than others, and most are multifaceted in their appeals to the unwary. Many are blatant in their activities, impacting our educational, judicial, medical, manufacturing, and social interests. Given the increased actions of the radical branches of the Moslem world, it is highly likely there will be more rather than less such organizations, covert or otherwise. "One might say that these methods are too crude to deceive anybody. Yet, the sheer scope of this deception is simply too huge for many to become suspicious. "An average Western man is not prepared to detect a colossal and audacious falsification."(1)..."But the West can remain free and secure only if it possesses the intelligence and courage to defend itself - and to recognize the utter futility of appeasement and cowardice."(2)

To name a few of the more established organizations: "World Peace Council", "Educators for Social Responsibility", "Physicians for Social Responsibility", "Institute for Policy Studies", "World Peacemakers", "Riverside Church Disarmament Program", "All People's Front", "Center for National Security Studies", "Institute for Defense and Disarmament Studies", Committee for National Security", "Lawyer's Committee on Nuclear Policy", "World Policy Institute", "Foundation for National Progress", "Transnational Institute", "National Center for Policy Alternatives", "Alliance to Rebuild America", "Americas Watch", "Center for Development Policy", "Coalition for a New Foreign and Military Policy", "Center for Defense Information", "World Peacemakers", "SANE/Freeze", "Rubin Foundation", "Center for Constitutional Rights", "CISPES", "Campaign for Political Rights", "ACLU," "Citizen's Project on National Security," "ACLU Political Surveillance Project", "Fund for Peace", "National Emergency Civil Liberties", Center for Constitutional Rights", "Center for International Policy" etc.,® and those linked to Mosques. But the danger comes more from those entities operating widely and freely within our society that seek to weaken and gain power in this country than militant Moslems, such as the Baathists and Jihadists who seek our country – and the world – to capitulate. Thus this chapter deals primarily with what we now have to fear in the U.S.

"Dezinformatsia"*, is described by Stanislav Levchenko, former KGB active measures specialist, as "the most complete study of active measures ever published in the West. It identifies the main lines of Soviet overt and covert propaganda, and discloses the structure and mechanisms of Soviet influence operations against the Free World", the following ten (international) 'front' organizations are listed as worthy of more than passing concern:

1.World Peace Council (WPC) – "...focus (is) on coordination of other international fronts."

2.World Federation of Trade Unions (WFCTU) – "Principal focus is coordination of Communist trade unions worldwide, publications, training, financing, educational programs, and seeking "unity" between Communist and non-Communist unions."

3.Afro-Asian People's Solidarity Organization (AAPSO) – "Principal function is to serve as channel for (Communist) influence in the Third World."

4.World Federation of Democratic Youth – "Principal function is to support (Communist) peace campaigns through publications, congresses, and World Youth Festivals."

5.International Union of Students (IUS) – "Principal functions are quite similar to those of the WFDY."

6.International Institute for Peace (IIP) – "Closely associated with the WPC. Principal function is to direct the Forum for East-West Discussions between Scientists."

7.International Organization of Journalists – "Principal function is to support Soviet (communist/socialist) peace campaigns, human rights campaigns, and other causes through publications, conferences, and similar activities."

8.Christian Peace Conference (CPC) – "Principal function is to develop Christian and theological support for (Communist) policies."

9. Women's International Democratic Federation – "Principal activities focus on publications, meetings, and seminars in support of general (Communist) foreign policy goals."

10. International Association of Democratic Lawyers (IADL) – Principal activities are publications and conferences in support of general (Communist) foreign policy goals."

And there are others, e.g., the "American Civil Liberties Union", the "National Education Association", and the United Nations General Assembly, which are dealt with separately in this chapter, whose activities are, at best, of questionable benefit to our Republic. In a booklet compiled and published by the editors of "Human Events" in 1986, there are five pages of organizations listed as "Major Liberal and Radical Forces. For example: American Civil Liberties Union; American Federation of Labor-Congress of Industrial Organizations; Americans for Democratic Action; American Friends Service Committee; Americas Watch Committee; Center for National Security Studies; Committee for An Effective Congress; Communist Party, U.S.A.; Democratic Socialists of America; Institute for Policy Studies; Mobilization for Survival; National Education Association; Riverside Church Disarmament Program; SANE; Socialist Workers Party; U.S. Peace Council; Youth Project; etc. •

The bottom line is that there are hundreds of organizations

alive and well that intend us harm and work hard to achieve these goals. Most are well financed, much of their money comes from abroad, and substantial amounts are collected from sources across our country, although it is probable that many contributors do not know the true purpose of the organization(s) seeking donations. Some donors, such as students, may simply believe they are contributing to a worthy cause. However, the majority who contribute to or labor for these organization know their true intention. That some of these organizations hide in the shadows of 'religion' should not be surprising. For example, the World Council of Churches has been an effective force in this regard for years, but is under the guidance of the World Peace Council. And 'EMILY's LIST, a radical feminist cause, gives 100% of its money to Democrats for that party's support of "liberal" causes.

How does one determine the true purpose of an organization? This is a no-brainer; read the brochures that accompany the pitch for money; go 'on-line' and bring up the organization in question; or go to the library and do a little research. Call the Better Business Bureau and inquire, or call your newspaper, or the FBI. And if it looks like a duck and makes noises like a duck, then it very likely is a duck. Do what you will, but know what you are doing.

If you are interested in supporting organizations having the best interests of our Republic in mind, there are many that would welcome and deserve your support, including: Judicial Watch; Mid-Atlantic Legal Foundation; Heritage Foundation; Eagle Forum; Cato Institute; Accuracy in Media; American Enterprise Institute; Accuracy in Academia; Lincoln Institute; Gulf & Great Plains Legal Foundation; Mountain States Legal Foundation; National Public Interest; Young Americans for Freedom; Thomas Moore Law Center; "The National Campaign Service Fund", "Center for Individual Rights", etc.

Of immediate concern as a "clear and present danger" to the well being of our country is the American Civil Liberties Union (ACLU).

The ACLU web page (of 01/01/2004) notes that it is "Defending the Bill of Rights", and states "There is a movement afoot in America – one aimed at reclaiming freedom that some were ready to count as lost. Because of this movement, the debate over civil liberties in American has reached a turning point. On one side are increasing numbers of people like you who recognize that, especially in difficult times, it is vitally important to stand up for rights like freedom of speech, freedom of association, due process and the right to privacy. On the other side are President Bush, Attorney General Ashcroft and their allies who are not only unrepentant about the ways in which they have weakened civil liberties, but are also determined to launch a new wave of assaults on fundamental freedoms." "If they get their way, America will end the coming year not with our lost freedoms restored, but with our civil liberties in even deeper jeopardy. Rest assured that the ACLU has no intention of letting that happen!" The article then states that with the support of "over 400,000 card-carrying members' the ACLU has stood up for freedom, "but the true test of our strength and determination is still before us." "To prevail in 2004, the ACLI must escalate the rally cry for liberty." The stated goal of the ACLU was to add 20,000 new card-carrying members by the beginning of 2004, and noted "As always, our success in this critical endeavor requires the generous financial support and personal involvement of people like you."

Might we ask the ACLU what freedoms have been lost? What real rights have been lost? What liberties have been threatened? What has the ACLU been doing to help our country counter the terrorist threats? Seems to me that the ACLU has done nothing but interfere with and oppose attempts to make this country more secure, and/or to undermine our basic religious values, and/or work against our principles of patriotism. I see no ACLU codes of honor; or clarion call to help our country against those who would harm us; no support of our Bill of Rights and Constitution; no support of or respect for our flag and heritage; no honest concern for the aspirations or views of the majority of Americans. On the other hand, I do see the ACLU (and the American Lawyer's Guild) trying to trample all "they" do not agree with, regardless of the views of others, including our religions, our Constitution, our flag, our cherished history, and our traditions.

The ACLU originated in 1920 "to protect the basic rights set forth in the Constitution". The founders were Jane Addams, a women's suffrage activist and pacifist, and Roger Baldwin, a pacifist and a civil libertarian, who in 1981 was awarded the 'Medal of Freedom' by Jimmy Carter. Baldwin readily explained that "The courts are the vehicles we use to assert *our interpretation (emphasis added)* of the U.S. Constitution", reflecting that the ACLU would attack our Constitution in order to achieve its goals. Norman Thomas joined the ACLU in 1926, noting he was supportive of "expression of opinion, equality before the law, and due process." Thomas then ran for president as the Socialist Party candidate in the next six presidential elections. Doesn't this provide a clue as to the true origin and nature of the ACLU? We might also remember that Russian leader Nikita Khruschev once promised "We will bury you!" When asked if he meant militarily, he said, "No, we will destroy you from inside out."...perhaps like the ACLU?

Judge Robert Bork explained the ACLU as: "There exists in this culture a significant disjunction in attitudes. The same people and organizations manage simultaneously to adopt positions of extreme moralism and extreme moral relativism. If one had to choose one organization to illustrate this feature of modern left-liberal culture, it would the American Civil Liberties Union. Its positions resemble those of many other public interest groups, and it is the primary litigating arm of the adversary culture. The ACLU's major asset with the general public is its claim that it is a nonpartisan civil liberties organization. That claim is demonstrably false. The ACLU is in fact a highly political organization and is determined to advance an agenda through the courts that often has little or nothing to do with civil liberties found in the Constitution or our statutes."(3)

"As a list of its positions demonstrates (e.g., the ACLU opposes laws outlawing homosexual conduct, pornography, abortion, suicide, etc. and seeks to have declared unconstitutional the tax exempt status of churches and synagogues; the employment of chaplains by Congress, prisons, and the armed services; all displays of nativity scenes on public property; the singing of "Silent Night" in the classroom; and the words "under God" in the pledge of allegiance), the ACLU favors intrusive governmental action in the service of morality in some areas but insists that the Constitution mandates moral relativism in others. Reviewing *Our Endangered Rights: the ACLU report on Civil Liberties Today,* Professor Jeremy Rabkin says, "Indeed, the most revealing thing in this book is the extra-ordinary scope the ACLU now gives to the term civil liberty." "It contains an essay on "Sexual Justice,' demanding special protection for homosexuals, legal recognition of homosexual marriages, and the rejection of any distinctions between men and women in the military. Then there is a neo-Marxist essay on 'Economic Justice,' demanding income redistribution and social control of corporate investment- in the name of enhancing liberty."(4)

We need to keep in mind that: "In general, why do those who boast of their determination to transform society through government action like to regard themselves as the greatest champions of liberty?" "Attention must be paid to the apparent inconsistency of positions that seem to demand, simultaneously, governmental coercion in the service of certain moral values and individual freedom from law in the service or moral relativism., Attention must be paid, because that syndrome, advanced in litigation by the ACLU, other public interest organizations, and law professors, has entered into our constitutional jurisprudence." "The group(s) under discussion hold strongly egalitarian social views. Whether that is a genuine commitment or simply a manifestation of hostility to bourgeois attitude, the fact remains that this stance may at least suggest coherence in the ideas advanced.". "An egalitarian morality naturally produces both extreme governmental restrictions of individual liberty and, simultaneously, the demand that government recognize the "right to be let alone."(5) "The egalitarian tendency is then to coerce equality of result by law. That tendency is, of course, frequently checked by public reaction, but it remains a tendency. Law may not be used, however, to enforce moral standards that are not egalitarian. This results in what may be seen as moral relativism or the privatization of morality."(6)

If the above is not sufficiently clear, Michael Savage had this to say about the ACLU: "If I had to point to the epicenter of this illiberal attack on our culture, it would have to be the hate-filled, socialist lawyers of the American Civil Liberties Union. These vultures have so picked apart the American legal system (that) American justice has been reduced to road-kill. All that remains of our courts is a rotting carcass of what was once the envy of the world."(7)..."The truth is that they are beasts of prey living off the down-trodden, and they are responsible for much of the lack of justice in our society and for the perpetuation of our social problems. America cries out for a strong leader to arise, disband, defund, and depose these lawyer-gangsters before we all end up in a new,

Rwanda-like reality." "Just as the Taliban cloaked themselves in the religion of Islam, the ACLU cloaks itself in the false premise that it speaks for the First Amendment. It's the exact same terrorism.! The ACLU is an enemy of the American people. It is an enemy of the American way."(7)

The ACLU, "The National Lawyers Guild", and "The Association of Trial Lawyers of America", all equally leftist organizations - and most other leftist Liberals – would have us believe that there is no 'right' nor 'wrong'; no absolute truths, just moral relativism; no 'Creator'; no 'unalienable rights'; and equality would be in the eyes of the beholder. "In October 1787, James Madison wrote to (Thomas) Jefferson about the debate at the Constitutional Convention over slavery: "Some contended for an unlimited power over trade including exports as well as imports, and over slaves as well as other imports; some for such power, provided the concurrence of two thirds of the states, with an exemption of exports only. The result is seen in the Constitution. S. Carolina & Georgia were inflexible on the point of slaves." "In autobiographical notes penned in 1821, Thomas Jefferson...recalled that when the Congress was considering the Declaration (of Independence), "the clause reprobating the enslaving of the inhabitants of Africa was struck out in [compliance] to South Carolina and Georgia, who had never attempted to restrain the importation of slaves, and who on the contrary still wished to continue it."(8)

The following quote summarizes our unique and grand history. "Without the principles enunciated in the Declaration of Independence and adopted by the colonies, however, there would have been no American Revolution, no United States, no Constitution, and no war to end slavery. The Declaration is a document that celebrates liberty. Its authors and signers deserve our praise and thanks, *not* our disdain, for they put in place the ability to correct wrong and injustices so that future generations would proudly want to say the Pledge of Allegiance."(9)

To balance the past with the present, "Are you beginning to understand why the radical left-wing agenda is so dangerous and why I've said extreme liberalism is a mental disease? I hope so, because if we don't reverse the spread of this cancerous growth, America will be cooked in a few short years."(10)

It is not because the ACLU is a secular organization that is irritating, but that it has chosen to attempt to force its aberrant views on all of us is the infuriating factor. The Supreme Court rulings in the early 1960s against prayer and patriotism in public schools were wrong then and they remain wrong now. Secular humanism is not part of America's landscape, but is simply an extension of moral relativism. The ACLU national board policy #81 states in part "The ACLU believed that any program of religious indoctrination – direct or indirect – in the public schools or by use of public resources is a violation of the constitution of separation church and state and must be opposed. (The ACLU) opposes the infusion of other types of religious practices and standards into the public schools. These include such practices as baccalaureate exercises in the form of religious services, prayer meetings, athletic events, the taking of religious census of pupils, and the profession of religious observances or belief as a consideration in the evaluation and promotion of teachers. The ACLU contends that public schools which allow religion would "indoctrinate impressionable young people into an officially endorsed religion. What is tyranny, if not that?"

Excuse me. What is your definition of tyranny? May I ask which, if any, religion is being officially endorsed? And tell me what harm there is in children learning-and practicing- the Ten Commandments? What civil rights of the children are being violated? Or are you representing the socialist oriented imprint of the "Children's Defense Fund"? And how can you so misread the first amendment which clearly states the federal government cannot establish one religion for all people, and that the government should do nothing to discourage religion? The Declaration of Independence and Constitution remain the twin beacons of our liberty and freedom, and attacks on them from abroad or from within should be dealt with as though enemy forces were attempting to wrest our country from us. You appear to me to be an enemy force.

Other questions that come to mind: Who are your members? Have you been audited, and if so, by whom and when? How do you justify a tax-free status? Who decides what issues the ACLU will defend or prosecute? What 'outside' financial resources does the ACLU have (both national and international), and how do you meet your annual $45 million budget? Who are your lobbyists? Who serves on your board? What is your agenda? How do you justify your defense of NAMBLA, certainly one of the most decadent and disgusting 'groups' in the country"? For shame —on all counts. Don't respond by citing the few times you have chosen to defend obviously outrageous law suits. Instead, defend your endless transgressions. Please answer the questions, or better still, let's have the Mainstream Media do the honors. Or doesn't Media have the will?

Next we have the NEA (and the American Federation of Teachers PAC):

"An organization called the National Education Association (NEA) has wrung control of the public schools from parents and local communities and placed it in its own offices. In 1976 this teacher's union got a political payoff for supporting President Carter – the Department of Education (DOE)."(11) "This does not mean some teachers don't take an interest in the children they teach. Nor does it mean that you have to father or mother a child to be interested in its upbringing. But it is true nevertheless that many teachers today have been driven by the NEA to think more about salary and politics than about children's education. It was easier for the NEA to steer teachers to this position than it would have been to make parents think this way. Teachers don't pay for education; they are paid by it. The NEA wants to get parents out of education. While sounding platitudes about helping facilitate parent-teacher relationships, getting parent involved, and so on, the NEA has nevertheless affirmed that final jurisdiction in education belongs to teachers, and because the NEA represents teachers, it belongs to the NEA. Parents should help out when they're needed, but they should butt out when they aren't wanted."(12)

"The NEA has consistently supported the doctrine of "compelling interest" on the part of states and the federal government. This doctrine says that since states and the federal government have a direct interest in the welfare of the child, they are as qualified as the child's parent in controlling the child's education. The Department of Education was a shameless giveaway by a politician with no sense of ethics. Jimmy Carter posed as a moralist but in fact he was an opportunist. He traded a cabinet department for a special interest endorsement, and a result, in the words of one NEA official, "We became the first union with a cabinet post of our own."(13) "The NEA's plan to control education consists of making all alternatives to NEA domination either illegal or extremely impractical."(14)

For example, the resistance to charter schools and vouchers is led by NEA spokespersons and lobbyists. "The NEA has for years termed and referred to parents as the conservative New Right, a grouping that the NEA sees as a distinct threat to their power. "Private education competes with public education and makes public education better. Also private education gives parents whose views are in a minority an alternative to the public schools."(15)

The NEA was founded in 1857 to advance the cause of education, promote professional excellence, and gain recognition for the basic importance of teachers in the educational process. For more than a century it did just this, gaining widespread respect for its dignified and diligent staff and positions. "Gone is the distinction between industrial workers and teachers as a class apart. This 'fall from grace" has in large part been due to the NEA, which has transformed itself from a sedate professional association into a militant union. In 1970 NEA President Fischer said, "The NEA was identified during a Congressional debate as the second most powerful lobby in Washington" In 1972 NEA President Barrett announced her target of a $10 million political war chest to fight elections on behalf of the Democratic Party. In 1974 President Wise declared the organization's goal: "To reverse the national leadership in Washington, to build NEA's force over the next two years to the point where the presidential candidates will seek NEA endorsement. We have the power to elect the next President." "In 1976, after the elections, NEA President Ryor claimed that his groups had become "the foremost political power in the nation."(16)

NEA now engages far more with political issues ranging from nuclear freeze, to supporting abortion to legalizing marijuana to opposing school vouchers to providing funds to willing liberal politicians, etc. than to improving the education of students. How sad.

"Communism began to infiltrate the NEA in the mid 1930s, as is clear from this statement by Richard Frank in *Communist*, a Party publication: "In the effort to organize the teachers, every care must be taken to bring together in united front actions all existing teacher organizations. Special attention must be paid to secure such action with the American Association of University Professors, the National Education Association and the Guild. Our party members in these organizations must work actively that end.". "A bit earlier, in 1932, Communist organizer William Z. Foster wrote in *Towards a Soviet America,* "The schools, colleges, and universities will be coordinated and grouped under the National Department of Education and its state and local branches." As former chairman of the Communist Party USA, Foster also wrote that the DOE should be revolutionized, cleaned of religious, patriotic and other features of the bourgeois ideology. The students will be taught on the basis of Marxian dialectical materialism, internationalism, and the general ethics of the new Socialist Society."(17)

The NEA has not supported testing, but branded the process as "biased, racist, and destructive." "These days the NEA functions as a trade union, pure and simple, as bellicose and demanding as the miners, the teamsters and the longshoremen. In the whole field of public employee unionism, few outfits are as militant."(18)

"How to use power?" The (Saul) Alinsky Report says: "After each meetings make a 3x5 card indicating the person's self interest, relationship with existing institutions, relationships with other people, and what type of 'action' they have been in previously. i.e. civil rights demonstrations." "Alinsky is interested not in working with people to advance their interests, and solve problems, but in generating confrontation. Alinsky would not recommend exchanges of letter or private discussions with the superintendent as a way of building the organization. Such meeting or such correspondence might solve the problem, but they would not promote the kind of action that is exciting and what makes your people want to get involved with the organization to participate. Instead, "You must seize the development of the other side's tactics. This corresponds to the idea of training people by making the other side insult or assault your people so that they can learn what the other side is really like". Concludes Alinsky, "It is hard to deal with an enemy with whom you have a personal relationship. You should not let your people fraternize with the enemy. Distance helps to polarize the issue. The organizer must not resolve issues even though he might be able to."(19) That the NEA is interested primarily in expanding its own bureaucratic power and not with educational issues other than to pay what the NEA considers necessary lip-service and cosmetics should be crystal clear by now. No? Read on.

"We are in a battle over a philosophical question regarding the rules of society, the concepts of right versus wrong, and good versus evil. One side supports what we call the "No Rules Society" and the other supports a "Law Abiding Society". "Some people seem to think that a society with no rules is a free society. The people and groups who support this thinking would say that man is his own authority; there is no higher spiritual authority than man, there are no absolutes, no right or wrong, sex is nothing more than a biological function, and morality is whatever you may determine it is at the moment. These people argue that Freedom equals no rules and no consequences."(20)

When the ACLU and the NEA are viewed side-by-side, the threat to our Republic is more obvious. "The American Civil Liberties Union is the strongest proponent of a "no rules society in America" "When we destroy our common set of rules we no longer will have common sense. It is the lack of common values that leads federal court judges, politicians, school administrators and other leader to make decisions that are completely void of any common sense."(21) The NEA has become much like the "no rules" group in that scholastic standards have retrogressed, testing is resisted, education in general has been "dumbed-down", school boards and parents are not in control of the school curriculum, deportment is all but a forgotten requirement, and graduation requirements continue to diminish, to note a few deficiencies. The "no rules" gang seems to be winning.

"Major attempts to pass a school prayer amendment occurred in 1964 (a reaction to the Engel and Abington Supreme Court rulings), 1980, 1983, and 1993. Unfortunately, each effort resulted in failure due to the concerted opposition from liberal Congressmen and anti-religious groups like the ACLU."(22)

Would the relatively large number of school shootings that occurred these past ten years been reduced had there been deportment standards, school prayers, and/or public displays of the Ten Commandments? Not difficult to answer, and certainly interesting to reflect upon.

"While the NEA says it is for quality education it also stands firmly against efforts to check teacher competency. Since 1994, the annual conventions have passed Resolution D-19 which says: The National Education Association believes that competency testing must not be used as a condition of employment, license retention, evaluation, placement, ranking, or promotion of licensed teachers."(23) "After 56% of prospective teachers failed a Massachusetts test for certification in 1998, Governor Paul Cellucci proposed that basic literacy and subject matter tests be given to current teachers before they were certified. The *Washington Times* in a July 2 editorial said: "The teacher's unions went ballistic, accusing the Governor of "finger pointing" and "teacher bashing" and "bashing the system". The newspaper added: Given that nearly 60% of prospective teachers failed these exams the first year they were required, it's a safe bet that Massachusetts' current teaching core is littered with teachers who are failing their students."(24)

"The NEA also leads the drive of education reformers and some politicians to get control of children from birth on. Since at least 1992, the annual convention has passed Resolution B-1 which reads: The National Education Association supports early childhood of education in public schools from birth through age eight. The Association further believes that early childhood education programs should include a full continuum of services for parent and children, including child care, child development, developmentally appropriate and diversity-based curricula, special education, and appropriate bias free screening devices. The Association believes that federal legislation should be enacted to assist in organizing the implementation of fully funded early childhood education programs offered through the public schools. These programs should be available to all on an equal basis and should include mandatory kindergarten with compulsory attendance." "For at least twenty years the NEA has promoted educator control of children at ever earlier ages. (In a January 1969 article titled "Forecast for the 70s", Teachers were told to expect and work for the day when educators will assume a formal responsibility for children when they reach the age of two."(25)

If these cockamamie ideas don't worry you I suggest you check your pulse. Also disturbing are the so-called "Governor's Schools and Academies (which) reflect the prevailing philosophy in the State Education Departments which establish them. The Arkansas

Governor's School, for example, was state funded and was conducted under the authority of the seven-person State Board of Education which was appointed by then Governor Clinton. Every American and parent should take note. The same Clinton who started the Arkansas Governor's School in 1979 was responsible for education "reform" in Arkansas. After twelve years of promoting education in Arkansas, the state's schools still ranked 45th or lower, in the nation. Then Governor Clinton was also a key figure in the 1989 education summit which produced Goals 2000, School-To-Work, outcome-based education and much of the education "reform" which then swept across the nation in the 1990s."(26) Parents have protested that after attending these sessions their children returned home with no regard for parental discipline or views, as though their children had been indoctrinated in a cult, and rejecting that there is a wrong or a right. Some attendees simply walked out of the programs, rejecting that which was being presented. There were also a few cases of "student" suicides during and/or after these special high pressure sessions. Some might call this "cult", if not political, indoctrination. Again, parents, be aware.

"The leftist ideological commitment of the NEA's leadership was demonstrated by endorsements of liberal presidential candidates since at least 1980. The support has been overwhelming. For example, in 1980 the NEA board of directors endorsed the Carter-Mondale presidential ticker over Reagan-Bush by a vote of 118-4. The leadership's leftist leaning do not necessarily represent the views of many NEA members. *Forbes* magazine reported that in spite of top-down leftist control, the NEA's own figures show that one-third of its members are Republicans and well over 40% voted for Ronald Reagan and George Bush." "NEA money goes almost totally to liberal candidates. In the 1996 elections a review of the Federal Election Commission's records showed that the NEA's Political Action Committee made contributions of $1,213,230 to candidates for Congress. Of this, some $1,202,880 went to 298 Democratic candidates while just $5,350 went to 11 Republicans. In the 1994 election cycle, the NEA-PAC gave candidates $2.26 million, 98.5% of it to Democrats. The American Federation of Teachers gave $1.29 million, 99.1% of it to Democrat congressional candidates. The *Washington Times* in an editorial titled "Educators or Lobbyists?" said: Even the $3.5 million donated to congressional candidates by the two national teacher association PAC's is dwarfed by the total contributions from the NEA's state and local affiliates."(27)

The NEA also developed a manual titled *How to Win: A practical Guide for Defeating the Radical Right in Your Community.* According to it, "all those who have a Christian World View – a view which holds that Biblical principles extend to every area of life and not just the matter of how to get to heaven someday - are enemies who must be stopped. Other education organizations which joined the NEA in sponsoring the *How to Win* manual included the National Associations of both Elementary and Secondary School Principals, the National Council of Teachers of English, The National Center for Science Education, the American Library Association, and the National PTA. Citizen protests labeled "extremist" or "radical" often result when two resolutions which NEA convention endorse annually are applied to local situations. Resolution E-9 demands total "academic freedom" for teachers. Resolution I-22 says it supports "freedom of expression in the creative arts and deplores any efforts to suppress, directly or indirectly, such expression."(28)

Of course, the "radicals" and "extremists" are those parents who want to know what is being taught in classes and that it is of benefit to their children. The NEA also labels those who do not support public schools and the NEA programs as the "New Right". Never mind the declining grades, the increasing number of functional illiterates, the failure rates, the drop-outs, the "tinkering" with testing and graduation requirements, and the obvious dumbing-down process that continues yearly. Yet the NEA doesn't understand our complaints?

"Over the years as the NEA developed its left-leaning, anti-traditional values political machine, the education of American children has suffered a tragic decline. Headlines have told that scores on the SAT and ACT college entrance exams were dropping. Functional illiteracy in cities has reached epidemic levels. The crisis in education has steadily worsened. The solution the NEA and it political supporters offer regularly is "Give us more money."(29) Karl Marx has to be one happy "Dude" as he views us today. He is winning!

"In 2002, the Josephson Institute of Ethics issued an update entitled "the Ethics of American Youth" The California think tank surveyed twelve thousand American high school students and found a serious deterioration in moral conduct over the past ten years.

Cheating: in the decade from 1992 to 2001, the number of high school student who admitted cheating on an exam increased from 61 percent to 74 percent – three quarters of the high school population!

Theft: 38 percent of high schoolers admitted shoplifting at least once in the previous twelve months; that was up from 33 percent in 1992.

Lying: An astounding 93 percent of teenagers admitted lying to their parents. In 1992, the number stood at 83 percent."(30)

"And in the rejection of vouchers, which would allow low-income American parents to choose schools based not only on educational standards but also upon the moral atmosphere, the politicians have sold out our most vulnerable families. Think about it: Are the anti-voucher people looking out for the kids? Or for the entrenched public school teachers unions and administrations? Roughly two-thirds of black American parents favor school vouchers. What say you, Bill Clinton, our nation's "first black president"? What say you, Edward Kennedy, often described as the champion of the poor?"(31) What say you Hillary, and your fellow CDFers?

"In April 1983, The National Commission on Excellence in Education issued its now historic report, *A Nation At Risk.* The (18-month study) said: The educational foundations of our society are presently being eroded by a rising tide of mediocrity that threatens our very future as a nation and as a people. The Presidential Commission added these chilling words: If an unfriendly foreign power had attempted to impose on America the mediocre educational performance that exists today, we might well have viewed it as an act of war. As it stands, we have allowed this to happen to ourselves." "What's been done since 1983? Multiplied billions in federal, state and local tax dollars have been poured into education, but in 1997 Brunno V. Manno, senior fellow at the Hudson Institute, said: "The fundamental indictment that came out the [Nation at Risk] report was that kids weren't learning enough. Fifteen years later we're still saying that kids in this country aren't learning enough and their academic performance is still mediocre." Assessments made in the intervening years confirmed Manno's charge. During those years Americans have become accustomed to seeing on-going headlines saying: City Schools Get Poor Grades In Survey; SAT Scores Fall To New Low; College Test Scores Lowest Since Mid 1950s; Students Today Have Vocabulary Problems, Educator Says; Schools Brainwashing Our Children; U.S. Students Lag In Math, Science."(32)

"Recent results from the National Assessment of Education Progress (NAEP) science exams showed that 40% of high school seniors and one-fourth of all 4th and 8th graders didn't meet minimum standards. Syndicated columnist James Kilpatrick reported: Recently, 7,812 11th graders took a test in history and literature. Of all the answers to the 141 questions on history, only 54.5 percent were correct. The 17- year olds demonstrated no sense of historical chronology. Almost one-fourth thought the Civil War was waged before 1800. One third thought Columbus had discovered the New World after 1750. Three or four could not place Lincoln within 20 years. Forty-three per cent guessed wrong on the half century that included World War I. The results in literature were even more depressing." "A 1997 op-ed article in the *St. Lois Dispatch* carried the bold headline, "Don't Fill College Classrooms With Unprepared Students." Paul Shore, assistant professor of education and American studies at St. Louis University, wrote: "Currently, literally million of American high school students are not receiving adequate preparation for college level work. Not only do many of these students lack the reading and math skills needed to succeed in college, they also have not learned the habits of self-discipline or study techniques that enable college student to do well. Before you discount this claim as just another complaint from a cranky college professor, consider how many colleges have already instituted remedial writing programs for entering freshmen."(33)

"President Clinton "pledged that if reelected he would work to insure that every student could read *by the time he finishes third grade!* Year after year, home schooled students and those enrolled in Christian schools which start training children with intensive systematic phonics consistently see them reading by January or

February of kindergarten year." John Agresto, president of St. Johns College in Santa Fe, New Mexico: "We've turned colleges into high schools. After the babysitting of high school, we've asked colleges to give students the education they should have gotten. This is disastrous for colleges. I can't teach kids physics, biology, and the history of the Roman Empire if I have to teach them the difference between "their" and "there". We get students – even here – who say, "What difference does it make if I spelled it wrong? You know what I meant!"(34)

"*U.S. News and World Report* columnist John Leo gave some insight on why teacher education colleges "routinely graduate demonstrable idiots" "In his August 3, 1998 column, Leo listed courses given to prospective teachers at the university of Massachusetts-Amherst. The teacher preparation courses include: "Leadership in Changing Times, four courses in "Social Diversity in Education," "Embracing Diversity," Diversity & Change," "Oppression & Education," "Introduction to Multicultural Education," "Black Identity," "Classism," "Racism," "Sexism," "Jewish Oppression," "Lesbian/Gay/Bisexual Oppression," "Oppression of the Disabled," and "Erroneous Beliefs." "In his column, Leo said of the college graduates who failed the grade-school-level test: "Apparently they went into ed school without knowing much about anything, then came out the same way. But at least they are prepared to drill children in separatism, oppression and erroneous beliefs."(35)

.....Laura Ingram notes: "Liberals can make great educators – as long as they keep their politics out of the classrooms and lecture halls. The problem is, they don't. Since the 1960s, they've turned our colleges and schools – many of which are funded by our tax money – into an elite-run fiefdom of political correctness and anti-Americanism.. But the elite Edu- Mafia is liberal/leftist to the core. The Ivory Tower is Red in tooth and claw."(36) "It's well known that in the 1960s, leftists conquered the Academy. They're still in charge. The faculties of virtually all colleges are overwhelmingly leftist/liberal Democrat, as survey after survey bears out."(37)

For example:

• "John McGinnis (of Northwestern University) and Matthew Schwartz (Columbia) reviewed all federal campaign contribution over $200 by professors at the country's top twenty-two law schools from 1994 to 2000. Of professors who contributed, only 16 percent donated to the GOP compared with 74 percent to the Democrats. At Yale, for instance, almost 50 percent of profs donated, and 95 percent gave primarily to the Democrats. At Georgetown, in the six years studied, the Democrats received $180,000, the Republicans to $2,000, and $1,500 went to the Greens."

• "At Ithaca College, according to Mark Finkelstein (chairman of the Tompkins County, New York, Republican Party) and Michelle Meredith (chairman of the Ithaca College Republicans) in a study published in www.frontpagemag.com, there are 117 registered Democrats or Greens (93.6 percent) versus only eight registered Republicans or Conservatives (6.4 percent). The Department of Politics has eight Democrats and no Republicans. The Department of History has seven Democrats and not a single Republican. The English Department has thirteen Democrats and no Republicans. The same fat "0" score for Republicans or conservatives goes for the departments of Psychology, Sociology, and Environmental Studies. We ought to be grateful that there's one Republican in the Department of writing (twenty-five Democrats). Even in the Department of Business, thirteen professors were Democrats and just two were Republicans."(38)

"The leftists of academia aren't the centrist, fair-minded liberals who populated the universities until the 1960s; they are the radicals who chanted "Hey, hey, LBJ, how many kids did you kill today?" and purged the centrist liberals, let alone conservatives. Now they run most colleges and universities with a rod of iron. Today, the Ivory Tower elites us their privileged – and unaccountable – positions to propagate extremist views designed to transform society into their own image."(39)

"The "tenured radicals" of our universities have little interest in actually enlightening students. And it shows. The National Association of Scholars, a higher-education reform group, conducted a survey in late 2002 that found that college seniors have slightly less cultural knowledge than *high school* graduates of the 1950s (i.e., before the radicals took charge.) On average, on a battery of 15 questions assessing general cultural knowledge, today's college senior gave 53.5 percent correct responses. In the 1950, high school graduates gave 54.5 percent correct answers to similar questions and college senior scored 77 percent."(40)

In an October/November, 2003, article in The American Enterprise by Lewis Andrews, executive director of the Yankee Public Institute for Public Policy in Hartford, Connecticut, he noted that: "According to the National Center for Education Statistics, per pupil spending in religious and independent school averages $4,600, versus $6,857 in public schools; (this) implies that there may be more than $100 billion in unnecessary spending for public schools. Reduce that, and the state and local budget deficits evaporate. *There would be no budget crisis* were public schools operating with the same efficiency as private schools." "Only half the people on the District of Columbia public school payroll are teachers. Fully 40 percent of per pupil spending in California goes to bureaucracy. Nationwide, public schools spend more than twice what other industrialized countries spend on administration." "In some public school districts, annual per pupil costs run five to ten times the cost of a local parochial school. Teachers' unions are well aware of these price advantages, and ferociously oppose any effort to take advantage of them. As a defensive measure, the American Federation of Teachers has launched a campaign to discredit the financial appeal of school choice." "More importantly, the graduates of private schools *at all tuition levels* score consistently higher on objective achievements tests and go on to college in significantly higher numbers."

In September 1, 2003 Human Events article, Robert Holland, Senior Fellow, Lexington Institute of Arlington, Virginia wrote: "Phyllis Schlafly accurately reports "NEA Gears Up To Elect Democrats," that the National Education Association brass declared its hostile intention toward No Child Left Behind (NCLB) at its recent New Orleans convention." "What bothers the teacher union bosses is the NCLB for the first time puts the power of the federal government on the side of parents instead of the education establishment. If children are stuck in chronically failing government-run schools, parents are entitled to move their children to a better performing public school (if one can be found) or to use their federal subsidy to purchase the services of a private contractor" "The NEA leaders are targeting those and other NCLB provision because they sense their monopoly to be shaky and in danger of failing. About that, they are right.".

Michael Savage, not one to obfuscate an issue, has written: "In the Democrat's America, Justice is not only blindfolded, she has been bound and prostituted. Justice has become a perverted masquerade where those in power are innocent even when proven guilty and where average Americans are guilty – with no money to prove their innocence. Their rights are mocked, and their values are the brunt of filthy humor." "A leading Democrat recently claimed before a radical "gay" group that the greatest problem today is hatred. But he himself is the source of this hatred. He and the others have done more than anyone to promulgate hostility and vision among the groups of our society. They push hatred, envy, and suspicion wherever they go, while feigning compassion and understanding." "The truth is, the Democrats are still up to their politics of personal destruction. Behind the smiles, finger wagging, and promises, there are only arrogance, corruption, and deceit." "This seismic shift in education is not an accident. It's a core strategy of liberalism. The libs know if it can inculcate young minds with their views of the world, they can change the direction of the American culture in. as little as one generation. In Plato's *Republic,* the philosopher warns of this brainwashing of youth. In classic Greek fashion, he first asks a question: "Shall we just carelessly allow children to hear any casual tales which may be devised by casual persons, and to receive into their minds for the most part the very opposite of those which we should wish them to have when they are grown up?" "Plato answers his own question this way: "We cannot. Anything received into the mind at that age is likely to become indelible and unalterable; and therefore it is most important that the tales which the young first hear should be models of virtuous thoughts. There can be no nobler training than that.". "Today, the Commu-Nazis in the liberal dominated education system resist such distinctions for fear that their socialist agenda would be exposed for what is: a fraud. A dangerous lie. Pure socialist propaganda. And a radical grab to redefine the American culture."(41)

Hopefully, you have the picture by now.

The efforts by the NEA, the ACLU, and the Liberal Leftist faculty members across our nation's colleges/universities have tried to create what would clearly be a new social order. Fortunately, they have failed to date, but not for want of trying. Although 86% of U.S. residents embrace and support religion, including school prayers, the Pledge of Allegiance, the Boy Scouts, our Flag, the Bill of Rights, and our Constitution, meaningful education for our youth, etc., it remains difficult to combat those who would have us a Socialist country, for both our Mainstream Media and our politicians have failed to recognize these threats to our Republic. Those working to undermine our treasures will not cease their efforts until they are declared illegal, outlawed or brought to reform. Unfortunately, the struggle for "Marxing" our youth extends onto nearly every college and university campus in the U.S.

In the week of September 29, 2003 Human Events article entitled "Dismal School Performances, Thomas Sowell wrote: "Some of the most heavily financed schools are doing miserably. Even after spending $17,000 per pupil, Cambridge, Mass., was still left with a huge gap between the test scores of its black and white students. In fact, black students in Cambridge scored lower than other black students in nearby communities with less than half as much spending per pupil." Those who believe that money is the answer are not going to be stopped by anything so mundane as facts. To many in politics and in the media – and to most in the teachers unions – "improving" the schools means spending more money on them. But what is called "investing" in better education could more accurately be called pouring money down a bottomless pit.

The NEA sees it differently. In its website (www.nea.orgesea/) article 'No Child Left Behind' Act/ESEA that I read on January 7, 2004, the NEA said: "The No Child Left Behind (NCLB) Act of 2001 presents real obstacles to helping students and strengthening public schools because it focuses on:

- punishments rather than assistance
- mandates rather than support for effective programs
- privatization rather than teacher-led, family oriented solutions

The Solution—

NEA will support that commitment (of the legislation's goals) through an unprecedented public awareness, legislative lobbying, and member empowerment campaign – working for the things children need to be successful, (and to) Provide parents and policymakers information that helps get at the causes of school failure, not just the effects." What malarkey! NEA is interested only in retaining and expanding its power, nothing else.

You may be less than convinced of the threats to our educational process, be they collective or individual. 'After all', some might say, 'what clear-headed American youth could be 'persuaded' on today's public school or college campus? Certainly our "students" aren't sheep! The "Cold war" is over. The Soviet Union is history. This is America!' Well, just for the hell of it, let's take a brief look at some our college and university campuses, starting with "The Dirty Dozen" as compiled by the Young America's Foundation as reported by NewsMax.com:

• "Students at Harvard University can take Feminist Biblical Interpretation, a class that will concentrate on "the significance of feminist Hermeneutics for contemporary theological reflection and education for ministry

• At Carnegie Mellon University, students will study Sex and Death. This course will address the question of "Whether we need to liberate death now that we have figured sex out."

• Brown University offer Black Lavender: A study of Black Gay and Lesbian plays and Dramatic Construction in the American Theatre. The course is described as "an interdisciplinary approach to the study of plays that address the identities and issues of black gay men and lesbians".

• Women's Studies students at Bowden can take Music and Gender. The main question addressed in the course: "Is Beethoven's ninth symphony a marvel of abstract architecture, culminating a gender-free paean to human solidarity, or does it model the process of rape?"

• The University of Texas offers Race and Sport in African American Life. The class will explore "how sports have been used to justify and promote antiquated, eugenic, and ultimately racist notion of blackness."

• At the University of Virginia, students can enroll in Marxism: What is to Be Learned From It? The description declares that Marx's work is the "standard against which all subsequent social thought must be judged" and that "it is worth devoting an entire semester to it"

• Students of Latino-Latin American Studies at Syracuse University will study Gay and Lesbian Caribbean Literature. These students will read "Caribbean poetry and short stories in Homosexual literature."

• Georgetown students can enroll in The Bible and Horror. The description states that the Bible "can be a scary book" and that it "often reads more like horror than religious literature." The class will examine the question of "What might religion and horror have in common?"

• Literature students at Cornell University can study Bodies Politic: Queer Theory and Literature of the Body. This course discusses such questions as "How do concepts of perversion and degeneration haunt the idea of the social body?" and "How are individual bodies stigmatized, encoded, and read within the social sphere?"

• Middlebury College offers Images of Blackness and Whiteness in American Literature in which students will study how American authors in the 20th century "construct images of blackness and whiteness" and examine how "culturally coded images both shape and are shaped by American attitudes about racial identity."

• Students at UCLA can enroll in Death, Suicide, and Trauma. Students will study "Definition and taxonomy of death; role of individual in his own demise; modes of death; development of ideas of death through life; partial death, megadeath; lethally psychologically autopsy; death of institutions and cultures."

• Bucknell University's sociology students will take Women and Social Inequality. They will explore "recent anthropological, historical, and sociological literature on gender inequality."

I have little doubt that those completing such challenging courses will make the adults paying for them proud and appreciative of the enlightenment offered. How did those attending school in decades past ever manage to succeed without such mentally rewarding exposures?

But there is more:

• California Democrat (former San Francisco Mayor) Willie Brown urged students to harass opponents of affirmative action. Brown told students to "enroll in [California affirmative action's chief opponent, Glynn Custred's} class, and believe me if you treat him correctly, he should really need therapy. You ought to do what you do best to terrorize professors you don't like, and I guarantee he will be a basket case by the end of the term. "

• An August 1998 article by John Leo, a senior editor with

U.S. News & World Report, included the following: "Many universities offer courses on television shows. The University of Wisconsin has one on soap operas, and Purdue offers one called "The Biology of ER." Other current or recent course include "Issues in rock Music and Black Culture" (Columbia); "Juggling" (Duke); "Star Trek", California State at Chico); "Vampires: The Undead," (University of Pennsylvania); "The Slavic Vampire," (University of Chicago), etc.

• From the September 1999 issue of "Campus Report, in an article entitled "The Dirty Dozen: The Worst College Courses in America" (10 are quoted below.) "Oberlin: Feminist Criticism of Shakespeare – "Student will explore how current feminist critics engage in critiques, resistant readings, and reinterpretation of these plays. Intersections of gender with other forms of social difference will emphasize how gender must be understood in relation to history, class, ethnicity, sexuality, and other significant social categories." UC Berkeley: Lesbian and Gay Detective Fiction – "We will be looking at the ways in which the outlaw sexuality of the protagonists plays out against the outlaw activities of the antagonists; at the social issues that tend to appear in gay and lesbian fiction; and at more mundane concerns, such as structure, plot, character, and style."

• Bucknell: Green Utopias - "Introduction to literary utopias and to the cultural writings of various ecological movements offering alternative concepts to the increasing destruction of nature."

• U. of Northern Arizona: Women, Gender Identity, and Ethnicity – "This course focuses on the experiences of women and the significance of gender in society. Topics will include gender, socialization, idealized female beauty and the body, violence, work and welfare, family, communication, witches, and fairytales."

• Indiana: Star trek and Religion - "We will explore two related areas of religious thinking in Trek. First, we will read critics who hope for the demise of religion, then explore those who think religion can be re-imagined in mystical or cosmic terms."

• San Francisco State: Kemet, Afrocentricity and the Dawn of Science – "Emphasis on the structured whole, the Nubian Vortex, Cartesian vs. Alchemical cosmologies; and the Afrocentric thought and achievement of African American scientist and engineers."

• U. of Pittsburgh: Readings in Critical Theory: Psychomarxism

– "Special attention will be given to Lacan's Seminar XVII of 1969-70, which deals with the relations between psychoanalytic theory and Marxism in the wake of May'68 and French Maoism."

• Michigan: Issues in Afro-American Development, Politics, Economics, and Development: Affirmative Action - "There is great concern that all the rights gained in the sixties are now being eroded by legal challenges to affirmative action rules. Indeed there is a hue and cry that there is now reverse discrimination and that preferential treatment is illegal. Some attention will be paid to Justice Clarence Thomas and Mr. Ward Connerly, two major figures against affirmative action. The objective is to begin the process of cogent action and to develop the language to articulate affirmative action as a right and not a benefit."

• U. of Pennsylvania: Feminist Revisions of Fairy Tales and Myths – "This course addresses feminist critiques and rewritings of patriarchal mythologies. The popular culture form of the fairy tale, in particular, has been a site of extensive feminist revision."

• UCLA: Gay and Lesbian Perspectives in Pop Music –"Survey of English-language popular music in the 20th century, with focus on lesbians, gay men, and members of other sexual minorities as creators, performer, and audience members."

Add to the above that half of Ivy League schools ban ROTC programs; a Marine recruiter seeking to sign-up law students for the judge advocate general program was shouted down last year by a group of faculty and administrators at the University of Nevada, Las Vegas; students, faculty, and staff at Hampshire College voted overwhelmingly to condemn the U.S. government's war on terrorism; the faculty of Washington University voted to officially discriminate against soldiers, sailors, airmen and Marines (Accuracy in Academia, April 2002); A professor at Rutgers said of September 11 (2001),"its ultimate cause is the fascism of U.S. foreign policy over the past decades." and a Purdue professor said "there is no ground to be certain that America and Israel weren't behind the 9/11 attacks." (The Heritage Foundation, February 7, 2003); "A course given by the U. of Connecticut's Dept. of History is titled "Hip-Hop, politics and Popular Culture in the Late 20th (century)." The stated objective is to "discuss, at a college-level proficiency, contributions of various artists on hip-hop and the significance of the art in the U.S. and abroad." and another syllabus, for an Afro-

American Studies course at Harvard, "Hip-Hop America: Power, Politics and the World," introduces scholars to the Hip-Hop Idealogy, i.e. "Representin, (sic)' , 'Coming' Correct, and Keep Real." and at Penn State, students were required to attend a "Mos Def Concert" and write up some Hip-Hop Literacies from the concert."(sic). (From Campus Report, February 2003)

With course loads like these, indeed "We have nothing to fear but fear itself". Both FDR and John Kennedy ("Ask not what your country can do for you, but what you can do for your country") must be asking one another "What in God's name went wrong?"

The answers to that question are self-evident, beginning with the influences of the Liberal Left, or worse, in our education system. From a column by nationally syndicated columnist Larry Elder in the February 2003 Campus Report:

- UCLA – of the 31 English professors with a registered political affiliation, 29 belong to a party of the left. Of the 56 history professors, 53 belong to party of the left. Of 13 journalism professors with an affiliation, 12 belong to a party of the left. Of 17 political science professors with a registration, 16 belong to party of the left. And of the 33 women's studies professors, 31 belong to a party of the left.
- Cornell – of the 12 anthropology professors, 11 are registered to a party of the left. Of 13 economics professors, 10 are with a party of the left. Also on the left, 35 out of 36 English professors. All 29 out of 29 history teachers are registered with a party of the left. Of the 17 political science teachers, 16 registered to the left. Psychology professors totaled 25 to the left out of 26. Sociology managed a registered 7 to the left out of 7; Women's studies was 33 to 0 to the left.
- Stanford – Anthropology, 15 of 16 to the left. Economics, 21 of 28. Of English, 31 of 33. History, 22 of 24. Political Science, 26 of 30. Psychology, 20 of 20. Sociology, 11 of 12. Women's studies, 5 of 5.
- A column by Christopher Chow in the October 2002 Campus Report noted that:

"Liberals outnumber conservatives 18 to one at Brown University. At Cornell, the number is even higher, with liberals outnumbering conservatives more than 26 times. Penn State displayed a bit more balance, with the ratio of liberals to conservatives being six to one. Even the smallest disparity, at the University of Houston, had a ration of three liberals to one conservative.

• Of the 166 professors examined at Cornell University, only six were conservative, with no conservatives at all in the fields of history and sociology. There were likewise no conservatives in these fields at Brown University.

• Some of the largest disparities were found in the University of California system. UCLA, for instance, has only 9 conservatives for 141 liberals. UC-Santa Barbara had only one conservative professor in the 73 asked. At the four UC schools surveyed, there were only five conservative political science professors compared to 90 liberals, and only seven of the 66 professors noted were conservatives, with none in the department of sociology, Berkeley.

• Extreme liberal bias was found even in states generally thought to be conservative, such as Colorado, with liberal professors outnumbering conservative professors by more than 23 to one.

It is not really surprising that at least 84 percent of college professors voted for Gore, and only 9 percent for George Bush. The sad part of this equation is that many college graduates go into government, or the media, or into other public oriented positions, including "teaching", bringing their liberally formed views with them. At the rate of graduating about 14 million every four years, the need to balance the scales is obvious. This, however, is not going to be easy, with the NEA 'controlling' the public school system as it has for nearly three decades, and the ridiculous imbalance at the college level between liberal and conservative professors. Serious and sustained parental involvement is the most logical answer for the immediate, with constant attention to this predicament by the Mainstream Media to achieve the needed long term antidotes. For example, why would any parent want to spend multiple thousands of dollars on such courses as noted above? Why would any college or university worthy of the name allow such courses to be offered? It is time to clean out the stables. Hopefully those parents who are products of the diminished education system of the past two plus decades can agree. A third rate nation? Look on yon horizon.

To put an exclamation point on the above sentence, Don Kaul, a former columnist with the Tribune Media Services, wrote a story last year titled "Life, liberty and the pursuit of ignorance", noting that a Washington group of conservatives dedicated to the promotion of liberal-arts study gave an American history exam to 556 randomly selected *seniors* at 55 universities and colleges of the better sort, and found out that the kids, for the most part, didn't know diddly-squat about American history. The average score on this 34 question, multiple-choice test was 53 percent, and only one student got them all right. This despite the fact that the questions were generously described as "high school level". (In the 1940s, they might have been asked of eighth-graders to determine whether they were ready for high school.) For example: "When was the Civil War?" And the choices given were 1750-1800, 1800-1850, 1850-1900, and 1900-1950 and after 1950. Sixty percent of the students got that one right. In other words, 40 percent of these hot-shots from elite universities and colleges could not place the American Civil War in the correct half-century. It gets worse. "Who was the American General at Yorktown?" the test inquired, the choices being William T. Sherman, Ulysses S. Grant, Douglas Mac Arthur and George Washington. Only 34 percent answered Washington, the only Revolutionary war general on the list. Two-thirds of the grads apparently didn't know Yorktown was a crucial battle of that war. You want to leave a full 70 percent of our Best and Brightest behind? Ask them which piece of social legislation was passed as part of Lyndon B. Johnson's Great Society program: the Sherman Antitrust Act, the Voting Rights Act, the Tennessee Valley Authority, or the Civilian Conservation Corps. Not knowing it was the Voting Rights Act means that 70 percent of graduates in our nation's best schools have virtually no knowledge of the history of perhaps the single most important social movement of the past half century. The two highest scores were achieved on questions that required knowing that Beavis and Butthead are television cartoon characters and that Snoopy Doggy Dog is a rap singer. Snoop got 98 percent and B&B got 99. This is the generation that's going to be running the country in 20 or 30 years, How can you run something when you don't know anything about its history? Democracy requires the intelligent participation of people who know something about the issues, who can place facts in historical context and make some sense of them; and you should know a little bit about how we got to where we are. The fact that we seem to be turning out a generation of youngsters who lack even that little bit is a national scandal – and tragedy." The redundancy of citing how sorry our education system has become is intentional – no apologies.

A final example of what has happened to it comes from an article by C. Bradley Thompson, chairman of the Department of History and Political Science at Ashland University and a senior writer for the Ayn Rand Institute that appeared in the February 18, 2003 issue of the Las Vegas Review Journal:

"This past year the U.S. Department of Education released its History Report Card and the results were predictably awful: 57 percent of high school seniors flunked even a basic knowledge of American history and 10 percent tested at grade level. What is less obvious – and more dangerous – is that the history they do know is utterly subversive of American culture and values. I recently attended the annual meeting of the American Historical Association, the nation's largest and most influential organization of academic historians was shocked by what I saw and heard." "Of the roughly 200 panels, there was virtually nothing on subjects such as the American Revolution, the Civil War, or America's involvement in the two World Wars. Instead, there were dozens of papers on subjects ranging from the banal to the bizarre and perverse. Participants were subjected to scintillating presentation on topics such as "Meditations on a Coffee Pot: Visual Culture and Spanish America, 1520-1820," or "The Joys of Cooking: Ideologies of Housework in Early Modern England," or "Body, Body, Burning Bright: Cremation in Victorian America."

"But without question the dominant theme of the conference was sex. Historians at America's best universities are obsessed with it."

"One historian from an Ivy League college delivered a paper on "Strong Hard Filth' and the "Aroma of Washington Square': Art, Homosexual Life, and Postal Censorship in the Ulysses Obscenity Trial of 1921." Another scholar from Berkeley spoke on "Solitary Self/Solitary sex." And one spoke on "Constructing Masculinity: Homosexual Sodomy, Ethnicity, and the Politics of Penetrative Manhood in Early Modern Spain."

"Academic history has become thoroughly egalitarian. It seeks to elevate the history of ordinary men and women doing ordinary things at the expense of great men and women doing great things.

Thus, the history department at Harvard University no longer offers a course on the American Revolution. In its place, it now offers a course on the history of midwives and quilting."

"Worse yet, mainstream historians are driven by a pernicious political agenda that seeks to elevate "group rights" over individual rights. But what of America's founding ideals, such as the principle of inalienable individual rights? There was a time, not too long ago, when students were required to study the great events, the magnanimous statesmen, the brave warriors, the brilliant inventors, and the ingenious industrialist of American history. There was a time when American students knew in intimate detail the heroic story of the American Revolution and the tragedy of the Civil War."

"American children once learned about honesty from George Washington, justice from Thomas Jefferson, integrity from John Adams, independence from Daniel Boone, oratory from Daniel Webster, ingenuity from Thomas Edison, perseverance from the Wright brothers, and courage from Sergeant York. They memorized and learned the principles of the Declaration of Independence, the Constitution, and the Gettysburg address. The history of America was the history of freedom."

"Today, our children are being taught to be ashamed of America. By denigrating the principles and great deeds of America's past and dethroning its heroes, today's college professors are destroying in our youth the proper reverence for the ideals this nation stands for. And a nation that hates itself cannot last."

We all badly need to remember these words and urge other to do so also.

I quoted most of Thompson's excellent article for the simple reason that it puts in perspective – and balance – what is basically wrong with our educational system, and, by extension, the sorry roles that the NEA and the campus liberals have played in taking what once was a system worthy of extensive praise and pride down to the shameful depths where it now resides. The Statue of Liberty may well shed tears, along with those from our extraordinary leaders commemorated on Mt. Rushmore, and those that framed and signed our Constitution and Bill or Rights. Our Congress? No heroes there; just greed and ostriches.

Were the ACLU, the college campuses, the NEA - and their affiliates - the only problems, it is quite likely that with a cooperative Media we might be able to set about making the orderly changes that are necessary to return this country to what it should be. Unfortunately, there are additional "union related" problems that need addressing as well. In a July 20, 2001 "Orange County Register:" *AP* article by Larry Margasak and John Solimon datelined Washington, they reported: "Documents that the Democratic Party and unions have sued to keep secret reveal a campaign strategy in which labor and party officials served side-by-side on committees that directed the Democrat's election activities in each state." ..."When the DNC and its National partners, including the AFL-CIO and the NEA (National Education Association) agree on the contents of a plan, each national partner will give their funding commitment to the state," and internal DNC memo titled "Rules of Engagement" said." "The documents detail extensive discussions between labor and party leaders on how to contact, register and influence voters to support Democrats, and they show where union in some instances got their money to accomplish the mission." "John Hiatt, AFL-CIO general counsel, acknowledged the union had veto power over Democratic activities it helped finance." "In North Carolina, the documents show, state AFL-CIO President Chris Scott and North Carolina NEA President John Wilson each served on the committee that handled day-to-day operations." "In Nebraska, the state party gave AFL-CIO and teachers union officials similar positions on its executive committee alongside official from Ben Nelson's Senate campaign and other candidates." Any questions?? Read on —

In a June 2002 NewsMax.com article by veteran broadcaster Wes Vernon: "For the 1999-2000 election cycle, the latest period for which the figures are complete, public employee unions spent just over $11 million in political contribution. Well over 90 percent of that money went to the Democrats." "The biggest players in the government employee political game are the teachers' union. Political Money Line figures show that between the National Education (NEA), the oldest teachers union, and its rival, the American Federation of Teachers (AFT), spent over $3.25 million in political donations during the two year leading up to the big election of 2000. Over 95 percent of that money went to Democrats.

Second in total political clout is the American Federation of State, County and Municipal Employees (AFSCME) at nearly $2.5 million. This union, it should be noted, has a history of leadership that has been candidly socialist in outlook. Close to 97 percent of its political largesse ended up in Democratic coffers." "What about a Hatch Act for (all) government unions and corporations that benefit from government contracts? The new idea is from one-time left-winger David Horowitz (who) notes that both government unions and government corporate suppliers have a conflict of interest in pushing political candidates who have to pay them back with government funds once they are elected."

Has anyone heard anything further on this excellent proposal? No is the answer, and unless the Mainstream Media pushes there is little likelihood you will, sad to say.

An *AP* article by Leigh Stroop datelined New York, titled "AFL-CIO ups dues to boost election fund" in the May 23, 2002 Las Vegas Review Journal said: "The AFL-CIO leadership, looking to November elections when control of Congress will be at stake, approved a member dues increase Wednesday that will bring $25 million into labor's war chest. The increase of 4 cents a month, voted by the federation's general board, will apply to members of all 66 unions that make up the AFL-CIO." "What this money does essentially is allow working families to compete in a political process that they would otherwise be completely shut out of," said Steve Rosenthal, the AFL-CIO's political director. "There is no other voice for workers in the legislative and political process." (Might I ask what do your many lobbyists do in Washington? If they don't provide a color-me-green voice in the process why are they there? Mr. Rosenthal, I think you think us all very gullible) "The $33 million the AFL-CIO plans to spend this election cycle is about the same that it spent in 1998 in the last mid-term elections, Rosenthal said. In the 2000 presidential election cycle, the federation spent about 42 million."

Stefan Gleason, vice president of the National Right to Work Legal Defense Foundation, in a September 3, 2003 article carried by Knight Rider/Tribune News Service, wrote: "Throughout the country, unchecked union coercive power has wreaked havoc in the lives of countless workers who have voiced dissent and therefore faced harassment, firings and even vicious, bloody beatings." "Meanwhile, Big Labor bosses on the board of directors of the Union Labor Life Insurance Co. (ULLICO), perpetrated an Enron-style insider trading scam that lined the pockets of Carpenters" said union chief Doug McCarron, among others, to the tune of nearly $300,000, money from the pension savings of rank-and-file workers. Other ULLICO board members, such as William Bernard of the Frost Insulators union, pulled well over $1 million in tainted profits." "Since 1975, the National Institute for Labor Relations Research has documented nearly 10,000 media-reported incidents of union violence and estimates than an additional 80,000 unreported incidents occurred within that same time period." "In United States v. Enmons (1973), the Supreme Court held that strike-related violence cannot be prosecuted under the Hobbs Act, which criminalizes the obstruction of interstate commerce through violence, threat or coercion." "This means that top union officials may order or condone violence against dissenting workers with little fear of punishment. As a result, thousands of acts of violence (usually directed against non-striking workers) have gone unpunished."

And this same date, same news service companion piece on 'labor' needs to be read and remembered as well. David Kendrick, director of the Organized Labor Accountability Project of the National Legal and Policy Center, wrote an article titled "Time to open Big Labor's books. "This Labor Day, America's unionized workers deserve a present: the information they need and want about the unions claiming to represent them." "All too often, union dues go toward funding lavish lifestyles of top union brass, rather than toward bread-and-butter issues workers care about." "And how are the top brass celebrating Labor Day 2003? By feverishly lobbying member of Congress to preserve the secrecy surrounding their financial affairs." "Updating the annual LM-2 disclosure form likely would cut down on union corruption. Had there been more openness, perhaps Barbara Bullick, the past president of the Washington affiliate of the American Federation of Teachers would not have stolen million of dollar from that union. Perhaps Jake West, former head of Ironworkers International, would not have embezzled funds from his union. Perhaps the board of directors of ULLICO, the labor focused finance company, would have been more hesitant to engage in their get-rich-quick stock manipulation scheme." "The Supreme Court ruled in Communications Workers v. Beck that unionized employees are required to pay dues only for collective bargaining-related, not for politics and other actions that are not bargaining-related. But union officials seek to hide the very information needed to enforce that right." "Union lobbyists are working furiously to keep the veil tied down." "It is far past time for the Senate to approve the Labor Department's budget without shutting down (Labor Secretary) Chao's proposed reform of organized labor's financial reporting requirements."

According to the week of February 16, 2004 edition of Human Events, the AFL-CIO plans to spend $52 million to defeat President Bush. Do all members support this financial drain on their union resources? 'Who' and 'What' is behind this amount? There are more questions should anyone care to dig a little. The article projects that even the NAACP will donate in excess of $10 million as well. Why oh why no Mainstream Media interest?

The NAACP is one of three other "national" political entities worthy of a brief review: the Black Caucus; and the Rainbow (PUSH) Coalition being the other two.

At one time the NAACP was an organization of merit and purpose, but slowly fell from grace after its members and supporters found that very large amounts of money had been taken from its coffers by irresponsible leaders. Riding to the rescue was Frizzell Gray, now known as Kweisi Mfume, Baltimore politician and, according to Christopher Caldwell's article in the April 22, 1996 edition of Weekly Standard, entitled "The National Association For the Advancement Of Liberal Guilt" the "father of five sons out of wedlock by four different women by the age of 22." At 47, after "a 10 year congressional career, Kweisi Mfume became president and CEO of the NAACP." "At his swearing in – held in the Department of Justice attended by President Clinton, Vice President Gore, Attorney General Janet Reno, and her civil rights deputy, Deval Patrick – Mfume condemned the "extreme ultra-right wing policies of the Republican Congress. He also attacked the "politics born of guilt or misdirected compassion of the ultra-left that seek to maintain the poor," There is no mystery about how rigid Mfume will be in defending federal programs as head of the NAACP. "This organization believes there's an important role for affirmative action in society", Mfume says." "In a sense the NAACP is a victim of its triumphs: The civil rights agenda has become government orthodoxy. Affirmative action has become so widespread that the black upper-middle classes who make up the NAACP's elite are now profoundly dependent on it." "Uniting the Farrakhanites and the Black Bourgeoisie is Mfume's 'Job One,' although no one at the NAACP will admit it publicly." "Mfume (said) he hoped to forge a "sacred covenant" with the Nation of Islam similar to the one the caucus had with the NAACP." It would appear that the "romance" has not gone too well, and that NAACP's Mfume, in the role of suitor, is being held at more than arm's length by Farrakhan, who understandably prefers the independence of his own well organized following and pursuits."

In a July 2001 article by Philip Terzian in The Providence Journal, titled "Bond's poisonous rhetoric suggest an organization unwilling to look forward" wrote "Julian Bond, chairman of the NAACP, stood before the delegates the other evening and, speaking generally about the Bush administration, declared that the president: has selected nominees from the Taliban wing of American politics, appeased the wretched appetites of the extreme right wing, and chosen Cabinet officials whose devotion to the Confederacy is nearly canine in its uncritical affection." "Bond suggests that those Americans who cast their ballots for George W. Bush last November – half the voting population – are the local equivalent of an Islamic terrorist movement." "To this is added the presumption that the conservative wing of the Republican Party has no legitimate interests in American political life, only "wretched appetites" that must be "appeased". "Yet Julian Bond's rhetoric cannot be so lightly dismissed. It is one thing to differ on the mission of government, or disagree on means to achieve a common end. But likening opponents to murderers and tyrants, or scorning intentions with excessive abuses, reduces public life to a poisonous well. The NAACP is habitually described in the press as "the venerable civil-rights organization." But at 95 years old, venerability is just about all it has going." "Largely devoid of substance in its work, it relies on celebrities (Myrlie Evers, Julian Bond) to raise corporate money, and serves as a vehicle for mobile politicians (Kweise Mfume) whose primary interest is garnering attention." "At a time in American history when the prospects for black people are increasingly bright, and progress is measured in levels of prosperity, the anger that animates the NAACP – invective as a substitute for reasoned debate – is nearly irrelevant as it is dangerously corrosive."

And for a look at the 39 Democrat members of the Black Caucus, in a November 30, 2003 Associate Press article by Jeffery McMurray titled "Congressional group tries to retain clout while not giving ground on liberal policies" he wrote "The agenda still is shaped by the liberal causes of urban black Americans, but it is no longer limited to them." "It just calls for the caucus to send a message out there that we have sent out in the past but maybe not as clearly: We not only represent African-Americans. We represent the vast majority of Americans," said the group's chairman, Rep. Elijah Cummings, D-Md." "The caucus was among the first to back President Clinton during the impeachment proceedings." "Vanderbilt University law professor Carol Swain, who wrote a book about blacks in Congress, said black Democrats are wise to broaden their vision beyond race. "When they organized 30 years ago, it was a different country in a lot of respects, less racially diverse and truly marginalized. When you have a race-based caucus that is as liberal as the black caucus has been in the past, it hinders in some cases the advancement of minority interests." "Rep. John Lewis, D-Ga., said "On campaign finance reform, there were members we literally had to beg, cajole and do everything else. Some members held their noses and voted for it." "Frederick Harris, director of the Center for the Study of African-American politics at the University of Rochester, says the caucus is undergoing a natural shift in identity. No longer will its clout come from high numbers of like-minded liberals, he says, but from assuring that the party's moderate wing and predominantly white leadership do not give too much ground to Republicans. It will lean toward a watchdog. They still promote issues that disproportionately affect the poor African-Americans. The question becomes, without the Black Caucus, would the Democrats be as forthcoming on issues involving civil rights?"

So where do these two entities go from here? Aren't they now little more than socio-political clubs? Certainly the 39 congresspersons won't disband, nor will the NAACP. Racist born and racist perpetuated, both will simply continue to huff and puff on issues real and imagined, or on behalf of the Democrats, regardless of the issue and whether it is right or not. Their existence depends on keeping the racist pot boiling. At least Rev. Farrakhan does not attempt to hide behind a political agenda or feed at the government trough. After all, if "racism" was not kept alive, and shrilled at every opportunity, how could those who prosper from the issue continue to live in the manner to which they have become accustomed? Racism has served as a handy excuse and cash cow for so many for so long that it is hard to imagine life without the issue being pushed daily by some advocate or highlighted somewhere in the Mainstream Media. Perhaps one of these days the "Abraham Lincoln Foundation" will gain the popular support it richly deserves for its balanced and moderate positions. ® But there are many Blacks that oppose the Foundation, for it is seen as threatening to their perpetuated livelihood derived from the race/diversity/quota issues.

The last 'organization' to be addressed is the United Nations, referred to by some as "Disney-World-on-the-Hudson", or "The Flat-Earth-Society".

In a December 2003 article by Fred Gedrich, a senior policy analyst at Freedom alliance and former State and Defense Department official titled "No UN in Iraq" appearing in "The American Enterprise", he wrote: "The fact is, the United Nations has a troubled history in peacekeeping, terror preventions, and nation-building. Since the founding of the world body, there have been 291 wars resulting in 22 million deaths. During this period the U.N. authorized military intervention to deter aggression only twice: in response to North Korea's invasion of South Korea in 1950, and Iraq's invasion of Kuwait in 1990 – both under strong U.S. leadership. Sadly, in both instances, the belligerent madmen who started the wars remained in power, due to U.N. restrictions, at great cost to the civilized world. Despite a dozen international treaties and scores of Security Council and General Assembly resolutions, the U.N. has done absolutely nothing to reduce global terror." "The U.S. Department of State lists 36 terrorist organizations operating with impunity in at least 60 U.N. member countries." "The U.N. human rights record is equally appalling." "Much of this nonsense occurs because 102 of the U.N.'s 191 members do not have completely free and democratic governments. Fully 47 members are notorious dictatorships, and the U.N. roster includes six terrorist states." "What is perhaps most relevant to today's debates over Iraq is that the U.N. record at rebuilding countries is poor. Its experiences in Haiti, Somalia, Cambodia, Kosovo, and elsewhere illustrate the body's ineptitude. In Cambodia, the U.N. allowed a despot to take over the government. When things get tough, the U.N. is more apt to cut and run than face a difficult challenge – as it did a few years ago in Somalia, and this summer in Baghdad (where it simply pulled out most of it workers rather than deal with security challenges. In Kosovo, four years and $2.4 billion in international aid haven't allowed the U.N. to turn over the reign of government to locals."

In a December, 2003 article by Allen Dowd, director of the Hudson Institute, written for the "American Institute", he wrote: "According to its charter, the U.N. is supposed to be an instrument for the "maintenance of international peace and security." But instead of functioning like a useful tool, the U.N. has come to think of itself as an end in itself." "The U.N.'s contempt for U.S. leadership and allergy to military action followed it into postwar Iraq. When U.S. troops offered to protect the U.N. headquarters in Baghdad, they were snubbed. To avoid being associated with the allies, the U.N. instead contracted with a private security firm. The foolishness of this course was demonstrated on August 19, when the weak protection allowed a truck bomber to kill the U.N. chief and scores of others. U.S. troops that had also been excluded from the scene of Iraq's other horrendous bombing, the assassination of Ayatollah Baquir Al-Hakim in Najaf, which killed more than eighty people." "Americans have been given a lesson. Bureaucrats at the U.N. must never be allowed to control and validate U.S. military action. President Bush's original position was the right one: "When it comes to our security, we don't need anybody's permission."

In a June 2003 issue of NewsMax.com article by Richard Perle titled "Perle Declares U.N. "Dead", he wrote: "During the Cold War, the Security Council was hopelessly paralyzed. The Soviet empire was wrestled to the ground, and Eastern Europe liberated, not by the United Nations but by the mother of all coalitions NATO". "Facing Milosevic's multiple aggressions, the U.N. could not stop the Balkan wars or even protect its victims. Remember Sarajevo? Remember Srebrenica? It took a coalition of the willing to save Bosnia from extinction. And when the war was over, peace was made in Dayton, Ohio, not in the United Nations. The rescue of Muslims in Kosovo was not a UN action: Their cause never gained Security Council approval. The United Kingdom, not the United Nations saved the Falklands." "The chronic failure of the Security Council to enforce its own resolution is unmistakable: It is simply not up to the task. We are left with 'coalitions of the willing'. Far from disparaging them as a threat to new world order, we should recognize that they are, by default, the best hope for that order, and the true alternative to anarchy on the abject failure of the United Nations."

• Also in the June 2003 issue of NewsMax.com, Dave Eberhart and Jarret Wellstein wrote in an article titled "The Failing of the U.N.": "The 57 years of the U.N.'s existence cover perhaps the bloodiest period in human history. The U.N.'s recent behavior regarding Iraq is just the latest in a long list of it failures to keep the peace. Other notable failures include the Korean War, China's theft of Tibet, the Vietnam War, genocide in Cambodia, war between Iran and Iraq, and between Indonesia and East Timor, the first Gulf War, the Bosnia war, war between Britain and Argentina over the Falkland Islands, and hundred of other conflicts. Tens of millions of innocent people have died in these conflicts. Rwanda provided an example of the abysmal failure of the U.N.'s "peacekeeping" operations. In 1994, the U.N. assured Tutsi tribesmen in Rwanda that the U.N. would protect them from rival Hutus after Hutus took control of the government and forced the Tutsis to turn in all weapons. As many warned, once the Tutsis surrendered their arms, Hutus massacred them. The U.N. did nothing. As many as 1 million Tutsis were killed and hundreds of thousands more, including children and pregnant women, were tortured, raped, crippled or had limbs hacked off." The U.N.'s human rights record can only be described as abysmal.

• The U.N did nothing to stop the Soviet Union's mass murder of dissidents.

• It did nothing to stop genocide by Pol Pot in Cambodia.

• It has done nothing to stop communist China's huge system of concentration camps and slave labor

• It has done nothing to stop the massacre of nearly 1 million or more people in Rwanda.

• It has done nothing to end the expropriation of property and mass murder still going on in Zimbabwe."

• It has done nothing to stop the ongoing practice of slavery in Africa and the mid-east.

• And now we learn that the U.N. was deeply involved at multiple levels in graft and corruption in Saddam Hussein's Iraq, ranging from illegal oil profiting, sales of foods and goods ostensibly provided the people of Iraq; contracts for prohibited materiel provided Iraq by European countries, etc.

"The U.N. plan for global disarmament would set the stage for a holocaust on a global scale and the imposition of a U.N.-led world government." "The U.S. is the U.N.'s largest financial backer. In 2001, the U.S. contributed $3.5 billion to the U.N. This included a $612 million assessment for the U.N.'s regular budget and U.N. specialized agencies, $716 million in assessments for U.N. "peacekeeping" and $2.2 billion in voluntary contribution to U.N. affiliated organizations and activities." "The reality is that the U.N. has been a failure and today does more to promote an international socialist agenda, which threatens U.S. sovereignty and peace, than to achieve any of its announced goals. There is simply no way that free states can compromise with tyrannies to achieve a peaceful world. Indeed, the U.N. is increasingly irrelevant in a world in which capitalist countries build a prosperous future while tyrannies remain mired in the poverty and oppression of the past."

Roger Scruton, a writer of many books on the west and terrorism, wrote in an October 13, 2003 National Review article titled "The U.N.: Now Less Than Ever "Ambassadors to the U.N. are sent by the people who have obtained power, by whatever means, in the territories recognized by that body as sovereign. In effect, the U.N. simply legitimizes whatever elites and tyrants have gained power over the particular nations named in its list." "Major threats to peace and stability in the modern world come from terrorist organizations that, by their very nature, can play no part in the dialogue of nations that the U.N. is supposed to represent." "Hence it is no longer important for terrorists to gain control of a sovereign territory. Power can be achieved and deployed more effectively without assuming the burdens of the nation state." "The

U.N., to put it bluntly, shows the error of optimism when addressing the real conflicts of human beings." "Hence the U.N. has acquired the habit of substituting liberal pieties for hard-headed judgments when faced with the serious threat of war." "The U.N. was invented by people steeped in the ethos of the nation-state, and it is designed to resolve conflicts between such states. It erects into a transnational goal the notion of a rule of law – a concept that rests on the particularly Western view of sovereignty as based on territory rather than faith or ideology." "But there is another, more dangerous effect of the United Nations, one that is insufficiently pondered by our politicians. Both the U.N. and many of its ancillary and subordinate institutions have legislative powers. They can use the original force of the Charter to bind national legislatures to measures that may be profoundly against the national interest. These measures will often be a huge burden to law-abiding states but no burden at all to dictatorships." "In effect, the lawless have acquired, through the U.N., the power to bind the law-abiding in chains that they themselves escape." "By leaving the club, Western nations will free themselves from the burden of transnational legislation. They can then rebuild through bilateral treaties the kind of international relations that protect their sovereignty." We seem to have forgotten that the U.N. is not exactly a "neutral" organization. The power is in the Security Council, where membership is correctly restricted. All other positions, including all members of the General Assembly and their staffs are the result of political plums and favors dispensed by the individual governments of respective members. In simple terms, it's a patronage system. The U.S. gives billions of dollars to countries around the world through U.S.A.I.D., and in return we are often sliced and diced even though we subsidize the U.N. at a 23 percentile for its budget. Madness! With very few exceptions, what has the U.N. done to earn its keep? The International Red Cross plus a modest sized international team of qualified arbitrators could likely handle 75% of what the U.N. claims to do at a small fraction of the cost. But I ramble.

The obvious conclusion? The Security Council and the General Assembly need to recognize and address these problems before they become so large there is no answer. Will this happen? Not likely, certainly not voluntarily in the UNGA, unless strongly led by the Security Council and/or forced by world-wide Media pressures, enumerating its many failures of both commission and omission over the pasta five plus decades.

None of the articles cited above came from Mainstream Media sources, yet the facts and figures are not new. Without question, the U.N. is antiquated both in structure and in philosophy. It no longer serves a meaningful purpose, high-lighted by its repeated failures as an effective "peace keeper" as well as its sorry Human Rights record. It is terribly out of whack, and comical in its sustained pretentiousness. Careers have been spent by many U.N. employees in the New York U.N. offices, including the present Secretary General Kofi Annan of Ghana who admits to some 25 or more years at the U.N. All salaries and benefits are inflated, nepotism is rampant, fraud and theft are commonplace, greed and graft are accepted as commonplace, illegal activity goes unpunished, and the rallying cry is always for more money. Even the U.N.F.P.A. (Population Fund), not content with having whitewashed China's forced abortion policy for years, now seeks more millions of U.S. taxpayer money. To repeat, the U.S. bears some 23% of the UN's cost. (Japan follows at 19.6%, with all others under 10%.) How is this justified, especially with the U.N.'s strong anti-U.S. voting record? We deservedly must be the laughing stock of the other 190 members!

If a meaningful and honest review is not in the cards, then the U.S. should withdraw, for in reality there is little hope for this costly, inefficient, straining bureaucratic monolith. Times have changed and the challenges of today are no longer those envisaged in the 1940's charter. The U.S. carries a big stick only because it contributes the most. With a "coalition of the willing" we might collectively be able to save that which is worth serving of the present U.N. – which is not much when examined closely.

A small example: When two colleagues and I traveled in Tibet in 1997, we were struck by the all but overwhelming disregard the Chinese had for the Tibetans wherever we went. Lhasa was teeming with Chinese Army and varied Chinese police forces, bureaucrats, and shop keepers. "New" and clean Lhasa was for the Chinese; while "old" and badly run-down Lhasa was where Tibetans lived in sub-standard hovels. The U.N. had an office in Lhasa, but care was given to the Chinese. The obvious needs of the Tibetans, especially the children, were simply ignored. Ironically, Ngapo

Ngwang Jigme, the turncoat Tibetan official who had signed the 'treaty' with China in 1950, was in charge of U.N. efforts in Tibet, although he was living in Beijing. The other locations we visited in central and westward Tibet were a microcosm of Lhasa – there was no concern or U.N. care for the Tibetans. During a meeting with the Dalai Lama in Dharamsala, India in 1999 I related our dismay about the U.N. performance in Tibet. He said he had tried to improve it, but had not yet been successful. Nor had the U.N. responded to any of the Dalai Lama's repeated requests for help in the early 1950s. Human Rights? China? Africa? The Mid-east? Cuba? Haiti? Or hadn't you noticed the U.N.'s record of absences, its silences, or its lip-services only? A U.N. military "peace keeping" force? Mind boggling, to say the least. Look at the record and tell me when the blue helmets were really effective under duress. Just as with their 'Resolutions', who pays attention to the U.N. dilettantes?

When there were expressions in the U.N. in the early 1980s suggesting that the United Nations leave the United Sates, our Ambassador to the U.N. at the time, Charles Lichenstein said "if, in the judicious determination of the members of the United Nations, they feel they were not welcome and not treated with the hostly consideration that is their due, the United States strongly encourages members states to seriously consider removing themselves and the organization from the soil of the United States. We will put no impediment in your way, and we will be at dockside bidding you a farewell as you set off into the sunset." That effectively stopped the gripers cold. Ambassador Lichenstein died in 2002, but surely his spirit - and words - lingers on.

A similar response is in order to any U.N./U.N.G.A. resistance to changing its soiled underwear, so to speak, after all these years of ineptitude, and should not be difficult to deliver. A favorable response from the U.N., followed by realistic revisions and changes reflecting a much revised role for the U.N. in the new millennium could benefit all. But don't hold your breath. The chances of any meaningful reform, operational or fiscal, are slim, for the U.N. wants more money and more power, not less. And certainly the U.S. does not need nor should allow U.N. mucking about in our Republic's business, or with our Constitution, or Bill of Rights, or parental rights, or any other internal matters, thank you very much!

Nor should we waste time thinking about any kind of a 'New World Order'. If anything, the United Nations itself provides the stark fallibilities of that idea! Ours is a world of different cultures, different languages, different customs, different religions, different beliefs, different mores, different values, different systems of government, differing dress, differing laws, differing social and economic standards, differing needs; different abilities, different educational standards and requirements, and, more importantly, different people. Thankfully, one size does not, nor should it, fit all, nor is there one so-called 'answer'. Think what a bureaucratic nightmare a "One World" concept would require, or how many pages of 'wiring diagrams' would be required as a blue print. There is no unanimity in the U.N.-or anything resembling effectiveness or lasting achievement. Multiply the U.N. 'struggles' by a factor of at least one hundred to begin to understand what a mess the 'One World' idea creates. Also consider the billions and billions of tax dollars the United States has "spent" in trying to influence countries around the world through our "Agency for International Development" and U.S. Information Office. Except in very few cases, probably less than ten total, A.I.D. efforts have been a failed boondoggle for years. But it's a bureaucrat's delight, much like the U.N. There may be a handful or programs worthy of support, but as presently constituted, A.I.D. too should be critically reviewed and its mission updated. It presently seems to respond more to U.N. requests than it does to serve U.S. policy interests.

If the U.N. example is not readily understood or convincing, let's review our own political system. Republicans and Democrats seldom agree on many subjects and largely pursue different agendas. Libertarians do not agree with either of the two major parties on most topics and the so-called Green Party identifies largely with Socialists. The Communist Party (U.S.A.) huddles unto itself, sulking, plotting and waiting. The lobbyists have major influences and impact on the politicians, especially on those in the top two parties, but cannot <u>totally</u> control them. Add religion and secularism, for each plays a role. This is not the case world-wide where many religions have extensive influences. So, given the problems within our own political system as a microcosm, how could a "Utopian" New World Order be achieved? Who would govern? Who would rule? Who would police it? How would it be financed? What of dissent? Etc. I repeat, forget it. I know it 'sounds' so nice, and Socialists drool about it, but it's *nonsense!*

That individual freedom is a universal right and need, and serves as a common bond and goal for all people, is one thing we all can agree upon and support. However, to attempt to codify any "New World Order" notion into a monolithic organization reigning over all people is simply stupidity in action and undeserving of a second or even a fleeting thought. It surely invites and would cause far more conflicts and disasters than it could possibly serve to resolve. Far better that the idea of "Coalitions of the Willing" be tried, financed by the 'willing', and, with successes, perhaps these could be expanded in time to help solve the myriad of problems we earthlings already face. Such coalitions could serve as a checks and balance system on its members as well, as they work together to resolve common problems.

In the interim, there is more than enough to keep all countries focused on their own respective problems. Let's keep in mind that Europe should be in position to best serve European needs; South America should serve South American needs; etc.; Central America and the United States should be able to resolve Central American problems, just as Canada and the United States should resolve North American problems. The Asian, Near East and Mid East problems are more complex, as are the problems in Africa, with each having factors that are common and not common to one another, thus requiring special attention by those concerned and able to assist. In the far Pacific, neither Australia nor New Zealand can be expected to carry the problem solving responsibility alone, just as Japan cannot be expected to resolve Korean and China related problems by herself, despite the threat that both China and North Korea represent in that area. Indonesia, Malaya, Thailand and the Philippines are equally unable to individually shoulder responsibility for that area, nor is Taiwan. But a 'Coalition of the Willing', similar to the original SEATO concept could well be the answer. Africa represents a challenge all by herself, given the wide range of problems that plague that continent. Harking back to the colonial days, European assistance would seem to be in order, should Belgium, the Dutch, the French, and the Germans be able to form their own 'Coalition of the Willing', assisted perhaps by others that might agree to assist in such an effort. Obviously, the United

Nations has chosen to ignore most of Africa's more serious problems, including genocide, and its actions in most parts of the world have been equally unimpressive for years.

For decades we have had an especially strong bond with England, Canada, Australia, New Zealand; and more recently with Japan, Taiwan, South Korea, Thailand, Israel, Saudi Arabia, etc. Hopefully these bonds will continue and grow stronger over the decades ahead. Unhealthy Liberalism and Socialism have weakened these bonds somewhat, but the world will be the loser if political ideologues or differing ideologies break these marvelous links that have served so well for so long.

The Mid East, beginning with Iraq, is on the path to its version of democracy, as is Libya and perhaps Iran in the near future, thanks to the 'Iraq Coalition' effort now more than a year old. There are a number of problems that will not be solved in the near term, for such is not the way of religious based wars that have waged for hundreds of years. However, what has transpired in Iraq could possibly be achieved in Palestine if the PLO fanatics ceased their terrorist and suicide activities, and gave the peace process a chance. The benefits of same for all are obvious. Their children need to be taught to learn, not just recite current dogma, including to respect the religious beliefs of others, rather than to hate - nor should their women be kept like mushrooms. Their men need employment and careers, for terrorist pursuits offer little in the way of bragging rights or longevity. The promise of Vestal Virgins may make an appealing commercial for aspiring terrorists, but like the Greek mythology of sirens luring mariners to their death by their singing, that's not exactly a winner! Arafat still has but one commodity to sell and has stayed in power these many years only because of his unceasing terrorist taunts. Syria will not endure as a friend, and then what will Arafat and the PLO do? Israel has been remarkably patient for years, but that seems to escape the Mainstream Media as well. This is 2004, or is there a different calendar involved? It's time for Arafat to ponder his next life and consider the future of all Palestinians. His 'vision' and terrorist practices have brought only false hopes, death and destruction, and his hate, delusions, and deceptions are not marketable pursuits. Arafat may have no conscience, but that certainly does not excuse the Mainstream Media's defense of him. If Media were to review its record on this subject the resulting blush could be seen on black and white televisions sets!

The United States is not the world's bank, or its general provider. Its role should be the Eagle Scout helping the disadvantaged to cross a busy street, or assisting like minded others to do the right thing in meeting our common interests and obligations. Teddy Roosevelt had it right about 'walking softly, but carrying a big stick'. Please note that dollars are not mentioned in this famous remark. Had he been of the 'liberal left- or a Socialist - money or special social programs surely would have been added, or political correctness, or affirmative action, or other such outmoded and negative nonsense. We need to recognize how fortunate we are to be Americans and be proud and thankful of it. We have freedoms that most other countries do not and we need to protect these precious gifts. The Declaration of Independence is not simply an idle term or an outdated document. Like our flag, it serves to remind us that we need always remain independent and free, and very thankful to our forefathers.

Next: Chapter 8 and the "Media".

® from the booklet, "National Committee to Restore Internal Security" regarding Gorbachev's Apparatus in the United States and (T)he "Moscow Spring", United States Senate, June 8, 1988.

* By Richard H. Shultz and Roy Goodson, published by Pergamon-Brassey's, 1984.

• Should you be interested in identifying more, I suggest you read Robert S. Meyer's "Peace Organizations Past and Present", published by McFarland and Company, 1988.

® Jesse Jackson and his Rainbow Coalition/Push organizations are not included. All that anyone needs to know about Jackson, his organizations, his scams, his extortions, etc. have been well documented in Kenneth R. Timmerman's "Shakedown" exposing the real Jesse Jackson", published by Regnery Publishing, Inc., 2002. The August 22, 2003 Wall Street Journal article by Dave Shiflett also summarizes Jackson's many deceits, as have other non-Mainstream Media articles. In sum, the man is described as a "fraud", and has lived his life as one. And should the reader be interested, the book "Scam" by Rev. Jesse Lee Peterson, (WND Books, 2003) is a superb read. It explains in stark terms how the "Black Leadership" has exploited and manipulated racism and Black America for decades. All adults should read it.

CHAPTER 8

MEDIA

(The Good, The Mediocre, & the Sorry)

Certainly the above three categories enable an accurate labeling of those in the trade. There is a wide diversity of talent and ethics within this vast field, but few in it pursue their profession with pride and objectivity. As with politicians, there are some on both ends of the spectrum, but there are far more to the 'left' than to the 'right', or in the handful in the so-called 'moderate' middle. As repeatedly noted, there's been very little objective, factual reporting on important issues by those in the category of Mainstream Media; this simply underscores the point that this country has been poorly served by the Media for far too long.

Sadly, there are many thousands across our land intent on destroying or undercutting the basic strengths of our country, yet they remain largely untouched by the Media. Why? One apparent answer is that a large majority of the members of the 4th Estate is on the Liberal Left. Thus the liberal Mainstream Media combined with the leftist trade unions, the leftist teachers' unions, the leftist campus liberals, the liberal leftist politicians (national and local), the leftist ACLU, and other like-minded lawyer's organizations, and the multiple sub-rosa leftist organizations that wish our Republic ill have knowingly provided 'them' a distinct advantage over 'us'. It's surprising that our Republic still has a pulse!

How many papers and magazines are published in this country? I suspect that no one has an accurate answer, but let's guess at least 80,000 to 100,000, with about 2,000 constituting the number of dailies published in our larger cities. Unfortunately, most all of these depend on wire services to provide them with national and international news. We used to see UPI, AP, Reuters, Knight-Ridder, etc., but now AP seems to be the wire service of choice. That its coverage is 'spotty' and more often than not less than accurate or objective seems not to matter. But since journalism and reporting seem to be a failing art, why should AP be different or have higher standards?

For instance, can you identify who said the following: ®

1."I think most newspapermen by definition have to be liberal; if they're not liberal, by my definition of it, then they can hardly be good newspapermen."

2."Gorbachev is helping the West by showing that the Soviet threat isn't what it used to be – and what's more, that it never was. The doves in the Great Debate of the past 40 years were right all along."

3."I think the House was right to turn down aid to the Contras."

4."I think that what is being proposed for expenditures on Star Wars, for example, is absolute nonsense."

5.In the days before the Iraq War "(this) will join the Bay of Pigs, Vietnam, Desert One, Beirut, and Somalia in the history of military consequences."

From a 1985 study, "the vast majority of former political activists currently working in the media have had professional ties to liberals and Democrats. In fact, every major national media outlet has reporters and executives who were previously professionally associated with liberals. A relatively small number of journalists have had associations with conservatives or Republicans."(1) "No less than 81 percent of the press interviewed voted for the Democratic presidential candidate in every election between 1964 and 1976."(2) "51 percent cast a ballot for President Jimmy Carter, 24 percent for liberal third party candidate John Anderson, and 25 percent for winner Ronald Reagan."(3) "Liberals outnumber conservatives in the newsroom by more than three-to-one: 55 percent considered themselves liberal, compared to just 17 percent who identified themselves as conservative." (4) In 1989, 62 percent of newsroom employees identified themselves as "Democrat or liberal" or "independent, but lean to Democrat/liberal. 22 percent called themselves "Republican or conservative" or "independent, but lean to Republican/conservative."(5) 85 percent of the journalism students called themselves liberal."(6)"As of early 1990 the

MRC (Media Research Center) had identified 235 reporters, editors, producers and news division executives who had passed through the "Revolving Door" (switching jobs/positions in Media) at least once. Far from representing a cross–section of the American political spectrum, virtually all belonged on the liberal side." "178 current or former Big Media reporters, editors, producers, and executives were connected at one time to liberal political groups, but only 57 had ties to Republicans or conservatives.(7)

For example: ®

• Douglas Bennet, President and Chief Executive Officer, National Public Radio, was Assistant to VP Humphrey, 1967-69; Administrative Assistant to Sen. Tom Eagleton (D-Mo.) 1969-1973; Admin. Assistant to Sen. Ribicoff (D-Ct), 1973-74; Assistant Secretary of State for congressional relations, 1977-79; Director, Agency for International Development, 1979-1981 (Jimmy Carter years)

• David Burke, President, CBS News. Executive Vice President, ABC News, 1986-1988; Chief of Staff, Sen. Ted Kennedy, 1965-1971; Chief of Staff to N.Y. Governor Carey, 1975-77.

• Wally Chalmers, Director of Broadcast Research, CBS News, 1984-86; CBS News Political Editor, 1984; Executive Director and Chief of Staff, Democratic National Committee, 1986-99; Executive Director of Nuclear Freeze Foundation and Fund for a Democratic Majority, early 1980s; Kennedy for President coordinator, 1980;

• Hedley Donovan, Editor-in-Chief, Time Inc., 1964-1979. Senior Adviser to Pres. Jimmy Carter, 1979-1981.

• Jeff Grainick, Vice President of ABC News; Press Secretary, Sen. Geo. McGovern, 1971; Executive Producer, *Jennings-Koppel Report,* 1987.

• Jane Pauley, Correspondent and anchor, NBC News, 1990. Administrative Assistant, Indiana Democratic State Central Committee.

• Tim Russert, NBC News Vice President and Washington Bureau Chief, 1989. NBC News Vice President, 1984-1988; Counselor and media strategist for N.Y. Governor Mario Cuomo, 1983-84.

• Jeff Greenfield, Media reporter, ABC News Nightline; political correspondent, ABC News, 1983-; Media critic, CBS News

Sunday Morning, 1979-1983; Speechwriter, Sen. Robert Kennedy, 1967-68.

• Walter Shapiro, Senior Writer, Time magazine, 1987-; Speechwriter, President Carter, 1979."

From the background information cited above, it is not difficult to note the liberal influences at work regarding who remain prominent in the Media. From the same source, there are additional examples:

• "Walter Cronkite, former CBS News anchorman: "I know liberalism isn't dead in this country. It simply has, temporarily we hope, lost its voice."..."We know that unilateral action in Grenada and Tripoli was wrong. We know that *Star Wars* means uncontrollable escalation of the arms race. We know that the real threat to democracy is the half of the nation in poverty. We know that no one should tell a woman she had to bear an unwanted child." "Gawd Almighty, we've got to shout these truths in which we believe from the housetops. Like that scene in the movie *Network*, we've got to throw open our windows and shout these truths to the streets and heavens. And I bet we'll find more windows are thrown open to join the chorus than we'd ever dreamed possible." Cronkite Speech at "People for the American Way" banquet, quoted in *Newsweek*, December 5, 1988. Even after all these years, Cronkite still has trouble discerning the truth and his errors; at least he finally admits being a liberal.

• Diane Sawyer, ABC's *Prime Time Live* co-anchor: "Once I got a card at *60 Minutes* that said, 'You are a brazen right-wing hussy,' she recalls, 'I was able to write back and say, I'm not right wing." How true!

• Sidney Blumenthal, *Washington Post* political and "Style" section writer, 1985-1989: Senior Editor, *The New Republic*, 1990 – (then joined the Clinton administration); Boston correspondent, *In These Times,* a socialist newspaper, late 1970s; writer for *The Progressive,* and *The Nation,* mid 1970s. And a bad joke even amongst most of his liberal peers.

• Christopher Matthews, MSNBC, formerly CBS *This Morning* political columnist; Chief of Staff for Speaker of the House Tip O'Neill, 1981-86; speechwriter for Pres. Carter; Campaign commentator/analyst, Mutual Broadcasting System, 1987-; former Peace Corps member. A mouthy, sarcastic, highly opinionated, liberal who usually has other liberals on his TV 'show'.

• Ted Koppel, *Nightline* anchor: "A testament to courage: the courage of some unabashed trade unionist and civil rights worker, Leftists and yes, American Communists, who fought for principles that we now take for granted." A Socialist 'wannabe', but prefers the comforts of his position.

• John Chancellor, NBC *Nightly News* commentator and former anchor: "The squeeze is on and a modest tax increase would help a lot, but the candidates rule it out. Bush's no new taxes position seems to mean that taxes would go up only if there were an economic crisis. That drastically limits the flexibility any government should have. Is it smart to wait for a crisis? Bush has accused Dukakis of being for unilateral disarmament, yet Bush has disarmed himself by promising no new taxes. Taxes are one of the most important economic weapons a government has." Nov. 2, 1988. Is he running for office or reporting the news?

• Bill Moyers, former CBS News commentator, now with PBS: "Behind the smiles, behind the one-liners, the pretty pictures, the manipulation of emotions, the government rots. Its costs soar, its failures mount, from the folly of Star Wars to the scandals of HUD. But on the bridge of the ship of state, no one is on watch and below the deck, none can see the iceberg. But everybody feels good." *The Public Mind,* Nov. 15, 1989. This man has a Media rap sheet similar to Seymour Hersch's.

• Jim Wooten, ABC News reporter: "Ah yes. The dreaded federal deficit, created, for the most part, by the most massive peacetime military buildup in America's history." *Nightline*, Jan. 29, 1990. No defense is better?

• Bryant Gumbel, NBC *Today* co-host: "[Except] for capital gains, it is certain the President won't mention the T-word, and yet taxes are very much at the heart of what all out potential solutions are. How long can both sides pretend that a hike's not needed?" Jan. 31, 1990, before Bush's State of the Union address. He, Carville, and Jesse Jackson ride the same horse.

• Sam Donaldson, ABC News correspondent: "I think the House was right to turn down aid to the Contras." *This Week with David Brinkley,* Feb. 7, 1987. His 'opinion' is objective commentary? A brassy Bill Moyers.

• Bill Moyers: "And in Nicaragua, the Contras used weapons

from the 'enterprise' against civilians. It's a terrorist war they're fighting. Old men, women and children are caught in the middle or killed deliberately as the Contras use violence against peasants to pressure their government. Thousand have died. Even when the hearings were taking place in Washington this summer, a Contra raid in Nicaragua killed three children and a pregnant woman. As the casualties mounted, the secret government in Washington knew that the Contra leaders were not such noble freedom fighters after all."—*The Secret Government, the Constitution in Crisis,* hosted by Bill Moyers, PBS, Sept. 14, 1988. 0bjective, on-the-spot news?

• Ted Turner, founder of CNN: "I absolutely trust them [Soviet leaders] with my life. They're not even an enemy anymore." *Washington Post,* Apr.13, 1989. "Gorbachev has probably moved more quickly than any person in the history of the world. Moving faster than Jesus Christ did. America is always lagging six months behind. I think we can get by easily with a $75 billion military budget. Those bombers and all of this stuff is an absolute waster of money and a joke." *Time,* Jan.22, 1990. No wonder Jane left!

• Eleanor Clift, *Newsweek* Washington correspondent: "Jackson is speaking to blue-collar voters for Wallace, who voted for Reagan. Jackson is right on South Africa, Jackson is right on Central America." *The McLaughlin Group*, weekend of April 9, 1988. Eleanor Clift's view is sought? Why?

• Walter Cronkite: "Jesse Jackson succeeded in conducting a brilliant presidential campaign in which he enlisted white support as well as black support, and conducted a mainstream campaign." Democratic National Convention, July 20, 1988. Hang in there Walter, you'll get it right yet.

• Eleanor Clift: "This is about racism, and not particularly about Boston. We were all had in this country because we all buy the racial stereotypes, and George Bush gave the green light with the Willie Horton campaign." *McLaughlin Group,* Jan. 14, 1990." Was Willie innocent and just not understood? Not exactly

• From his comments and questions, John McLaughlin in recent months impresses as leaning further and further to the Liberal Left and unwilling to entertain logic. Perhaps the (understandable) cause is the shrillness of Eleanor Clift, who still staunchly defends the indefensible Clinton record.

Bernard Goldberg's said in his book *Bias (Regnery Press),* "The problem is that so many TV journalists simply don't know what to think about certain issues until the *New York Times* and the *Washington Post* tell them what to think. Those big, important newspapers set the agenda that network news people follow." "Liberals have an uneasy feeling about tax cuts in general, and are downright hostile to the kinds of cuts that benefit the wealthy in particular, even if they also help a lot of other Americans."(8) "I wrote about the liberal bias that permeated the national press, and I dissected Engberg's "Reality Check" point by point, showing how slanted it was. Then I wrote, "One thing to remember about network news is that it steals just about everything from print. So if the *New York Times* is against the flat tax, and the *Washington Post* is against the flat tax, the networks can't and won't be far behind."(9)

Goldberg also noted in his book: "I noticed that we pointedly identified conservatives but didn't bother to identify liberals as liberals." "The Christian Coalition is identified as a conservative organization – so far, so good – but we don't identify the National Organization for Women as a liberal organization, which it surely is." "Robert Bork is the "conservative" judge. But Laurence Tribe, who must have been on the *CBS Evening News* ten million times in the 1980s (and who during the contested presidential election in 2000 was a leading member of Team Gore, arguing the vice president's case before the Supreme Court) is identified simply as a "Harvard law professor." But Tribe is not *simply* a law professor. He's easily as liberal as Bork is a conservative." "If we do a Hollywood story, it's not unusual to identify certain actors, like Tom Selleck or Bruce Willis, as conservatives. But Barbara Streisand or Rob Reiner, no matter how active they are in liberal Democratic politics, are just Barbara Streisand and Rob Reiner."

It follows that Liberals think of themselves as 'Mainstream' and therefore should not be labeled, whereas 'others' need to be identified as 'conservatives'. You may detect this same flawed thinking in the way the Mainstream Media and certainly most of its spokespersons practice this 'rule-of-thumb'. As the House Managers commented about the senate Democrats, and some of their Republican colleagues during the Clinton impeachment 'trial': "I'm glad they weren't at Valley Forge". So are we. We don't need more Political 'cowards'. Media coverage remained sympathetic to Clinton, of course.

In commenting on the distortions in both the print and TV Media, Goldberg noted that "The Gannett newspaper chain's policy that *requires* reporters at all seventy-five Gannett papers to include minority sources in all their stories."(10) Think about that for a moment - this means that irrespective of the subject matter there must be a minority source quoted, regardless if the story is about Asians, Jews, Russians, Hanukah food, organized crime, or the weather. And obituaries?

Dan Rather said on Tom Snyder's late-night TV show of February 8, 1995: *"Most reporters don't know if they're Republican or Democrat and vote every which way."* Goldberg's comment was: "I don't think so. They vote overwhelmingly Democratic. Could Rather really be the only person in the entire United States of American who doesn't know this?"(11) Yes, Rather could be!

Goldberg also cited data from 'The Freedom Forum', an independent foundation that examines media issues, and 'The Roper Center', a reputable opinion research firm, that found "Washington journalists are far more liberal and far more Democratic than the typical American voter:

• 89 percent of the journalists said they voted for Bill Clinton in 1992, compared with just 43 percent of the non-journalist voters.

• 7 percent of the journalists voted for George Bush; 37 percent of the voters did.

• 2 percent of the news people voted for Ross Perot while 19 percent of the electorate did.

• 89 percent voted for Bill Clinton There's hardly a candidate in the entire United States of America who carries his or her district with 89 percent of the vote. The only politicians who get numbers like that are called Fidel Castro or Saddam Hussein."(12)

Goldberg also offered the following quotes in his book:

• "The man is on the Court. You know, I hope his wife feeds him lots of eggs and butter and he dies early, like many black men do, of heart disease. Well, that's how I feel. He is an absolutely reprehensible person." USA Today columnist and Pacifica Radio talk show host Julianne Malveaux on Justice Clarence Thomas, November 4, 1994, PBS, 'To the Contrary'.

• *Inside Washington* (TV) host Gulland: "I don't think I have any Jesse Helms defenders here, Nina?" Nina Totenberg: "Not me.

I think he ought to be worried about what's going on in the Good Lord's mind, because if there is retributive justice, he'll get AIDS from a transfusion, or one of his grandchildren will get it." National Public Radio and ABC reporter Nina Totenberg reacting to Senator Jesse Helm's claim that the government spends too much on AIDS research, July 8, 1995. Nina, all-American girl!

• *Newsweek's* Eleanor Clift, after Bill Clinton was impeached, said, "That herd of managers from the House, I mean, frankly, all they were missing was the white sheets." (McLaughlin Group.) Clementine nearly at her best.

"And that's why the media stars can so easily talk about "right-wing" Republicans and "right-wing" Christians and "right-wing Miami Cubans" and "right-wing" radio talk show hosts. But the only time they utter the words "left-wing" is when they're talking about an airplane."(13)

But Goldberg is not the only one who has presented the obvious liberal bias of the Media. For example:

• "There is no evidence that Bill Clinton has lied. He's done nothing illegal. He has what I would call the politician's disease. He has tailored the truth to adapt to the reality of running in a conservative southern state." Eleanor Clift on *The McLaughlin Group.* Tailoring the truth?

• "Clinton told Phil Donahue "I didn't have anything to do with the (Vietnam Moratorium Committee). But in his 1969 letter to his ROTC commander he wrote that "After I left Arkansas last summer, I went to Washington to work in the national headquarters of the Moratorium, then to England to organize the Americans here for demonstrations." Clinton also claimed on *Donahue* that "I was not a big organizer of anti-war activities." Of course not, Bill, nor were you a philanderer, or smoke pot, etc., etc.

• "On *The McLaughlin Group, Newsweek* reporter Eleanor Clift protested: "This is in the finest McCarthy tradition, to suggest there was something suspicious with no evidence. I don't see how George Bush sleeps at night after stooping this low." Eleanor, McCarthy was right!

• "Time headlined its story "Anatomy of a Smear." *Newsweek's* Joe Klein wrote an article titled "Bush's Desperate Game", and wrote "Never before has a candidate for President been so directly, *personally* mean-spirited and derogatory toward his opponents, so willing to indulge in demagogic innuendo." (Reference was to Pres. Bush asking Clinton about the draft.) With *Time,* everything is pro Clinton.

• Admiral James Stockdale (Ross Perot's running mate) "blasted Clinton's involvement in protests against the Vietnam War. Stockdale, who spent eight years in Vietnamese prison camps, declared: Those comrades of mine who died, the extra 10, 15, 20 thousand, that blood is on your hands, you war protesters" "While the *Washington Times* and *USA Today* put the story on the front page and spotlighted in on page 2, *The Washington Post buried it in the 23rd paragraph of a Perot story on Page A-11 (emphasis added) The New York Times* buried it at the very end of a Perot story on Page A-17. CBS and NBC read brief stories late in their newscast. ABC totally ignored it."(14) Our very even-handed and balanced press hard at work.

• Dan Rather in a satellite interview with Clinton on May 27, 1993: "If we could be one-hundreth as great as you and Hillary Rodham Clinton have been in the White House, we'd take it right now and walk away winners. Thank you very much and tell Mrs. Clinton we respect her and we're pulling for her."(15) Golly gee, Dan, that's great reporting!

• *Time's* Margaret Carlson, January 4, 1993. "She is the duty-bound, disciplined Methodist, carrying her favorite Scriptures around in her briefcase and holding herself and others to a higher standard. Perhaps a First Lady who consults law-books rather than astrologers doesn't look so frightening after all. And perhaps Bill Clinton, rather than seeming weak by comparison with his wife, has proved that it takes a solid, secure man to marry a strong woman."(16) Isn't it amazing how Margaret put it all together for us?

• *CNN* talk show host, Larry King in his *USA Today* column October 4, 1993: "I saw a Hillary Clinton that I'd never seen before. She was funny, charming, sexy – yes, gang, sexy. We are both Scorpios, which tells you a lot. Meanwhile, she's earned the respect of everyone, except the wackos, with her handling of the health care issue. Indeed she has gotten everyone (except the wackos) to agree that we need health care for everyone. This is a very formidable idea, ladies and gentlemen."(17) Larry, all of us aren't Scorpios, or wacko, or liberal. Your Puberty days are long gone.

• "NPR's Nina Totenberg, the leaker of (Anita) Hill's unproven stories, had the audacity to decry the (Arkansas) trooper story (about Clinton's Arkansas dalliances). "You get allegations that are printed in a fringe magazine, or at least a magazine with a very definite political agenda, and then you see how long it takes the rest of the press to come and bite."(18) Nina, is it true that you may run on the Socialist ticket soon?

• *Newsweek's* Evan Thomas on *Inside Washington,* May 7, 1994, regarding Paula Jones: "Yes, the case is being fomented by right-wing nuts, and yes, she is not very credible witness, and it's really not a law case at all, some sleazy woman with long hair coming out of the trailer parks."(19) Thomas has a wonderful way with words and conclusions.

• On 17 July, 1995, *CBS's* Dan Rather reported the Whitewater investigation: "From another offensive wave on Whitewater to a sweeping rollback of federal regulations on Health, safety, and the environment, it's a political carpet-bombing attack, wall to wall, House and Senate." Then "When the House joined the Senate in hearings," Rather described the committees as "the Whitewater tag team."(20) And what a marvel was Rather's investigation of Whitewater!

A "modified" 1986 Gallup Survey tracked closely with the findings of a 1985 survey by the American Society of Newspaper Editors, both of which concluded that no 'print' outlet enjoyed a favorable rating by the majority of those contacted. The PBS MacNeil-Lehrer rating of 43 percent favorable, and 57 percent unfavorable, received the highest rating, followed by Cable News Network at 38 percent, and 62 percent unfavorable. Of the 'news' personalities, only Walter Cronkite at 57 percent favorable, and 43 percent unfavorable, received a positive rating. He was followed by Dan Rather at 44 percent favorable, and 56 percent unfavorable. Phil Donahue and Jack Anderson were at the bottom, with 77 and 83 percent unfavorable ratings. A subsequent *New York Times* telephone poll of 1,502 adults in 1987 provided more favorable ratings, but these figures were "widely challenged". After all, how can we expect those in the Media that have difficulty getting their news reports balanced and factual to measure themselves accurately? "And in recent years, in depth surveys of the personal opinions of American journalists and careful quantitative analyses of their work-product have established beyond serious dispute what common sense has told objective observers for decades: that the members of the media elite are, in personal terms, vastly more liberal, politically, than the American people as a whole; *and* that they very frequently permit their personal views to bias and distort the "news" they produce."(21)

"One final point: There is one political group in America that almost never protests over unfair treatment by the media, and this is the liberals themselves. It is true that from time to time some liberal organization or individual – most often a politician or a political candidate – will complain about distortions in a particular story. On the contrary, liberals are usually quick to praise the media, deny that the media are biased, and defend the media's claimed prerogatives against all challengers."(22). That's like "Run to the roundhouse, Pa, they can't corner us there!" Or let's play in my sand box. But recently the Liberals are whining that Conservatives are starting to monopolize radio and TV news audiences and blame this unfair development on the right wing. Of course.

In *The News Manipulators*, by Reed Irvine, Joseph C. Goulden and Cliff Kincad, there are multiple examples of Media bias, errors of commission and omissions, etc., such as:

• "Despite media and governmental inattention, over the years much credible evidence accumulated indicated that Vietnam was holding an unknown number of live American prisoners. In 1979, an intelligence officer named Le Dinh who had defected from the Vietnamese People's Army reported that he had heard from army generals that they were holding 700 American POWs as "a Strategic asset." "Defense Intelligence Agency agents interviewed Le Dinh in Paris in November 1979. The DIA. was skeptical. General Tran Van Quang's secret 1972 report verified Le Dinh's story, and Le Dinh's account confirms the validity of the Quang report. Independently, they both put the number of American POWs at about 700. Le Dinh provided important additional information – that the prisoners were still alive in 1975 and were being held for ransom. These reports not only confirm each other, but they also are confirmed by aerial photography and satellite imagery indicating the POWs have been signaling their presence over the years. The Quang report itself was given little attention by the three TV broadcast networks, they mentioned it a total of only six times in the three weeks after it was released. The print media have given it more attention, but they have focused more on those who are trying, in vain, to discredit it than on the evidence that it tells the awful truth that we abandoned 700 of our men. Almost none of the reporting brought up the evidence that many of those men are still alive today, praying to be rescued. (May 4, 1993)."(23) Hell'o Sen. Kerry, a little pucker factor, perhaps? Justified, say I.

• "*The Washington Post* is widely regarded as one of the most liberal papers in the United States. Katharine Graham, the principal owner and chairman of the board of the Washington Post company, persists in denying that there is any liberal bias at the *Post.* Mrs. Graham says the paper is "independent," not liberal. Kay Graham's vision was rudely shattered when *The City Paper*, a Washington weekly, decided to check the voter registration of 49 top Post editors, reporters and columnists. The results, published in its October 26, 1990 edition, revealed that only one registered Republican could be found: sports columnist Tony Kornheiser. An embarrassed Kornheiser explained that he wasn't a real Republican."(24)

• "Are our media letting a $2,000 a day private eye hired by the Clinton campaign decide what is fit to print about "bimbos" the Arkansas governor enjoyed during his days as an admitted adulterer? *The Washington Post* ran an extraordinary story on July 26 about investigator Jack Palladino. Clinton campaign aide and longtime intimate advisor Betsey Wright told the Post she was using Palladino to investigate women causing "bimbo eruptions". "The eruptions are many – no less than 17 since Clinton's nomination, according to Wright, to go along with five during the primaries, including Gennifer Flowers, who had tapes to back her story of a lengthy affair with Clinton. By Betsey Wright's account, Palladino is digging up dirt (she used the euphemism "deflect allegations") on women bold enough to confess intimacies with Clinton. The dirt is then fed to journalists to discourage them from reporting on the women. (August 7, 1992)" (25)

• "In an editorial thundering with righteousness, *The New York Times* on April 20 denounced "Republican critics" and "Republican hostility" which it claimed threatens the liberals' cherished icon, public television. "Their complaints that public broadcasting is marred by liberal bias have yet to be substantiated," the editorial said. "Senate Republicans say public television is too liberal, but they have failed to make their case." "Do *Times* editorial writers read their own newspaper? One of the most biting mainstream critics of public TV in recent years has been none other than the *Times'* own Walter Goodman. We suggest that *Times* editorialists might spend profitable time reading Goodman reviews before giving another knee-jerk endorsement of public TV." "Here is a sampling:

- "May 6, 1991, headline, "Documentaries That Lean Left." "When it comes to documentaries, it is difficult to argue that [PBS's] drift is anywhere but to port side. How come?"
- "March 22, 1992, headline, "Pull the Plug on PBS?" "Neither the Corporation for Public Broadcasting Service nor the Public Broadcasting Service has ever come clean about the chronic bent to the left of their public affairs programs."
- "May 18, 1992, headline, "A Running Debate About Public TV." Goodman took issue with Bill Moyers' earnest claim in an ABC *Nightline* debate with George Will that PBS was not biased. Despite these denials, Goodman wrote, "at least in prime time, PBS documentaries on subjects like foreign policy, welfare, social issues and the environment have long had a pronounced slant." "Goodman made it plain that he felt that the "slant" was to the left. (May 29, 1992)" The projected 1996 budget for our government – taxpayer, that is - subsidizing the Corporation for Public Broadcasting was $425 million." (26) Today's subsidy? Ask your Congressman for an accurate figure.
- "Bob Woodward is another Bradlees protégé who has found lying no obstacle to career advancement at the *Post.* Woodward and Bernstein's Watergate reporting earned them dishonorable mention in Sisselas Bok's book, "*Lying*". Their book, "*The Final Days*", was so riddled with mendacious reporting that John Osborne, who covered the White House for the liberal *New Republic*, described it as "the worst job of nationally noted reporting that I've observed during the 49 years in the business." In his book "*Veil*", Woodward "earned the sobriquet "Mortuary Bob" with his tale about interviewing the late William J. Casey shortly before he died. .Woodward claimed to have conversed with Casey, who was paralyzed and unable to speak intelligibly after his brain surgery."

"Woodward also told three conflicting versions how he got into Casey's heavily guarded hospital room." (27) I would simply add that there is no way Woodward's 'version' could be true: * 1.) C.I.A.'s excellent security would not have allowed Woodward into Casey's room without either Mrs. Casey's approval or 2.) via a request by the Deputy Director to Mrs. Casey that Woodward be allowed in, with her then agreeing, and being present. Neither of these happened. Mrs. Casey, and family, kept a vigil at the hospital and strongly deny that Woodward ever had such an interview. Woodward's 'creative' writing has continued, however. Blended with other stories and data available to most, if not all, serves him well as a masterful spinner of yarns. How or why the *Post* continues to protect him even though it has not carried many of Woodward's fanciful tales, is a mystery. Bradlee's comment is self explanatory: "I'm no longer interested in news, I'm interested in causes,' you've got a whole kooky constituency to respond to, which you can waste a lot of time on." "The Bradlee's message: it's okay to fudge the facts, but don't tell your readers and viewers."(28) Unfortunately, Woodward is not alone.

- "On November 9, 1990, *The Wall Street Journal* reported that researchers preparing Dr. King's papers for publication had discovered that substantial portions of his doctoral dissertation and other academic studies had been plagiarized. The "borrowing" was described as "heavy". "The *Globe* (Boston) defended King, quoting scholars who did not feel plagiarism would harm King's reputation. One justification offered was that King was bright enough to have written his dissertation without plagiarism. Perhaps – but had the plagiarism been caught in time, King would not have received a doctorate. The story quickly vanished from the *Globe*." (July 19, 1991)(29)
- "*ABC News* (and others) did not tarnish its image of Nelson Mandela's wife by quoting her zest for "necklacing," in which African National Congress goons cut off an opponent's hands, drape a tire filled with gasoline around his neck, and set afire. Winnie Mandela said on April 13, 1986, "Together, hand in hand, with our boxes of matches and our necklaces, we shall liberate this country." Some 172 "lynchings by fire" occurred in three months that spring. .The media cover-up for Mandela and his ANC is twofold. The media are keeping quiet about control of the ANC by

the South African Communist Party (SACP) and about ANC's publicly stated intentions of using violence to convert South Africa into a one-party Marxist-Leninst dictatorship. On February 11, communists blatantly draped their hammer-and-cycle flag over the podium from which he (Mandela) spoke. Network commentators did not mention its presence; CBS aired mostly tight shots of Mandela's face, keeping the red flag out of viewer range. .Mandela saluted the Communist Party for its "sterling contribution to the struggle for democracy", praising its leaders as patriots. *The New York Times* devoted one sentence to this, deep in its story. *The Washington Post* story ignored it altogether. Neither paper ran a picture of the communist flag. At least 16 of the 30 members of the ANC executive committee were identified as communists in biographical data released by William J. Casey, Director of Central Intelligence on July 30, 1986, that was published in the *Congressional Record* the next day; hence it has been available to the media for years. Numerous policy statements over the years have linked the ANC and South African Communist Party. ANC head Oliver Tambo said in 1981, "Ours (the ANC and the SACO) is not merely a paper alliance, it is a living organism that has grown out of struggle."(February 17, 1990).(30) How can the Media overlook so much so often? Maybe it just takes practice.

- "With the media panting with eagerness to play its volunteered role as publicist, professional environmentalists are girding for "Earth Day 1990," a major propaganda event aimed at convincing Americans we live in a messy, dangerous world. .It will begin at sunrise on April 22, with church bells pealing for the health of the planet," *Time* magazine breathlessly informed readers on December 18. "Perhaps unkindly, we recently ran down some of the predictions made as a scary buildup to the first Earth Day in 1970. *Life* magazine, a sister publication of *Time*, did a lengthy layout it its Jan. 30, 1970 issue entitled "ECOLOGY" A Cause as Movement." A preface page stated, "Unless something is done to reverse environmental deterioration, say many qualified experts, horrors lie in wait. Others disagree, but scientists have solid experimental and theoretical evidence to support each of the following predictions:
- "In a decade, urban dweller will have to wear gas masks to survive air pollution."

• "In the early 1980s air pollution combined with a temperature inversion will kill thousands in some U.S. city."

• "By 1985 air pollution will have reduced the amount of sunlight reaching earth by one half."

• "In the 1980s a major ecological system –soil – or water - will break down somewhere in the U.S. New diseases that humans cannot resist will reach plague proportions."

• "Increased carbon dioxide in the atmosphere will affect the earth's temperature, leading to mass flooding or a new ice age."

• "Rising noise levels will cause more heart disease and hearing loss. Sonic booms from SSTs will damage children before birth."

• "Residual DDT collecting in the human liver will make the use of certain common drugs dangerous and increase live cancer."

• "Let's face it: The environmental movement is big business, with some groups putting more emphasis on lawyers, publicity and fund-raising than science. Consider the Natural Resources Defense Council, which in 1989 huckstered the bogus Alar/apple scare. NRDC's most recent annual report lists as staff members 29 lawyers; four press agents, and nine persons working "development," a euphemism for fundraising. NRDC's $10 million budget supports lavish offices in Washington and New York – but only 9 scientists." (January 5, 1990.)(31) The ecologist tubas keep playing the same old notes, with the Mainstream Media marching along, all but oblivious. To be added to the above list of great scientific findings can be added the recent report by "The Union of Concerned Scientists". In an *Associate Press* article by Randolph E. Schmid that appeared in the Las Vegas Review Journal on February 19, 2004, titled "Scientists criticize administration", the august 'Union' cherry-picked some items of interest to them, and tried to convert them into a political attack. That the National Academy of Sciences had released its study commending the administration apparently had escaped the Union group and the *AP* as well.

• Jane Fonda, who in 1970 told the students at Michigan State University, "I would think that if you understood what communism was, you would hope, you would pray on your knees, that we would someday become communists." Jane, you should have studied under Mother Teresa, not Hillary.

• "Walter Cronkite, who was quoted in the *Moscow Literary*

Gazette of May 1979, seven months before the invasion of Afghanistan, as saying that "an honest person cannot believe that the USSR is preparing for war, that it might attack someone."(32) Wrong again, Walter!

Hopefully the above examples serve to make the reader aware of the vagaries of the Media, and their failure to serve us honestly for a long time. Little has changed regardless of the subject matter or the way Mainstream Media passes it to the public. This in itself is less than encouraging, but also gives rise to the speculation that there are underground forces at work, much akin to termites busily chomping on wooden support beams. Secularist thinking seems to have invaded not only our Media world, but our schools and campuses, our politicians, our legal system, and in effect is moving towards replacing our basic religious foundations. Has the Media addressed any of these rather vital points? Seldom and only superficially is the answer.

"Consider the following list: " Social justice; "hate" crimes; living constitution; sensitivity training; globalism; global governance; global security; world citizenship; coming New World Order; participatory democracy; 'new' Great Society; moral equivalence; diversity; democratic socialism; cultural imperialism; patriarchy; socialized medicine; universal healthcare; political correctness; progressive left; progressive constitutionalism; cultural hegemony; feminism; feminization of the military; tolerance; partial-birth abortion; free love; quality time; homophobic; sexism; animal rights; gay rights; facilitators; progressive education; women's studies; gay studies; critical-race theorists; eco-theology; evolving morality; victimization; lesbian rights; power to the people, redistribution of wealth; liberated art; domestic partnerships; alternative lifestyle; "white male oligarchy"; agents of change; vast right-wing conspiracy;" (33) etc. And of course we also have: extreme/ radical right; conservative right; outside the 'mainstream of America"; bigoted right; "for the good of the children"; equal opportunity; gravitas; uncaring; upward mobility; quota preference; disadvantaged; affirmative action; and similar labels that are often used as part of the lexicon of the Liberal Left. Isn't it strange that seldom does one hear such labels from Republicans? And most Media pundits invariably use "Conservative" whenever mentioning a Republican as opposed to not using a label or an adjective with a Democrat.

Of considerable concern is that "the Web site of the Democratic Socialists of America (DSA) identifies humanists in high office. Fifty-five members of Congress are currently listed on the DSA Web site. Some of the leading voices of the Democratic Party are on that list. And how a hard-core, procommunist humanist could head the House Armed Services Committee is nearly impossible to comprehend". "Not all of the (elected government officials) are humanists, of course, but as a group they have supported humanistic goals often enough to bring us to the threshold of (a) secular, amoral society ready to collapse. Humanists do not mind voting for outsiders as long they agree with a number of key secularist doctrines. Humanists will vote for and protect Bill Clinton, for example, because he is pro-abortion, pro-homosexual, and promiscuous. He supports placing American servicemen under UN command. He supports women in combat even though he avoided the draft. He loves to rule by executive order. He loves to bomb foreign countries despite the U.S. Constitution (a document which all leftists consider confining and reactionary). Some politicians, and only God knows how many, are truly evil. (The average citizen has no idea how much vote trading goes on behind legislative doors.)"(35) Is there a picture emerging here? Yes is the answer, and not an encouraging one..

"If the humanists and their powerful homosexualist friends succeed in forcing the pro-moral majority to accept homosexuality as normal behavior, two thousand years of Western culture and more than three hundred years of American history will go up in smoke." "But perhaps it's even later than we think. Already the humanists are seeking to bring pro-homosexual education to the earliest grades. According to "*Just the Facts About Sexual Orientation and Youth*", a twelve-page booklet produced by a group of psychologists, counselors, educators, and doctors, states "homosexuality is normal and not mentally unhealthy." That booklet was mailed to all of the Nation's public school districts by "Just the Facts Coalition". The coalition consists of the American Psychological Association, the American Academy of Pediatrics, and the Gay, Lesbian and Straight Education Network. The Coalition also included the American Federation of Teachers and the National Education Association. "...the booklet cautions school officials that "endorsement or promotion" of religious ministries aimed at overcoming homosexuality by officials of school employees of a public school district in a school-related context could raise constitutional problems."(36) What has the Mainstream Media said about any of this? Is the NEA seamless? You don't suppose the Mainstream Media has a very large number of humanist members, do you? What's next?

"The media elite are a homogeneous and cosmopolitan group, who were raised at some distance from the social and cultural tradition of small-town middle America. Drawn mainly from big cities in the northeast and north central states, their parents tended to be well off, highly educated member of the upper middle class. In short, the typical leading journalist is the very model of the modern eastern urbanite. Today's leading journalists are politically liberal and alienated from traditional norms and institutions. Most place themselves to the left of center and regularly vote the Democratic Ticket. Yet theirs is not the New Deal liberalism of the underprivileged, but the contemporary social liberalism of the urban sophisticate. They favor a strong welfare state within a capitalist framework. They differ most from the general public, however, on the divisive social issues that have emerged since the 1960s – abortion, gay rights, affirmative action, etc. Many are alienated from the "system" and quite critical of America's world role. Journalists from the post-Watergate generation place themselves well to the left of their older colleagues, and take a quantum leap beyond their elders in criticizing America's economic and political institutions, while sharing their rejection of traditional social and cultural norms. They also profess admiration for liberal public figures and media organs, but disapprove of conservative newsmakers and news outlets."(37)

The most interesting point in the above is the concentration on the orientation of the journalist, and the total omission of comment on the need for journalists to provide "fair and balanced" news, or "objectivity", or whatever else journalism once stood for that apparently no longer exists. How does the Media explain the absence of any of these rather basic requisites? Perhaps Media simply doesn't care.

"The Newspaper Association of America is a publisher's organization, concerned with advertising, circulation, and other business-related issues. The American Society of Newspaper Editors has a different brief, concerning itself with the broad issue of news coverage and the news-gathering process – with journalism proper. The nature of their relationship is often likened to that between Church and State; when the two sides are in agreement, it is often a cause for anxiety, at least among the journalists who are always fretting about the perception that they are sacrificing editorial integrity for the sake of ad sales or circulation figures." (38) This strikes me as a good definition of the respective positions and responsibilities of the two organizations. However, the topic being discussed was not in regards to how best to reform Mainstream Media. The author described the two organizations meeting to embrace the need for "diversity", a mystical word that has crept into our parlance that means much in a social or political discussion, but next to nothing in assessing reporting talents and practical application. The topic these two organizations next agreed on was "pluralism"—which boiled down to how best to achieve "diversity". The solution reached apparently required the use of "quotas". To watch and hear the respective giants of the Media industry arrive at such conclusions must have been inspiring to all who were in attendance, and without doubt the assembled moguls were certain that together they had reinvented the wheel. Exhausting work, indeed, even for empty suits.

The result? Now the leading newspapers have internship programs devoted exclusively to non-whites, programs for "multicultural management" programs, and "Senior Vice-President, Diversity" positions. All this is wonderful, of course, but none of it begins to heal the serious flaws that have grown over the years, including overwhelming mediocrity in the Mainstream Media as it continues its leftist pursuits. Media's current problems have nothing to do with "diversity", or quotas, or multiculturalism. Quality, not quantity is needed. For example, look to Thomas Sowell, Alan Paige, Dr. Walter Williams, Tony Snow, Britt Hume, etc. as the Pied Piper examples needed to bring Media back from the black holes created. Forget the Jesse Jackson "wannabes", or the Al Sharpton/Charles Rangel models. Media needs talented journalists, not just a fishing net filled largely by quotas, or Minstrel Show try-outs. Why is this so hard to understand? Perhaps the view from the top in the Media World is more clouded than suspected.

Contrast the excellent writing of Thomas Sowell (a senior fellow at the Hoover Institution) in a January 22, 2004 article (Creators Syndicate) with what other Media pundits have said about the same subject. "The headline story out of the Iowa caucuses is Sen. John Kerry's surprising victory and the collapse of Gov. Howard Dean's political bubble. This is all very well for those in the media who treat politics as the personal stories of politicians. But, for those who see politics as being about the fate of this country, rather than the careers of candidates, the story out of Iowa is much more sobering. How did Howard Dean become the early front runner in the first place? And what does that say about the current attitudes and future prospects of America? Strident leftist rhetoric was enough to propel him to the front of the pack with Democratic primary voters. Sen. Kerry caught up and overtook Dean largely by doing the same thing, but without the gaffes that revealed Dean's immaturity. What all the Democratic candidates seem to offer are higher taxes – and blaming Pres. Bush for all problems from hang nails to not having a smooth time in Iraq. This has been the liberal formula for decades. Solve problems by throwing the taxpayers money at them and blame others for problems. And when that doesn't work, then throw some more money and more blame. Sen. John Kerry is not only in the classic liberal tradition, he got to the front of the pack by being on all sides of the issue of war in Iraq. He did not win by showing leadership but by following the same rhetoric of Howard Dean while maintaining an air of dignity – but only an air. Perhaps most ominously of all, Sen. Kerry also followed the other liberal candidate's practice of playing the race card. When President Bush appointed Judge Charles Pickering to the 5th U.S. Circuit Court of Appeals, Kerry denounced Judge Pickering as an advocate for a cross-burner. What makes Sen. Kerry's race card truly despicable is that Judge Pickering not only fought against the Ku Klux Klan in Mississippi, back when it was dangerous to do so, he sent his own child to a newly integrated school with black children – which is more than most liberal politicians will do, even today. From the standpoint of the future of this country, Sen. John Kerry's political victory in Iowa means far less than the character flaws he revealed in getting that victory." The entire article is not quoted, but I think you get the gist.

And in a February, 2004NewsMax.com article titled "Courts

Without Law", Thomas Sowell (Creators Syndicate, Inc.) wrote: "One of the tragedies of our time, and a harbinger of future tragedies, is that court decision at all levels have come to be judged by whether we agree or disagree with the policy that is upheld or overturned. Recent controversies over gay marriage have been a classic example of failing to see the woods for the trees. The most fundamental issue is not gay marriage. The most fundamental issue is who is to decide whether or not to legalize gay marriage – and all the other decisions that define a free, self-governing people, as distinguished from people living under dictators in black robes. The political left is all for judicial activism, because courts can impose much of the liberal agenda that most elected officials are afraid to impose, such as racial quotas, gay marriage and driving religious expression underground. Bitter and ugly fights over judicial nominees are one consequence of liberals' heavy dependence on judges to impose policies elected officials dare not impose. Decent, honorable and highly qualified people like California Janice Rogers Brown are smeared and lied about because they insist that what the Constitution says still matters." Succinct, and to the root of the problem; be gone activist judges!

And a March 16, 2004 article by Dr. Walter E. Williams (Creators Syndicate) titled "Politicized teachers push trivia misinformation" reads: "What passes for educational enlightenment these days boggles the mind. Matt Gouras of the Associated Press, writing in the Jan. 5 Seattle Times, tells a story about Tennessee schools. The success of some students has made other students feel bad about themselves. What're the schools responses? Public schools in Nashville have stopped posting honor rolls. Some are considering a ban on posting exemplary schoolwork on bulletin boards. Others have cancelled academic pep rallies, while others might eliminate spelling bees. Nashville's Julia Green Elementary School principal Steven Baum agrees, thinking that spelling bees and graded events are leftovers from the days of ranking and sorting students. He says: "I discourage competitive games at school. They just don't fit my worldview of what a school should be." This is a vision all too common among today's educationalists, but there's a good reason for it: too large a percentage of teachers represent the very bottom of the academic barrel and as such fall easy prey to mindless and destructive fads. There are other causes for the sorry state of today's primary and secondary education. There's been the politicizing of education. Teachers have recruited student to write letters to the president protesting the war and participate in demonstrations against school budget cuts. Very often, good teachers and principals are faced with the impossible task of having to deal with administrators and school boards who are intellectual inferiors and motivated by political considerations rather that what's best for children. One of the very best things that can be done for education is to eliminate schools of education. There's little in the curriculum that contributes directly to the development of the mind. To improve teaching, we must attract people of higher intellectual ability and we must make teacher salaries related to ability and effectiveness. We must ensure that teachers have more academic freedom, better working conditions and a suitable environment for teaching. An important component of that environment is the capacity to remove students who are alien and hostile to the education process. Finally, we should consider curriculum changes that eliminate courses that have little, if anything, to do with reading, writing and arithmetic. The low academic quality of many of our teachers is neither flattering nor comfortable to confront, but confront it we must if we're to do anything about our sorry state of education." Again, the full article is not quoted, but enough for all to chew on.

The college campuses are overwhelmed with diversity related topics and certainly news-rooms don't need to repeat that crapulous indoctrination. But will the Mainstream Media just indulge itself further as it pursues its self-righteous airs and moral primping? "As one perceptive reporter at the *San Francisco Journal* reflected: "The ultimate goal is a society with as much racial and ethnic fairness and harmony as possible, but we can't get there unless we in the press are ready to talk about it in full."[x] (I would hope this means discussing the wide spectrum of Media's problems, not just Media's pursuing its version of L.B.J.'s 'Great Society', plus the inclusion of ability and qualifications being equally important.)

"The journalistic establishments never tire of celebrating the Civil Rights Movement itself, yet its own reporting would make it seem (that) the struggle's gains have been minimal."(40) Newspapers, including the *Times, ABC News, The Washington Post, Los*

Angeles Times, NBC, NPR, Times Magazine, etc., and reporters such as Ted Koppel, Bryant Gumbel, Dorothy Gilliam, Michael Marriott, Richard Cohen, etc. – like Jesse Jackson & Company - have often further exacerbated or contributed to pushing racism than to soothing or solving it, knowingly or otherwise. The Media coverage of other national issues such as secularist activities, the attacks on the Boy Scouts, the surge in homosexual demands, the Bureau of Prisons (Justice) system, welfare, activist judges, NEA attributable failures, litigious lawyers, illegal immigration, etc., are usually distorted and slanted well to the left. Why? William McGowan's "Coloring the News" contains numerous examples of these varying types of perfidy.

Perhaps one of the most egregious examples was "NBC *Dateline's* July 1996 report by correspondent Gary Matsumoto, who repeated the feminist line that there was a double standard in the Navy cutting against pioneer female fliers like (Carey) Lohrenz. .Matsumoto did not report that in Lohrenz's particular case, six separate panels had upheld the decision to ground her. He also failed to note that the "special considerations" she had been given were not minor. "Equally one-sided were the reports of NPR's Pentagon reporter Martha Radditz, who gave Lohrenz a platform to say that those opposed to her flying because of lowered standards were individuals "whose sole purpose is not to have women in the military or women in combat aviation." (41) or that Admiral Bien's description or Lohrenz's flying performance had been "unsafe, undisciplined and unpredictable".

Academic disparities involving "reverse" discrimination have long been on the table, but largely untouched by the Mainstream Media, including, for example, not only Michigan, but Georgetown and Rutgers. Lowered SAT and GPA scores are not "new" in academia, but it takes a brave reporter to shine light on the process or a law suit of some magnitude to get the Media's attention on the process. "The gap in test scores that was at the root of the controversy (the average combined SAT score for blacks at Rutgers was 750, nearly 350 points lower than the average for whites) was important for understanding the story. At no point, though, did *Times* or any other mainstream news organization report it."(42) In 1999, Bob Herbert of the *Times* said (a bid to scale back remediation programs) was equivalent to "ethnic cleansing."(43) Had he expanded on this conclusion to include 'how', it might have been an interesting story, but his 'objectivity' did not reach that far – obviously.

The Mainstream Media has also been tardy, if not remiss, in covering such nation-wide anomalies as "lobbyists" and their specific activities; the politicians courted by the lobbyists, noting voting records of politicians paralleling their lobbyist 'friends'. And why isn't there reporting on the qualifications of members of Congress and the "committees" on which they serve; or the virtual political lock-step and parroting by politicians in their public speeches and voting records? Where is the reporting on the political games and trade-offs that are played in both houses and the pressures that are brought to bear by "outside" special interest groups? Is there no congressional interest in the increase and organization of "gang activities" across the country? Is Media not bothered by congressional disinterest in the growth of "gaming" across the country, including Indian gaming? What about congressional absenteeism? Or the extensive ills and adverse impact of illegal immigration? Or nation-wide organized crime, including "gang" involvement in drug activities? Why is there no coverage of the disgraceful decline in the accuracy of school text books, and how they are selected; or school drop-out rates and the consequential costs? Media has failed to report regularly on welfare abuses and the cost that accrue to federal, state and local governments. And where is coverage of abuses and failures in mandated "Special Education" programs; the nation's extensive "homeless" organizations/problems; the abuses of the tax-free status of many if not most of the "activist" organizations (such as the NEA, NOW, ACLU, etc,)? The role of shyster lawyers in our increasingly litigious society and the resulting costs are largely ignored. Why aren't ethics violations by local, city, county, state and national figures newsworthy? Has Media focused on the politics and finances involved in the election and/or selection of our Judges at all levels and the effectiveness of our courts? Are the rants of environmentalists who have little or no regard for what havoc they may cause because of their exaggerated/distorted claims and pursuits publicized? Why is there no in depth coverage of the dangers and possible abuses in adopting pre-school 'requirements' as proposed by some activists; or the product and problems that result from (further) 'dumbing down' of scholastic requirements? Has a worthy report been made on the impact resulting from the lack of discipline in our secondary schools? Or on the sorry conduct of athletes - and spectators - in this era of 'trash talk' which has resulted in the near total abandonment of 'sportsmanship' standards in high school and college athletic programs? Is there a parallel to the increasing lack of civility that is present across our country, be it on the road or in public gathering places? The list of subject matter is all but endless. Where are you, Mainstream Media? Never mind where have you been – we know the answer to that one. We want to know where you are now, and if you will join in remedying our Republic, or are you intent on watching its destruction? And then what do you plan to do, or who will you blame? Finger pointing is not yet an Olympic sport.

Investigative reporting requires more effort, intellect, and talent than doing simple stories and interviews, or reacting to daily events, or relying on wire services to help fill in the gaps, but surely there must be some journalists who are willing to break out of the mould, or is this just asking too much? Hopefully we have had more than enough of the Seymour Hershs, the Daniel Schorrs, the Peter Arnetts, and those hordes of others on the radical left – or worse – who have deliberately served both the public and our country poorly. As Eldridge Cleaver ® said, "there *are* communists under the bed", yet that threat goes unheeded. Remember, the favorite liberal epithet "McCarthyism" was first used many years ago in *The Daily Worker*, the Communist newspaper.

"According to Michael Barone, senior write for *U.S. News & World Report*, "As managers seek a more superficial facial diversity of women, blacks, Hispanics, Asians, Native Americans, etc., rather than a true variety of viewpoints, newsroom cultures move farther left every year. But in most newsrooms those few Republicans tend to keep their mouths shut, while those who vote Democratic smugly continue to assume that every decent, thinking person does the same."(44) This should serve to unsettle most readers.

"Former *Washington Post* ombudsman Richard Harwood said it well: Diversity may have brought attitudes, perspectives, political values, and interpretations of reality that challenged or diluted the predominant worldview of white males, but newsroom diversity has done little to change its dominant social political values. Journalists for the most part are children of the professional and managerial classes and attend colleges and universities staffed by leftist intellectuals. The best and brightest move on to the most influential newspapers, all of them located in metropolitan centers dominated by the urban liberal bourgeoisie."(45) "The most serious consequence of diversity-obsessed journalism is the deepening credibility crisis of the entire news profession. Letting its own preconceived view of the world interfere with its reporting, the press has simply gotten the story wrong too many time."(46)

The failure of the Mainstream Media to "get it right" is exemplified by the following two examples from the October, 2003 issue of NEWSMAX: "Quoting inside sources at the CIA, Christopher Ruddy reported: "Clinton's anti-intelligence plants implemented a universal 'human rights scrub' of all assets, virtually shutting down operations for 6 months to a year. What this did was to put off limits, in effect, terrorists, criminals, and anyone else who would have info on these kind of people. NewsMax was first with that major story." "Carl Limbacher attended a speech given by Bill Clinton and recorded the ex-president himself admitting he blew a chance to take bin Laden into custody. "Mr. bin Laden used to live in Sudan," Clinton explained to a New York business group in February 2001. "He was expelled from Saudi Arabia in 1991, and then he went to Sudan. And we'd been hearing that the Sudanese wanted America to start meeting with them again. They released him. At the time, in 1996, he had committed no crime against America, so I did not bring him here because we had no basis on which to hold him, though we knew he wanted to commit crimes against America." (NewsMax) Did the Mainstream Media publicize this or did it simply ignore it? You know the answer.

To review the leading lights in today's Media, we find the following:

- "On NBC's June (2003) 15 edition of Meet the Press", Tim Russert interviewed retired Gen. Wesley Clark, who might be a presidential candidate in 2004. Clark criticized President Bush's tax cuts. That's fine, but Clark demonstrated gross ignorance when he said, "I thought this country was founded on a principle of progressive taxation. In other words, it's not only that the more you make, the more you give. Tim Russert, apparently just as ignorant, passed over the statement." From an article entitled "Media Ignorance", by Walter E. Williams, Human Events, week of July 21, 2003. (Dean fell by the wayside in his bid but is being touted at Kerry's choice for Secretary of State.)

• "Lawrence O'Donnell, a former aide to Senator Pat Moynihan, Hollywood reporter and occasional TV pundit, was tapped by TV writer/producer Aaron Sorkin to serve as a consultant on NBC's "The West Wing". Interviewed on Howard Kurtz's CNN Media Show, "Reliable Sources" on May 25, Kurtz asked O'Donnell why there aren't any Republicans on TV shows such as "The West Wing" or "Mr. Sterling." O'Donnell replied: "you will never get that TV show. You'll never, ever get the Republican TV shows. The Writers Guild of American, my union, is at a minimum, 99 percent leftist liberal, and like me, socialist. And we don't know how to write it. We don't." (The American Enterprise, September 2003.) Yet O'Donnell often appears as a 'political analyst' on many of the major networks. Certainly he's fair and objective; why would you think otherwise?

• Bernie Goldberg, author of *Bias,* wrote "the major media, specifically CBS News, led by Dan Rather, had veered so far left they didn't even know they were biased!" "During the '90s, a Roper study found that 89 percent of the Washington press corps voted for Bill Clinton." (A Christopher Ruddy article, NewsMax.Com, February, 2003)

• "Sen. Ted Kennedy called President Bush's judicial nominees "Neanderthals", a group that includes Hispanic lawyer Miguel Estrada and African-American Judge Janice Rogers Brown. Boasting of his party's resolve in the face of the GOP attempt in November to stop the Democrats' filibuster, Kennedy told the Senate, "What has not tended is the resolution and the determination of the members of the United States Senate to continue to resist any Neanderthal that is nominated by this president of the United States for any court, federal court, in the United States." Kennedy's overtly racist language stunned even liberal CNN correspondent Jonathan Karl, who reported, "Strong words from Ted Kennedy suggesting that some of these nominees are Neanderthals." NewsMax, January 2004. Otherwise the Massachusetts Dorian Gray of the Senate remains a paragon of virtue, protected by the Liberal Media elite, of course, and his fellow Liberal Senators.

• "It took only 48 hours for the museum to be destroyed, with at least 170,000 artifacts carried away by looters." New York Times, April 13 (2003). Charles Krauthammer wrote on August 16, 2003, "Turns out the Iraqui National Museum lost not 170,000 treasures, but 33. Frank Rich best captured the spirit of antiwar vindication when he wrote (New York Times, April 27) that "the pillaging of the Baghdad museum has become more of a symbol of Baghdad's fall than the toppling of a less exalted asset, the Saddam statues. The narcissism, the sheer snobbery of this statement, is staggering. The toppling of Saddam freed 25 million people from 30 years of torture, murder, war, starvation and impoverishment at the hands of a psychotic family that matched Stalin for cruelty but took far more pleasure in it. For Upper West Side liberalism, this matters less than the destruction of a museum. .The WMD-hyping charge is nothing more than the Iraqui museum story Part II: A way for opponents of the war – deeply embarrassed by the mass graves, torture chambers and grotesque palaces discovered after the war – to change the subject and relieve themselves of the shame of having opposed the liberation of 25 million people."

• August 6, 2003, AIM Report. Chris Matthews of MSNBC asked "was it the White House's fault or the CIA's fault that the President spoke inappropriately in the State of the Union, claiming there was a threat of nuclear war with Saddam Hussein because of the so-called traffic with Niger that never in fact occurred" "He also alleged that "the main point of their [the White House] argument that the Saddam Hussein regime was trying to buy nuclear weapons so they could attack us or someone else in the world, is unfounded." "After Republican official Ralph Reed appeared on the show to dismiss the controversy as a "feeding frenzy" by Democrats and the media, Matthews declared, "I think Ralph Reed has been proven wrong, because it isn't just a frenzy. It's a fact." "Even liberals were shocked by his callous disregard for the facts." (Cliff Kincaid article, AIM Report, August 6, 1963,) Matthew's constant – bordering on desperate - pursuit of any derogatory information on Vice President Cheney, especially anything connecting him to Halliburton, or oil interests, qualifies as disequilibration. The fast talking liberal feasts on anything negative about the Bush administration, including the Iraq War, which Matthews so badly called.

• "NBC's Tim Russert highlighted the claim by John Deutch,

CIA director under Clinton from 1995-1996, that the failure to find chemical or biological weapons in Iraq would represent "an intelligence failure of massive proportions." "We suggest writing to Russert about his omission of the fact that Deutch had warned of Iraqui WMD in the past, as late as 2002, and was guilty of his own intelligence failure. The use of Deutch as an expert is an affront to our intelligence and common sense." (Cliff Kincaid, AIM Report, August 6, 2003.) That Deutch had totally 'bombed' as DCI was not mentioned by Russert either. Not surprising. Russert more often than not cites articles in the *Washington Post* and *the New York Times* when interviewing guests, or has a panel of liberal guests to exchange less than subtle negatives about non-liberals, or to denigrate Pres. Bush and/or his administration..

• "One must ask where this kind of negativity toward soldiers operating in difficult circumstances come from. The answer is: from the inchoate anti-military suspicion that is permanently in the air in most editorial suites. Most of the pressies I ran into in Iraq were just typical reporters – meaning they were disproportionately left-wing, cynical, wise-guy Ivy League types, with a high *prima donna* quotient." (Karl Zinmeister article in The American Enterprise, July, 2003)

• Peter "Jennings, ABC's anchor, got a flat F. "More than any other anchor" reports the MRC, he "zeroed in on purported weaknesses and failings in the U.S. effort to win support among liberated Iraquis." On the March 22 edition of "World News Tonight", for instance, Jennings disparaged Iraqui pro-American celebrations on the day before: "Yesterday we saw images of a jubilant reception in the southern Iraqui town of Safwan where soldiers stood by as people tore down a picture of Saddam Hussein and jumped in the streets, at least for the cameras." "For the cameras?" (Article by Karina Rollins and the Media Research Center, July/August edition of The American Institute.) Shame on you, Peter; but of course you must have been taken out of context!

• "And then there's Al Gore, who declared last September, presumably based on what he had learned as vice president, that Saddam had "stored secret supplies of biological and chemical weapons throughout his country." "Finally we get to Bill Clinton. In a speech delivered at the Pentagon in February 1998, Clinton described what he called Iraq's "offensive biological warfare capability, notably 5,000 gallons of botulinum, which causes botulism; 2,000 gallons of anthrax; 25 biological-filled Scud warheads, and 157 aerial bombs, and that Iraq still has stockpiles of chemical and biological munitions, a small force of Scud-type missiles, and the capacity to restart its production program and build many more weapons." "It was essential", Clinton declared, that the world address "the very kind of threat Iraq poses a rogue state with weapons of mass destruction, ready to use them or provide them to terrorists who travel the world among us unnoticed." (article by Robert Kagan, June 12, 2003, Special to The Washington Post.)

• "(Bill) Moyers, an old Lyndon Johnson acolyte, received the tongue-in-cheek "Dishonors" in two categories: The "I hate You Conservatives Award" and "Quote of the Year." "Moyers, whose Public Television salary is subsidized by the American taxpayer, bemoaned "the Republican Party's monopoly control of the American government" that was "united by a right-wing agenda." "Moyers accused Bush of trying to "force women to give up control over their own lives," "using taxing power to transfer wealth from working people to the rich" and of bringing America "judges with a political agenda appointed for life." "Others receiving an award included Barbara Walters (for saying Cuba was "one of the freest nations on earth", and Jessica Lange, who said this is "an embarrassing time to be an American." (Article by Wes Vernon, NewsMax.com, June 2003.)

• "On June 9, Peter Jennings' network announced it had hired Rick Kaplan, the former President of CNN, to be the number-two man in its news operation. Kaplan will oversee *World News Tonight, Nightline, This Week*, and the ABC News Political Unit. The Clintons may not be running the White House, but the Clintonites are still running the networks. A long-time friend of Bill and Hillary, Kaplan is as partisan as any journalist in America and has never been shy about using his position and influence to support liberal politicians. (Article by L. Brent Bozell III, Media Research Center, June/July 2003.)

• "The modern GOP is built on a legacy of racism," *Newsweek* columnist Eleanor Clift screamed." (MSNBC.com. piece on December 13, 2002.) "Capital Gang's Eleanor Clift belittles the coalition of the willing: "It's really a coalition of the bullied and bribed." (Media Research Center, March/April, 2003) Hasn't Sen. Kerry adopted your line, Eleanor?

• "On the left, he [CBS reporter Wyatt Andrews] said, the liberal group 'MoveOn.org' is using soft money against President Bush, while on the right the 'Club for Growth' is using soft money against Howard Dean. Andrews didn't mention that, so far at least, the political left has a massive advantage in this area, primarily because some billionaire liberals are putting millions of dollars into MoveOn.org, a group created to save President Clinton from impeachment, and 'Americans Coming Together' (ACT). One, George Soros, is a billionaire liberal with global ambitions who has pledged $25 million (minimum) to defeat Bush, and who has compared the President to the Nazis. Another billionaire leftist, Peter Lewis, has promised another $12 million to this effort." (AIM Report, January 4, 2004.)

• "While the Communist Party USA and Socialist Workers Party played key roles in the anti-Vietnam protests, it was the once-obscure communist Workers World Party (WWP) that took the leading role in the demonstrations against U.S. policy on Iraq. Iraq officials frequently encouraged them to continue. Yet the major U.S. media consistently refused to explain the WWP role, its use of front groups such as International ANSWER and International Action Center, and the fact their top leaders had traveled to Havana, Baghdad, and Pyongyang. At best, the WWP was described as a "radical" or "socialist" group having a secondary role. "Village Voice" Nat Hentoff called the proposal (to investigate the funding sources behind the WWP and the protests) "constitutionally un-American", but acknowledged to radio host Bob Grant that many of the Iraq protests were being organized by "old line communists."(Aim Report, March, 2003.) Didn't know you cared, Nat.

• "Between 1969 and 1974, Nixon: Raised the civil rights enforcement budget 800%; doubled the budget for black colleges; appointed more blacks to federal post and high positions than any president, including LBJ; adopted the Philadelphia Plan mandating quotas for blacks and union and for black scholars in colleges and universities; invented "Black Capitalism" (the Office of Minority Business Enterprise), raised U.S. purchases for black businesses from $9 million to $153 million, increased small business loans to minorities 1,000%, increased U.S. deposits in minority-owned banks 4,000%; Raised the share of Southern schools that were desegregated from 10% to 70%, wrote the U.S. Commission on Civil Rights in 1975. "It has only been since 1968 that substantial reduction of racial segregation has taken place in the South." "The charge that we built our Republican party on race is a lie. Nixon routed the left because it had shown itself incompetent to win or end a war into which it had plunged the United State and too befuddled or cowardly to denounce the rioters burning our cities or the brats rampaging on our campuses. Nixon led America out of a dismal decade and was rewarded with a 49-state landslide." (article entitled 'Neocons Distort Nixon's "68 Southern Strategy", by Pat Buchanan, Human Events, week of January 13, 2003) But Pat, Eleanor Clift and James Carville insist the legacy of racism belongs to the G.O.P.!

The examples cited should be enough to convey the idea that all is not well in the land of the Liberal Left and the Mainstream Media. An article in the 2003 July/ August *American Enterprise* edited by Kathryn Bowman ("A Not-So-Great Media Report Card") underscores what the above makes clear: "58% of those polled believe news organizations' stories and reports are often inaccurate; 58% believe the news media get in the way of society solving problems; 72% believe the news media have too much power and influence in Washington; 50% believe the media protects democracy; 67% believe the media tries to cover up mistakes; that 49% of the media stands up for America." ® This should not strike the Mainstream Media as a very comforting report card, but liberals have a way of interpreting things very differently from most of us.

To the above should be added Media's updates on the Iraq War, the capture of Saddam Hussein, and the less that admirable reporting on the incredible accomplishments of our military in the year plus that has transpired since that war began. To summarize:

• The Mainstream and liberal political pundits were woefully wrong in their predictions of disaster, and few have had the courage to admit that they were mistaken. Lumped in this group are Richard Cohen, Eleanor Clift, Susan Estrich, John Brummett, Ted Koppel, Peter Jennings, Al Gore, Al Hunt, Ted Kennedy, Tom Daschle, Chris Dodds, Jesse Jackson, Chris Matthews, Mike Wallace, Walter Cronkite, Dan Rather, Bill Moyers, Jimmy Carter, Peter Arnett, Margaret Carlson, most of the Democrat candidates for president, Ramsey Clark, Nina Totenberg, Tom Brokaw, Andrew Cockburn, our liberal congresspersons, the liberal left in Hollywood, etc. among others. They should pose for a group Post Office picture.

• Even though there were scores of Media 'imbedded' with the various strike forces committed to Iraq, one might have concluded that there had been major disasters, setbacks, heavy civilian casualties, inept command structures, multiple successes by Iraq forces, ineffective air strikes, heavy losses by Coalition forces, and very poor planning at all levels. Fortunately, Fox News reporting, and a few of the other services, presented a well balanced picture of the Coalition effort, and made it clear that the faulty reporting of other imbedded Media, and TV anchors, was far less than accurate. That the Pentagon had gambled on letting Media ride along and report daily was in itself not noted as significant, but had there been less than an amazing success by the Coalition Forces it was abundantly clear that the nay-sayers were predisposed to strike and condemn. Their editors must have been upset with the way the military took care of business.

• The multiple mass graves, the torture chambers, a terrified populace, the well hidden weapons caches, the opulent living by the Hussein family and followers - as opposed to the standards of living by the general civilian population - became engraved pictures on our TV screens, as did the lack of medicines in the hospitals, the lack of schools, the lack of power/water, etc., even neglected farm lands. The challenges facing the Coalition Forces were monumental in all aspects of nation-building and in overcoming nearly 30 years of "hate America" indoctrination. Incredibly, the Coalition Forces met those multiple challenges and brought a standard of living in every category to Iraq's people they had not enjoyed in 40 years, ranging from medical care, to schools, to water, to electricity, to restoring markets, and, most importantly, freedom for 35 million people. A similar success story is taking place in Afghanistan, again defying the negatively inclined in our Media. That both countries remain a work in progress effort seems to mean only failure to most on the liberal left. There has been – and will be - a cost in lives, of course, but the numbers have been impressively low, especially given what might have been had the Hussein forces chosen to fight, using the vast stores of ordnance they had available. How long did it take to accomplish something similar in Japan, Korea, England, the Philippines, Malaysia and mainland Europe, including Bosnia??

• Some of the imbedded Media did outstanding work throughout the conflict, such as Lt. Col Oliver North (ret.) and Geraldo Rivera (a recovering Liberal, but doing quite well given the many challenges involved). Sadly, some of the imbedded died before the end of hostilities, but not before they had fearlessly brought clarity to the listening audiences. (There are others that deserve recognition for their balanced reporting, but that is more appropriately left to their respective media affiliations and peers.)

• Has the Media done any comparable studies, such as the number of Israelis killed by PLO terrorist bombings in a year's time? Or the number of civilians that Moslem radicals have killed around the world in a year's time? Why not? Or is their interest only in counting the Americans or other Coalition forces killed in Iraq? Just for the sake of comparisons, how many murders occur daily across our country? How many are killed in auto accidents daily in the U.S. How many in America die daily of AIDs? How many Africans die daily in Africa from attacks by their neighboring countries, or from starvation, or from AIDs? How many are in slavery in Africa? How many Tibetans are being held in prisons, and how many die there in a year's time?

• Or doesn't the U.N. have such figures? Or doesn't the U.N. really care beyond pontificating about 'Human Rights'? Are you enjoying your Human Rights in New York?

• *The Atlantic Monthly*, January/February, 2004, in an article by James Fallows, describes the "U.S. occupation of Iraq [as] a debacle not because the government did no planning but because a vast amount of expert planning was willfully ignored by the people in charge. The inside story of a historic failure." Fallows obviously managed to find malcontents in the U.S. State Department, the Agency for International Development, the C.I.A., the Pentagon, the Iraq Working Group, the NGOs (non-governmental organizations), The National Intelligence Council, the Office of Management and Budget, the Council on Foreign Relations, the Center for Strategic and International Studies, Army War College, the International Rescue Committee, Sam Gardiner (retired air Force), Inter Action, the Brookings Institute, the Iraq Survey Group, the

United Nations (weapons inspectors and UNSCOM), etc. that led him to also write: "The exaggerated estimates of Iraq's WMD will loom over future diplomatic discussions. No foreigner trusts U.S. intelligence to get it right anymore, or trust the Bush Administration to tell the truth." After 22 plus pages, Fallows would have us believe that there had been little or no planning not only for the war itself, but especially , and more importantly, for the challenges the Coalition Forces would face after the combat phase of operations was completed. He faults the office of the Secretary of Defense for not seeking the pearls of wisdom that old, experienced hands had to offer, or not seeking the insights to be gleaned from humanitarian and liberal organizations on how to manage the post-war problems, or for the military planners not having sufficiently "war-gamed" the post war period as had been done by the C.I.A. (reportedly a two day effort at Langley) and by U.S.A.I.D. etc., and not seeking the council of the U.N. gurus on what was needed or "how to" in post combat Iraq. (Didn't the first U.N. 'team' cut and run shortly after it reached Iraq?, but then the U.N, gang has run before.) The criticisms include too few troops, lack of detailed planning, not stockpiling supplies, uncoordinated troop movements, use of old military doctrines, troops not being trained to meet the multiplicity of problems they would experience, short-sighted leadership at most levels in the Bush administration, etc. Given the liberal rhetoric and blatant distortions the article included, Noam Chomsky or James Carville could have been co-authors! I don't subscribe to *The Atlantic Monthly*; It was at my dentist's office and taken with his blessing after he noted he doesn't read such nonsense either. The thought has occurred that a more fitting reading room for that article might have been a proctologist's office - no offense intended to proctologists. As for Fallows, don't quit your day job! - perhaps you can help with the *Doonesbury* cartoons.

• Certainly there have been and will be problems as the transition process continues. This is a far different setting and situation than heretofore experienced. Academic solutions do not necessarily apply or are practical. Radical sects and their militias bolstered by fanatics from neighboring countries will try to stop the process of democratization, for it means previous influences and power will be reduced. The faint hearts will continue to preach appeasement (remember "Better Red than Dead" slogans by yesterday's faint hearts?), or try to push the U.N. peanut as the answer. No war or contested change of government has gone according to the originally drawn plan for the simple reason there is no simple solution to difficult challenges. But the liberal, often over the top, blather that has belched forth from our liberal politicians and many of the Mainstream Media makes most of us ill because of its inaccuracies and its denigration of the outstanding efforts of our U.S. forces and Coalition partners.

• The capture of Saddam Hussein was an exceptional find, the culmination of efforts by determined U.S. forces intent on success. Sadly, it did not gain the full plaudits it deserved from all sources. Those on the Liberal Left almost seemed saddened. It was ridiculed by some, rejected as false by some, challenged as violating his rights by some, and described as inconsequential by others. Even the Mainstream Media displayed skeptics and doubters, and conjecture that his capture had any meaning at this point in time. How can those liberal left suits get more empty? Marxism is their standard answer, but Marx was no tailor.

• The Democrat candidates for president enjoyed months and months of "Bush bashing", parading across our nation in McAuliffe's orchestrated effort to achieve the passions of hate and division the Liberals hope to ride into the fall elections of 2004. The central theme is to blame Bush for all terrorist related problems, past and present, the military interventions in Afghanistan and Iraq, failing to have 100 percent accurate intelligence coverage on all subjects since taking office, for not knowing that terrorists intended to hit the Twin Towers and the Pentagon, not giving "diplomacy" or the U.N. inspectors in Iraq more time even after 12 years, and not having a huge Coalition Force join us in attacking Saddam Hussein's forces, plus not gaining specific U.N. blessing to strike. Political garbage is one thing, but to undercut our President and our troops in the field solely for political gain is a bit much. In the minds of most such attacks gives aid and comfort to the "enemy". In the minds of some such attacks constitute "hate crimes". Facts have been ignored, truths stretched beyond breaking, and innuendos repeatedly made that disgust most Americans. The baneful chatter continues as though there were no consequences involved. Some of the criticisms border on the psychotic in presentations, while others have been less virulent but similarly unmindful or uncaring of the truth.

• The hazing and intimidation of some prisoners taken in Iraq, including the indignities that the more suspect prisoners underwent, has been blown far out of perspective. Much of what photos showed could be identical to fraternity initiation practices around the world. Excesses – and deaths – occur in both. In Iraq, interrogation is not a game, but in pursuit of gaining information. This is not to excuse abuse or torture, but the reality is that what has been seen and reported does not begin to approach the torture and abuse that has bee documented as practiced by North Vietnam, North Korea, Cambodia, the Pathet Lao, Russia, China, Cuba. Haiti, Iran, Africa, or Saddam Hussein's Iraq – to name a few -, Geneva Accords not withstanding.

• Some Liberal Democrats have tried to deny their financial benefactor connections (e.g., George Soros), while praising their own record of dedicated Puritanism. From joining and leading the war protestors of yesterday to conveniently remembering (after many years of silence) "heroics" from service in Vietnam to accusing the current administration of dirty politics, the Kerry parade continues. That this is more a panoply of science fiction than a majestic display doesn't seem to bother the Liberal Left. But the silence of most of the Mainstream Media on this near-Shakespearean tragedy is difficult to stomach. Too many Falstaff characters do not a play make.

Most of the Media are already in an amazing variety of costumes, are they not? And, sadly, how can we expect anything different from the clownish but dangerous left? But the considerable amount of egg that has accumulated on Mainstream Media's multiple faces has been put there by themselves, individually, collectively, and deservedly. It neither looks nor smells good.

® 1. Walter Cronkite, former CBS News Anchor. 2.Strobe Talbott, N.Y. Times, and Clinton appointee to State Dep't. 3. Sam Donaldson, ABC News Correspondent. 4. Ted Koppel, ABC Nightline Anchor/ 4. 5. N.B.C.'s Chris Matthews,

® from *That's the Way it Is(n't)*, by L. Brent Bozell, III, and Brent H. Baker, Media Research Center.

* The brain tumor that finally caused Casey's death was causing him speech problem as early as 1981. Words were beginning to be slurred, and accompanying spittle was a problem that prompted many to seat themselves out of range during meetings. I remember on two occasions being the last to enter a conference room before Casey's arrival and having to take the only empty chair to his left, to the amusement of my fellow senior colleagues. The rule-of-thumb was to be sure to arrive early. Casey was an enthusiastic speaker, and one who preferred to 'sell' his point of view, rather than mandate it, which at times extended discussions…and discomfort. Woodward's claim of having a "deathbed conversation with Casey is simply 'horse hockey'.

x From "Coloring the News", by William McGowan.

® Eldridge Cleaver, author of "Soul On Fire" and "Soul On Ice" was originally a member of the Black Panthers in California but came to believe strongly in our Republic.

® Note that all examples in the foregoing portion are dated 2003 or later.

EPILOGUE

(Journey's End, or a Beginning?)

Despite the many and varied destructive forces that impacted our Republic these past fifty years, hopefully it will survive more millenniums. But to do so, she needs our help.

The previous eight chapters clearly related a number of messages, lessons and factual examples intended to impact the reader relating the many and all but unbelievable ways inept leaders of our government have poorly led this country, how pitifully most of our politicians have served us in Washington, failed to identify organizations in America that intend harm to our Republic, and how and why our youth are increasingly poorly educated. The main culprit? Mainstream Media. Had, during this five decade period of review, our Mainstream Media met even its basic responsibilities to provide the public with factual, balanced, and honest coverage, most if not all of what occurred during this period of review clearly could and would have been avoided.

Indo China, especially Vietnam would have had a far different ending; LBJ and McNamara would have been forced into operating overtly and honestly; Kissinger would have been either summarily dismissed or formally charged; Senator Kerry would not have been able to manipulate the Senate Committee hearings regarding POW/MIAs nor dismiss/destroy the evidence presented; Pres. Carter might well have been impeached; William Jefferson Blythe and Hillary Rodham Clinton would not have occupied the White House for eight years, and might well be serving time for their many questionable if not criminal acts; Congress would be a much cleaner stable; the ACLU, the NEA., and the U.N. would be much reformed or likely gone. Simple, but unfortunately such is not the case. Sadly, dooh-dooh did and does happen, and the past is just that.

But numerous changes remain paramount requirements, ranging from revising and reforming our educational system, to reigning in those organizations that seek to convert our Republic to a Socialist (or Communist) system of government, while we begin to reform and restructure Congress, and realign our Judicial system to mirror our Constitution both as the basis of our government and to serve the needs and true interests of our citizens. Ambitious but unattainable goals? No, but certainly challenging ones. Not in my time, but hopefully, in the decades ahead. Time is of the essence for our Republic badly needs these transfusions.

As to the leadership qualities needed in the years ahead, the tasks involved are not for the timid, or the compliant, or the appeasers. These types have had their days of trying to crew our ship of state, and rate a mediocre rating at best. Thanks to political partisanship, often too little was accomplished within our borders that stands the test of time. To the contrary, it has been a downhill slide in most major categories. For those that may not agree, I suggest you re-read some of the chapters herein, for apparently you missed some key points, especially in regards to responsibilities, shortcomings, education, threats, and consequences.

In simple terms, our country cannot afford any more Lyndon Johnsons, James Carters, Gerald Fords, William Clintons, Hillary Clintons, Al Gores, Walter Mondales, Mario Cuomos, Henry Kissingers, Gens. Alexander Haigs or Brent Scowcrofts, or Janet Renos, nor Intelligence Community amateurs such as John Deutschs, Stansfield Turners, and George Tenets. The main strengths of all the foregoing were not necessarily of basic character but simply in pursuing the art of equivocation and political self-aggrandizement. And we definitely don't need more Ted Kennedys, Patrick Leahys, Charles Schumers, John Kerrys, Barbara Boxers, Thomas Daschles; Robert Byrds, Chris Dodds, John Conyers, Dennis Kucinichs, Ron Emanuels, Charles Rangels, Gerald Nadlers, Henry Waxmans, David Boniors, Nancy Pelosis; Bella Abzugs, James Jeffords, Barney Franks, Madeline Albrights, Sandy Bergers, Maxine Waters, Warren Christophers, Hazel O'Learys, Ronald Browns, Jesse Jacksons, etc., all of whom strived more to weaken our Republic than to strengthen it while claiming they were pursuing role(s) of the righteous, and whining for "compassion" when uncovered. The era of the hypocrite, the deceivers, the two – or more – faced is hopefully over, but without doubt there are more of the same ilk loitering in the wings hoping to replace those named, just as others similar are behind them. Our Republic has been too badly damaged to survive another five decades of the regressive forces just experienced.

The preceding eight chapters outlined considerable data and many examples of what our country has managed to endure and survive. Much more could have been added to each chapter, but that is left for others to do. If the needed reforms are not made, there will be further decline. Only the names will change from those cited herein, plus perhaps there will be signs urging the last to leave to turn off the lights.

The urgent goal at this point is to persuade the Mainstream Media to don the mantle of responsibility and objectivity, rather than continuing to pursue the liberal left agenda it has chosen to adopt over the years. It has been a needlessly shameful time for and within the Mainstream Media, and has served our country badly. With an honest Media, other problems can be remedied as well.

There is a glimmer of hope for there is a stirring across the country that bodes ill for the leftist liberals, especially those who choose to continue to pursue their partisan goals at the expense of truth and objectivity. The swing of the pendulum is encouraging, although barely measurable at this point in time. The appeal of Fox's "Fair and Balanced" and "The No Spin Zone", and MSNBC's "The Real Deal", and CNBC's Dennis Miller's news show will sustain if not increase their popularity and probably the adoption of their balanced format by others. Hopefully, Rush Limbaugh, Michael Savage, Michael Reagan, Sean Hannity, etc. will hopefully be on the radio for years as well. (Can't you feel the Liberals cringe?)

The liberal left news stations will continue their downward spiral, for they have naught to offer of substance to a viewing audience. Why the optimism? The major networks require lots of money and large viewing audiences; without both ingredients their bottom lines will continue to decline, which in turn means unhappy stockholders. Voila! To remain competitive, many will be forced to switch and go with the flow, meaning the viewer wins as the liberal left loses. The print media will not be as affected over the near term, but eventually the Los Angeles Times, The Washington Post and the New York Times, not necessarily in that order, will also have to adjust. After all, is it asking too much that the "news" be factual, balanced and objective? Even the news services can and will understand that! There will be grumbling and denials and harsh commentary about these blasphemous expressions, but like the imaginative 7-Up commercials pitch, 7-Up Yours! Competition means it's time to get real.

The Liberal Left TV News tried to feature Al Franken and his unbalanced views (does he have other?). He failed, but others from the Left will follow and falter as well. Even Cronkite, no longer a closet Liberal leftist, or curmudgeon Liberal Andy Rooney, or irascible liberal Dan Rather, can't staunch or reverse the flow to objective and honest reporting. But the parade of willing and eager "guests", or hosts, such as James Carville, Paul Begala, Eleanor Clift, David Gergen (Media's equivalent of opportunistic political analyst Dick Morris), Al Gore, Michael Moore, or agnostic "comedian" Bill Maher, etc., are always available, but unless the Democratic Party or George Soros is willing to fund such an effort and pay the luminaries, who would invest? Sponsors are not that easy to find or keep. George Stephanopoulos might agree to work for Soros following the Franken flop; Paul Begala could easily be a Chris Matthews impersonator; or Michael Moore seek the liberal spotlight, but even cows can chew their 'cuds' but one way. The parade of "guests" would certainly include Bill Maher, Al "Gone Wild" Gore, Howard "Screams" Dean, Michael Moore and the usual Hollywood Liberal Left wannabe-socialists. The ACLU and NEA could provide guest speakers as well. The AFL/CIO might provide some funds, but it's unlikely the industrial unions would risk such public exposure. Trial lawyers, or the U.N. spokespersons? Probably not. Bill and Hillary Clinton? Sure, if the price was right. Start digging, Mr. Soros!

But remember that Soros is already contributing millions to Hillary Clinton's "America Coming Together" (ACT) effort in New York, and has his own agenda regarding 'open' drug policies and his "Open Society Institute". That Soros has also contributed heavily to John Kerry's campaign underscores the old cliché that there is no such thing as a free lunch, thus Soros wins if Kerry wins. Also involved are the following who serve as the ACT Executive Committee:* Ellen Malcolm (founder of Emily's list, the nation's largest political action committee; Steve Rosenthal (former political director at the AFL-CIO); Gina Glantz (30 years in grass-roots political campaigns); Minyon Moore (formerly Chief Operatins Officer of the D.N.C. and Director of White House Public Affairs for the Clintons); Carl Pope (executive director of the 700,000 member Sierra Club); Cecile Richards (President of America Votes, a coalition of 17 national organizations including environment, civil rights, women's rights, abortion, education and labor issues); Andrew Stern (President of the 1.6 million members Service Employees International Union, with a budget of about $150 million). This isn't just an aspiring 'fife and drum corps' outfit, it's a well financed 'mutual fund' strutting down the side streets!

The Mainstream Media can and should play a key role in sorting out and exposing the various spider webs in all the above; whether or not it chooses to will tell us a lot. Some may not be afraid of the challenges involved. Some may even seize current opportunities to leave the "Liberal Left" nest, and jump ahead of those who prefer to remain in the comfort zones they have created for themselves as "liberal" journalists. Herd instincts are not easily broken, as lemmings have regularly demonstrated.

The extensive NEA problems have been well documented, but unless the Mainstream Media pursues them little will change. Congress will do little if anything, for the NEA lobbyists have heavy leverage. However, with Media's help, the subject can be addressed and the necessary corrections eventually made. Between feature stories, investigative reporting, and regular updates the picture can be changed, much to the benefit of all – except of course to the socialists entrenched in the NEA system. And identifying them will eventually lead to their exposure and dismissal. The NEA problem is such that the remedial effort is worthy of a modern crusade involving parents, the Media, P.T.A.s, School Boards, Regents, ranging from pre-grade and secondary schools to college and university curriculums. Without Mainstream Media's help the necessary cures won't happen.

Activist judges are more than just isolated threats to our Republic. The ACLU is the drummer in this unfolding drama and needs to be identified as such across the country. The activist judges have usurped the role of Congress by establishing precedents that are then 'accepted' as law. (School prayer, for example.) Such temerity and foolhardy contempt should not be tolerated. Judges are impeachable and should be when they choose to make the law, thereby exceeding their authority, or if they fail to meet judicial standards. There are cases in city after city where judges have failed their oath of office and the justice system. Willfully or unwittingly makes little difference in the final analysis, for people suffer from the unjust/illegal decisions rendered at all levels by rogue or inept judges. ® That this nation is all but overrun by litigious lawyers is part and symptomatic of the general problem. For example, how can any judge seriously entertain a law suit against a tobacco company at this point in time? Warning labels on cigarette packs began in the 1960s, yet "victims" continue to blame 'others' for their own irresponsible actions. 'Obesity' is becoming a favorite lawsuit pursuit for lawyers. Where is personal responsibility? People still stage 'falling' in a casino and then try to sue, thanks to equally greedy lawyers. The true merit of cases matters little to the "ambulance chasing" lawyers, and some judges, for they prosper from fees and heavy commissions regardless. And don't forget "hate" crimes, or not being "politically correct", and "reparations" are still part of the horizon. The point is that frivolous law suits need to be tossed out summarily by judges, and those bringing such suits, including lawyers, should be fined enough so that at least others will hesitate to bring deceptive cases to court.

Electing judges is not the best of options, but unless the Media presents the voters with the respective records and ratings of those seeking an opening, the process is subject to manipulation. Politicians nominating federal judgeships is obviously flawed, but could be improved if such nominations were supported by the majority of congresspersons from both parties, with the ABA again being the final arbitrator. Just because a process in use repeats itself doesn't mean it can't be improved. Think about it.

Federal judgeships need to be re-thought and the appointment system involved revised so that politicians are excluded from the process. The current sickening display of partisanship in the U.S.

Senate is evidence enough to justify this change. A responsible ABA rendering should be sufficient, including appointments to the Supreme Court. An administration could seek and receive a list of "well qualified" judges from the ABA, and vacancies then filled by administrations choosing from the ABA list. This would eliminate the sorry politically inspired performances – and delays caused - by recalcitrant Senators. I can hear the Liberal Left screaming now, but other than wounded feelings and real or imagined slights by the current crop of liberal turkeys on the Judicial Committee, why should such a vital process be left to obviously partisan politicians? Wealthy and very ambitious Sen. John Edwards, (D, S.C.) a former trial lawyer, has readily joined with fellow Democrat Senators Leahy, Kennedy, Biden, Schumer, Kohl, Feinstein, etc., on the Judicial Committee in degrading and rejecting most nominations – despite favorable ratings by the ABA – submitted by the Bush Administration. Although a junior senator, his ego is such that he sees himself eminently qualified to be president. But thank you, Mr. Edwards, for underscoring the point that you made during your run that the selection system needs to be improved. What really qualifies a Senate committee to pass judgment on any federal judge nominee to begin with? We need more "activist" judges? Surely you jest! Is this truly part of Advise and Consent, or is it merely Arrogance and Pomposity? Come on, Media, let's speak out!

Term limits is another subject that badly needs to be revisited and approved. Why is it we think we need to keep any one of our elected officials in office for more than three terms? Why couldn't House members be elected to terms of three or four years, with a total of 12 years being the limit? Why can't the term for a Senator be set at 12 years? The cost of running for either the House or Senate is already far out of control, and "campaign finance reform" remains a bad joke. Why aren't the tax payers allowed a voice in this? Why do we allow presidential 'campaigns' to go beyond two or three months? Certainly two or three months are sufficient time for the respective parties to choose their candidate at all levels of offices. Five or six months total for the entire selection process, including conventions, is more than enough time. Those who make substantial money from the carnival-like "debates" and "campaigning", and the parochial interests involved, would oppose, but the tax payer would realize savings as well. The lengthy campaigns/ debates conducted by the aspiring Democrat candidates across the country wasn't exactly entertainment, but did surface instabilities and faults in some who were eventually pegged as unworthy. Those who have tried to disclaim questionable past records but couldn't then quickly have to be able to adroitly put on new makeup, or face defeat. The frightening aspect of all this stems from the multiple "buys" and "sells" involved behind the scenes, as touched upon previously. Unfortunately, the "sale" or buying of political offices has become part of our political landscape.

Democracy isn't synonymous with Democrats, or vice versa, just as Republicans isn't a term synonymous with Republic. However, 'Tax and Spend', especially in support of social programs, has long been synonymous with Democrats, just as 'States Rights', a strong central government, and strong defense is the labarum of the Republicans. The rest of the labeling of and on both parties is the result of political pundits on both sides, and by the Media, highlighted by Hillary Clinton's fatuous claim of a "vast right-wing conspiracy" as being the cause of Bill Clinton's many problems. But the Democrats decided they had come up with something exploitable and soon the Liberal Mainstream Media willfully joined in the branding and extensive name calling that followed, and remains in vogue. That this will continue for another generation is a good bet. We'll see what labels stick and where.

I have engaged in 'labeling' because I want to make clear that Liberal Left is, politely, a shade short of socialism, but only a bit less at best. And a Liberal is just a nudge from being Liberal Left. We need to remember that our forefathers defended their political beliefs, at times heatedly so, but there were not differences regarding basic beliefs, goals, dedication and "patriotism". The basic structure and needs of the new government had to be outlined, including treasury needs, postal system, laws, diplomatic, including treaties, and policy considerations, as well as the rights of states resolved. The daunting challenges and tasks were all but overwhelming, yet the Articles of Confederation were agreed to in 1777 and finally ratified on March 1, 1781 thanks to the dedicated efforts of the delegates from the 13 states. The state's commissioners began their proceedings in May 1787 in Philadelphia and by summer's end they had hammered out a majority agreement. A year later nine states had ratified it. The Federalist Papers detail the original intent and purpose of the Constitutional Convention, and the Articles of Confederation provide the blue print resulting. We now have 535 egos wearing expensive suits most of whom seem to care not for our history, and precious little for our future. But the speeches drone on "Americans want (blah, blah, blah)", or "Americans need (blah, blah, blah)". Who are they to tell us about what we want or need? Who writes those silly scripts that are repeated endlessly until a 'new' one finally appears? The answer is probably "staffs", but that merely compounds the resulting insult, for most of the 435 politicians have trouble reading that which, purportedly, are their own lines! Is this Robotics 101, or the DUH factor?

Yes, America has survived to date, but how many more rebounds can she make, and how many more serious domestic and foreign threats ahead can she successfully overcome if we have more LBJs, Carters, and Clintons at our helm? Certainly the advisory team that Sen. Kerry has chosen of former Sec. State Albright, former U.N. Ambassadors Holbrooke and Richardson, former NSC advisor Sandy Berger, former Sec. Def. Perry, former Joint Chiefs of Staff Chairman Shalikashvili, former NATO Commander Wesley Clark, and Sen. Biden qualifies more as the "Gang That Couldn't Shoot Straight" than as a reassuring, constructive brain trust. It appears that Kerry's intent is to repeat the Clinton years of turmoil and policy by polls, but perhaps with fewer cigars. Not visible in their inner sanctum, at this point at least, are the Chinese and Indonesian deep pockets, and Hillary. But stay tuned.

Iraq has served as a focal point for the negatives, be they from politicians, reporters, or talking heads. Balanced reporting is obviously not a requirement and is indeed a rarity. Is there recall that it was in March 2003 when U.S. and Coalition forces first went into Iraq? The military successes were outstanding! However, because of many years of neglect, Iraq's entire infrastructure needed to be rebuilt, and has been. Schools and Universities, hospitals, power plants, irrigation systems, telephone systems, police forces, small businesses, etc. are operating. Political agreements and a governing system have been agreed to by the Moslem tribes. That the U.N. did little in all this complicated process beyond urging the Iraquis to agree to an orderly, not a hasty, process, is recognized and applauded, but the Iraquis themselves have been the main movers toward assuming responsibility for governing, a new constitution, and elections, despite the sustained efforts by Hussein loyalists and foreign terrorists. Yet all these achievements seem not to satisfy the vocal critics on the left. Amazing! Reminds me of the "on call" (spontaneous) demonstrations and demonstrators the "left" magically assembles for Media coverage, and about as persuasive. C-Span should be especially embarrassed – again! But that's their problem. Sure, it "sells", but so do drugs, pornography, and countless scams.

I don't want the future of our country, or my children, or my grandchildren, or their children, etc., under any kind of control by any segment or appendage of the United Nations. Nor do I want the United States to become "socialist" in any manner, much less communist, the next sequential step. I don't want our country under the leadership of Liberals, thinly veiled Leftist Liberals or leaning-left moderates. There has never been a "threat" of any kind from Conservatives, or the "Right", despite the extensive name-calling and groundless charges by Liberals. However, there have been, are, and probably will be threats from the "Left". Why? It's the nature of the beast and the rabble rousers are always eager, for small stipends, to display their limited capabilities. The Left's modus operandi? Bluster and promises over substance, and whining demands for more money and centralized power

Allow me to make my views in very simple terms: I love my country, its flag, its unique history, its freedoms, and am proud of its many and extraordinary accomplishments. I get teary eyed when the National Anthem or 'America' is played, or when an Honor Guard passes by with our flag. I am proud to have served our country both in uniform and as a civilian. With the exception of William of Orange, I cherish our British Isles heritage, but am thankful the brothers McCarthy left southern Ireland in the 1800s to settle here. I much admire and respect our Australian colleagues who have done so much for us over the years in so many ways and places - and New Zealand, which has usually been a reliable friend as well. Canada is a special neighbor. Apparently she has abandoned the socialist path she unwisely chose to pursue earlier, and is returning to her former and wiser ways. Great Britain? Hopefully the close historical bonds will continue. Certainly – and thankfully - Winston Churchill, Margaret Thatcher, and Tony Blair have become an integral part of our history as well.

However, Mt. Rushmore remains our special sacred grove, for it appropriately celebrates those courageous leaders who led this country through dangerous and difficult times, making it possible that America achieve what no other country on earth has been able to match, despite the formidable odds and challenges experienced along the way.

North America is God's work at its best. This does not mean that there are not other beautiful places in this world; it simply means that we Americans appreciate having been born in the United States, and appreciate what our country has and offers. Although still a young country, we're proud of our history and the many and unique advances we've made - and then shared them around the world. We have no colonies, nor do we seek any, but we've willingly come to the assistance of many countries, be it to help repel invaders or bring relief where disaster or famines have struck. Many countries have called for our assistance and we have always responded. America receives little recognition or thanks for this, and often memories are short, such as in France and Germany. That's fine, but in turn let's play the game fair and up-front, not unfairly and dishonestly. The Liberal Left does not seem to grasp this point. We Conservatives think of the glass being half full, not half empty, and appreciate having both a glass and the water. The Liberals whine for and demand a larger glass and more water.

These recent decades have been interesting and challenging. Our country has experienced a lot, done a lot, and given a lot. For instance: from the Korean War in the early 1950s, to the incredible shootings in the House of Representatives by Puerto Ricans in 1954, to developing the St. Lawrence Seaway with Canada, to the development of the Hydrogen bomb, to discovering DNA, to developing nuclear powered submarines and S.D.I. In the middle 1950s, SEATO was formed; Pakistan became the first Islamic Republic, and finally again became a friend of the U.S., as did India. The U.S. and the U.S.S.R. developed intercontinental missiles; and both sides described the relationship as a "cold war". In the late 1950s the Soviets launched "Sputnik"; NORAD (North American Air Defense system) was formed with Canada; U.S. Marines went into Lebanon; the European Common Market was formed; the first U.S. satellite was launched; Castro came to power in Cuba;

Alaska and Hawaii became our 49th and 50th states; China invaded Tibet and the 14th Dalai Lama fled to India in 1959; and the 400 mile St. Lawrence Seaway opened.

In the early 1960s, Gary Francis Powers was forced to land his U-2 spy plane in Russia; Dr. Sabin improved the Salk vaccine; John Kennedy won the presidency over Richard Nixon; the Peace Corps was formed; Castro declared Cuba a socialist nation; the border between East and West Berlin was blocked and walled; Pres. Kennedy increased the U.S. advisers in South Vietnam to 16,000; and the U.S. Supreme Court ruled that the reciting of an official prayer in New York public schools was unconstitutional. In the mid-60s, the USSR announced it would not pay its share of U.N. expenses; a nuclear test ban treaty was signed; President Diem and his brother were killed in Saigon; China attacked India across Tibetan borders; President Kennedy was assassinated in Dallas; Castro's sister defected to Mexico; racial tensions were high in New York City, New Jersey, Chicago, and Philadelphia; the 888 page Warren Commission report on the Kennedy assassination was released; China detonated a nuclear device; the U.S. Surgeon General issued warning that smoking was injurious to your health; Canada adopted a new flag; Malcolm X was killed by Black Muslims; Syria nationalized nine oil companies; Indonesia seized 3 U.S. oil corporations and a Goodyear Tire Co. facility in Indonesia; U.S. Marines and Airborne forces deployed to the Dominican Republic; Australia sent an 800 man battalion to S. Vietnam and New Zealand sent a smaller group; there were race riots in Watts, California; thousands of Cuban refugees left Cuba; Anti-Vietnam War demonstrations began in Berkeley, Oakland, NYC, Boston, Philadelphia, Ann Arbor, Washington; there were race riots in Chicago, Cleveland; Brooklyn, Omaha; Baltimore, San Francisco, Jacksonville, Newark, and Detroit; Suharto ousted Sukarno in Indonesia; National Organization of Women was founded; Six Day War occurred, and 53 nations sign the General Agreement on Tariffs and Trade. In the late '60s, Che Guevara was killed in Bolivia; the "Pueblo" was seized with her 83 man crew; North Vietnamese forces attack South Vietnam forces during the 1968 Tet holiday; Martin Luther King was murdered; Robert Kennedy was killed in Los Angeles; Students for Democratic Society staged riots nation wide; Iraq publicly hanged 25 "Israeli and CIA spies" in Baghdad;

China and USSR engaged in border clashes; Ted Kennedy scandal involving the drowning of Mary Jo Kopechne in Massachusetts; Neil Armstrong is first man to step onto the moon; Ho Chih Minh dies at age 79; Washington and San Francisco "Moratorium" demonstrations; 89 Indians seized Alcatraz Island; Gay Rights movement starts; and Woodstock attended by 300,000.

In the early 1970s, Jordan and Palestinian Liberation forces clashed; 4 died at Kent State when some of the 100 National Guard fired on 500-600 demonstrators; Palestinian commandos hi-jacked three western airlines bound for New York; the 10 Day War and King Hussein crushed PLO forces in Jordan; Environmental Protection Agency established; civil war in Pakistan created ten million refugees; China and U.S. engage in tennis table diplomacy; Apollo 14 and 15 flights to the moon; 200,000 anti-war demonstrators in Washington, D.C., and 156,000 in San Francisco; Jordan captures 2,000 PLO commandos; China seated at the U.N.; Soviets signed 15 year friendship pact with Iraq; Pres. Nixon visits China; five men arrested for Watergate break-in; Pres. Sadat terminates the Soviet military mission; U.S. ended its ground combat role in Vietnam; eight members of the Arab Black September terrorist group kill two Israeli Olympic athletes and seize nine hostages; Pres. Nixon and Gromyko sign SALT accords, U.S. military draft ends; 11,380 foot tall Twin Tower World Trade Center completed; Life Magazine ceases publication; American POW release begins; last U.S. troops leave Vietnam (8,500 civilians remain); Senate began hearings on Watergate; Libya nationalized foreign oil companies; Yom Kippur War began; V.P. Spiro Agnew resigns, replaced by Gerald Ford; Arab oil producing nations imposee total ban on export of oil to U.S.; and vaccine to prevent rabies is developed

In the mid to late 1970s, China seizes the Paracel Islands; Iran and Iraq conflict begins; Arab guerrillas raid Israeli border town killing 18, mostly women and children, with the Lebanese Marxist Popular Front for the Liberation of Palestine claiming credit. President Sadat announces Egypt ending reliance on Soviet arms aid; President Nixon resigns after House of Representatives opened impeachment hearings; Sandinista guerrillas seize some 30 hostages and demand one million dollar ransom, free passage to Cuba and release of 14 political prisoners; war between Moslem and

Christians in Beirut, Lebanon, resulting in 100 killed and 200 wounded; Lon Nol surrenders to Khmer Rouge; Saigon falls; U.S. cargo ship Maiaguez seized by Khmer rouge, and 38 U.S. servicemen killed in rescue attempt; Pathet Lao take control of Laos; Mao Tse-tung dies; Castro sends aid to revolutions in Latin America, Angola, Ethiopia and Grenada; six Palestinian terrorists raid the Vienna meeting of the Organization of Oil Exporting countries, killing 3, wounding 6 and taking 81 hostages (Austria provides the terrorist with a plane to fly to Algeria); Soyuz 19 docks with Apollo 18; Chou En-lai dies; Senate discloses Lockheed Aircraft Corporation has illegally paid 22 million to Japan, Turkey, Italy and Netherland representatives; racial violence in South Africa results in 128 dead, 1,112 injured and thirty five million in damage; Israeli commandos rescue 103 hostages in Entebbe, Uganda, being held by Palestinian terrorists; the U.S. celebrates its bicentennial; Pres. Carter ends restrictions on travel to Cuba, Vietnam, North Korea, and Cambodia; Spain rules that Communist party can participate in parliament elections; 1,717 Clamshell Alliance protestors arrested in Seabrook, N.H. for civil disobedience protest against nuclear plant construction; 150 U.S. politicians accused by F.B.I. of taking several million dollars from South Korean agents; Canada and U.S. build 2,700 mile oil pipeline in Alaska; Pres. Sadat breaks diplomatic relations with Syria, Iraq, Libya, Algeria, and South Yemen; Israel invades southern Lebanon to root out Palestinian guerrilla bases; Afghanistan Communist party leader creates Marxist state with Soviet help, despite Moslem resistance; thousands of ethnic Chinese flee Vietnam seeking refuge in Malaysia, Indonesia, Thailand and Hong Kong; Moslem anti-government riots in Iran; in Zaire, 1,750 French paratroopers and 1,750 Belgian troops flown by U.S. Air Force, rescue 2,500 Europeans; Jim Jones and 911 of his followers commit suicide in Jonestown, Guyana; U.S. and China establish diplomatic relations (China promises to pay 41 cents on the dollar in settlement of $196.6 million in claims by the U.S. for seizures in 1949); the Shah of Iran leaves after 37 years of rule; Ayatollah Khomeni returns to Iran after 15 years in exile; 200,000-300,000 Chinese invade Vietnam along 480 mile border, but withdraws after one month; Panamanians take control of Panama Canal (with full control to be given to Panama on last day of 1999; and the U.S. Embassy is seized in Iran by Khomeni followers.

In the early 1980s, the U.S. Olympic Committee votes to boycott the summer Olympics in Moscow; six U.S. embassy employees who had hidden for five months in the Canadian Embassy in Iran were flown out with Canadians; 31 public officials, including a U.S. Senator and seven representatives were charged by the FBI in connection with ABSCAM, resulting in five Congressmen being indicted; 10,000 Cubans seek exit visas from Peruvian Embassy and are flown to Costa Rica, leading to the Mariel boat lift which resulted in 125,000 Cubans leaving Cuba; Iran and Iraq go to war; Mt. St. Helens exploded, killing 26, scattering ash 120 miles; U.N. has 140 members, which is slightly 100 more than at inception.

In the early 1980s, Iran releases all 53 hostages; Great Britain has race riots; Socialist Party leader Francois Mitterand defeats incumbent Giscard D'Estering in France; Israeli warplanes destroy a nearly completed atomic reactor near Baghdad; President Sadat is killed by Moslem extremist soldiers; Egypt expels 1,000 Russians. In the mid to late 1980s, England, Paris, Brussels, Potsdam, East Germany, and Amsterdam have mass demonstrations for nuclear disarmament; pneumonia, meningitis, rheumatic and scarlet fever have been nearly eradicated by antibiotics; vaccines have overcome diphtheria, whooping cough, typhoid fever, polio and measles. Smallpox was wiped out in 1977; malaria, tuberculosis and influenza are under control; President Reagan proposes states assume responsibility for food stamps, aid to dependent children and highway construction, but Democrats attack his proposal; Mitterand assumes ownership of companies with combined work force of 650,000 including five of France's largest industrial groups, 18 commercial banks and two leading investment banks- France will reimburse eight billion total; El Salvador has first free elections in 50 years, with centrist Christian Democrats and five right wing parties forming National Unity Party; Falkland Island conflict begins, with Secretary of State (Gen.) Haig attempting to negotiate, but resigns on 25 June after failing; all Palesinian forced out of Lebanon; France expels 47 Soviet diplomats and U.N. officials on charges of conducting espionage; ANC (African National Congress) car bomb explodes outside Pretoria air force base killing 18 and injuring 200; Korean Airline flight 007, a Boeing 747, is shot down by Soviet MIG, killing 240 passengers and 28 crew members (Soviets admit "stopping" the plane eight days later.); In Beirut, 237 Marines die, and 80 wounded from Islamic Holy War, a pro Iranian Moslem group, terrorist bombing of the Marine Barracks; a French military post was also attacked, killing 58 and injuring 15; an Israeli post was also attacked, killing 29; U.S. troops invade Grenada to rescue American student hostages from 600 Cubans; U.N. Security Council then votes resolution "deploring U.S. action in Grenada.

In the early 1990s, President Bush authorizes invasion of Panama to seize Panama's President Noriega who was heavily engaged in international drug trafficking and for increasingly threatening both the U.S. civilians living in Panama and our military personnel assigned to Southern Command, Panama. President Bush also authorized U.S. and Coalition forces to free Kuwait from Iraq forces that had invaded Kuwait in an action that became known as the 100 hours war. Unfortunately, U.N. restrictions did not allow Coalition forces to seize Baghdad or capture Saddam Hussein. Hopefully, you already know the summarized "rest of the story" into the 1990s.

Obviously our country has not been found wanting, although many of our key officials and politicians can't claim that same distinction.

The choice by many countries, including England, Australia, Poland, Italy and Spain to join the U.S. led Coalition forces in attacking the Taliban in Afghanistan and Saddam Hussein's Iraq forces in March 2003, deserves special praise. Steadfast friends are hard to find when the going gets tough, but they shine all the more brightly for their courage in being at our side.

Those who are faint of heart and unwilling, for whatever reason, need to reexamine their core values and weigh their decisions, for cowardice and pacifism solve no problems, but merely encourages the enemy, whoever and wherever he may be, to recognize and exploit the timid for years to come. Strength of beliefs and values are not commodities worth trading, especially when the future of one's country is at stake, but such decisions are only yours to make. We Americans do not pretend to be anyone's conscience. But understandably, we may express our disappointment and views, for freedom of speech is an integral part of our heritage. Chairman

Mao noted that strength came from the barrel of a gun. He spoke truthfully, but crudely. No one has ever proposed or said that strength came from the mouth alone. Beware the glib, for words alone do not make the man.

To you Americans who aspire to lead this country let me say that unless you have the strength of character and moral standards necessary to do so, don't bother to try. Slickness and good looks and a smooth line are no longer enough. If you are without probity you'll be found wanting and receive the ridicule you richly deserve. If you have enriched yourself as a politician from special interests or lobbyists, or just plain graft, you'll be exposed. The "attack dog" concept employed by the McAuliffe D.N.C. gang will not save you, for those acting in that capacity will be exposed as well, with or without the Mainstream Media. Al Gore and his "win at any cost" lawyers failed, and so will any like-minded "liberals" who try to follow that shameful effort, whether supported by the Mainstream Media or not. A Pox on your disgusting, collective efforts, say I, just as the Mainstream Media should have so exclaimed at the time. And if the Mainstream Media does not change its ways, I say a Pox on your house as well.

The Liberal Left has no conscience, but does possess a well developed capacity to lie. Bill and Hillary Clinton's administration gave us eight years of uninterrupted lies. As shameful as that period was, it makes one wonder if there isn't a course that Liberals have access to that provides them with basic instruction, including "some assembly needed" tips. Or do these people just develop the requisite attributes on their own, perhaps beginning with successfully cheating in school, or telling whoppers to one another, practicing and testing their skills. By this process they would soon recognize that such talent has potential in law, or in politics, or in journalistic pursuits, wherein the one who tells the most convincing lies may eventually get to Washington, or gain reporting fame. Once there, it is like pursuing a Master's degree, and of course, finally a Doctorate's, or perhaps like gaining "tenure" as have some of our longest serving politicians. Most Liberals seem to be from the same tree, so to speak, in appearance, mannerisms, bluster, and the ability to whine and, on cue, shout – yet avoiding the questions asked. But come to think about it, ACLU lawyers have the same talents and characteristics, as do Mainstream Media news anchors. We may finally have discovered the formula involved in these professions. Now we need o find one or more anecdotes for the afflicted. While not all lawyers or politicians or Media types are in this category, obviously far too many are.

Inexplicably, Richard Clarke, former anti-terror Czar for Pres. Clinton, (and having also previously served in the Reagan and Bush 41 administrations in lower capacities) held a counterterrorism coordinator position under Pres. Bush and Condoleeza Rice until he was released. He now claims Pres. Bush chose not to heed any of his (Clarke's) many warnings about al-Qaida's impending attacks on the U.S. He – and others - also charges the Bush administration with having misled the American public about Iraq, etc., etc. He has encapsulated these charges, and his evident anger, in a self-serving book, "Against All Enemies", released in March 2004, following his March 21 bombastic appearance on *Sixty Minutes.* His having a cabinet level position under Clinton but a much reduced one under Pres. Bush apparently has taken its toll on Clarke. His rants against Bush, like Gore's and Dean's, do not include any explanations as to why next to nothing was done about the al-Qaida problem during Clinton's eight years. According to Clark, there had been hundreds of conversations and meetings about Osama bin Laden and al-Qaida. Perhaps so, but the overall impression is that Mr. Clarke now seeks to capitalize on his years with Clinton, profit from his obvious angst, gain financial rewards from speaking engagements, and endear himself to the liberal minded at Harvard University where he presently holds forth. He and former Gov. Howard Dean share much in common, and both will likely be featured in Sen. Kerry's campaign, as will Bill Clinton and others from that mindless era. Why not? One, or more, can – and will – lie, while others of the like-minded swear to it! How convenient. Wasn't it Howard Newton who said "When a man blames others for his failures, it's a good idea to credit others with his successes"? But surely other liberal left whiners will surface not only in the months ahead, but in the coming decades as well, trying to sell similar pitiful tales. It's their life's blood. Like labels on cigarette packages, warnings of their dangers should be mandatory and constant. Buyer beware remains good advice.

Although she's in peril, may God save our Republic, as he has seen fit to do since the days of our forefathers! But Mainstream

Media, your help is sorely needed. You have failed us for far too long. It's time for you to make amends.

We all need to remember what Abraham Lincoln said about America: "This country is the last great hope of the human race." This is worth repeating often, even by Liberals.

* Human Events, the week of March 1, 2004

®For example, a malpractice suit filed nearly ten years ago against a doctor for flawed treatment has dragged on in Las Vegas courts, (and the Nevada Supreme Court) with different judges presiding and rendering decisions that finally prompted local papers to question the process and the very apparent injustices. One judge even disallowed the plaintiff to present evidence and supporting testimony to a jury…then summarily dismissed the jury, finding for the 'defense'. Included in the plaintiff's testimony was evidence of defense lawyers attempting to interfere with testimonies, etc., on behalf of the plaintiff. That the legal office representing the doctor's insurance company is large, prestigious and influential is exemplified by one of Nevada's U.S. Congresspersons, a former senior partner in the firm, writing on official U.S. Congressional letterhead, speculating that the plaintiff apparently could not find a lawyer willing to take the case, thus likely did not have a legitimate claim. Wrong! Despite the sustained and often questionable efforts by the law firm in question to deny the plaintiff a fair hearing, the case remains ongoing as this is written. Because of the stacked deck(s) encountered, the plaintiff has largely done her own legal work, although now legally blind. "Hear ye, hear ye, the kangaroo court is now in session!" A very sorry travesty, to understate it.

End Notes

Chapter 1: “Indo China”

1.Stephen Pan; Daniel Lyons, “Vietnam Crisis”, (Robert Speller & Sons, 1966) pg.12
2. *Ibid* pg. 13
3. Ibid pg. 15
4. George Kahin, John W. Lewis, “The U.S. in Vietnam” (The Dial Press, 1967) pg. 32
5. *Ibid* pg. 17
6. *Ibid...*pg. 20
7. Stephen Pan, Daniel Lyons, “Vietnam Crisis”, pg.24, 25
8. *Ibid...*pg. 30
9. *Ibid...*pg. 42
10. *Ibid...*pg. 67
11. *Ibid* pg 90
12. *Ibid...*pg. 95, 96
13. *Ibid...*pg. 86
14. *Ibid* pg. 86
15. *Ibid* pg. 97
16. *Ibid* pg. 99
17. *Ibid* pg. 104
18. *Ibid* pg. 116
19. *Ibid* pg. 116
20. *Ibid* pg. 104
21. *Ibid...*pg. 105
22. *Ibid* pg. 108
23. *Ibid* pg. 115
24. *Ibid* pg. 115

25. *Ibid* pg. 115
26. *Ibid* pg. 118
27. *Ibid* pg. 118
28. *Ibid* pg. 118
29. *Ibid* pg. 119
30. *Ibid* pg. 125
31. *Ibid* pg. 125
32. *Ibid* pg. 125
33. *Ibid* pg. 125
34. *Ibid* pg. 125
35. *Ibid* pg. 126
36. *Ibid* pg. 131
37. David Halberstam, "Making of a Quagmire" (Random House, 1964) pg. 32
38. *Ibid* pg. 30
39. *Ibid* pg. 305
40. "The U.S. in Vietnam", pg. 145
41. "Vietnam Crisis", pg. 133
42. Chester A. Bain, "Vietnam, The Roots of Conflict" (Prentice Hall) pg. 132
43. *Ibid* pg. 133
44. Richard Philip Jennett, "A Sadness That Was Vietnam", (The Boggastow Book Company, 1987) pg. 112
45. "Vietnam, The Roots of Conflict", pg. 137
46. *Ibid* pg. 139
47. *Ibid* pg. 141
48. *Ibid* pg. 168
49. *Ibid* pg. 172
50. "A Sadness That Was Vietnam", pg. 114
51. *Ibid* pg. 66
52. *Ibid* pg. 67
53. *Ibid* pg. 82
54. *Ibid* pg. 118
55. *Ibid* pg. 161
56. *Ibid* pg. 174
57. *Ibid* pg. 177
58. *Ibid* pg. 178
59. *Ibid* pg. 182
60. *Ibid* pg. 184

61. *Ibid* pg. 267
62. *Ibid* pg. 321
63. *Ibid* pg. 320
64. *Ibid* pg. 321
65. John Clark Pratt, "Vietnam Voices" (Viking, 1984) pg.268
66. *Ibid* pg. 286
67. *Ibid* pg. 453, 454
68. *Ibid* pg. 499
69. *Ibid* pg. 587
70. Arthur J. Dommen, "Conflict in Laos" (Praeger Publishers, 1971) pg. 170
71. *Ibid* pg. 379
72. *Ibid* pg. 302
73. *Ibid* pg. 190, 191
74. Henry A. Kissinger, "Years of Renewal" (Simon & Schuster, 2000) pg. 501
75. *Ibid* pg. 546

Chapter 2. "Lyndon Baines Johnson" (& Roberts S. McNamara)

1.David C. Whitney, "American Presidents" (Reader's Digest, 1996) pg. 316
2. *Ibid* pg. 317
3. *Ibid* pg. 318
4.Robert Caro, "The Years of Lyndon Johnson, The Path to Power" (Knopf, Inc. 1982) pg. 3
5. *Ibid* pg. 6
6. *Ibid* pg. 10, 13 (brief quotes)
7. *Ibid* pg. 21
8. *Ibid* pg. 22
9. *Ibid* pg. 20
10. *Ibid* pg. 16
11. Ronald Kessler, "Inside the White House" (Pocket Books, 1995) pg. 16
12. *Ibid* pg. 19
13. George C. Herring, "Shadow On The White House" (University of Kansas, 1993) pg. 115
14. Rowland Evans and Robert Novak, "The Exercise of Power" (New America Library, 1966) pg. 531
15. Henry L. Trewhit, "McNamara – His Ordeal in the Pentagon" (Harper &

Row, 1971) pg.195
16. *Ibid* pg. 197, 201-202 (tied brief quotes)
17. *Ibid* pg. 201
18. Michael R. Beschloss, "Taking Charge", (Simon & Schuster, 1997) Review by David Garner carried in Las Vegas Review Journal of "Taking Charge" .
19. "The Exercise of Power", pg. 426, 427
20. *Ibid* pg. 428, 429
21. *Ibid*...pg. 433, 434
22. *Ibid*...pg. 411, 413
23. "Taking Charge" pgs.256, 257
24. *Ibid* pg. 266, 267
25. *Ibid* pg. 366
26. *Ibid* pg. 390
27. *Ibid* pg. 400
28. *Ibid* pg. 415
29 Robert S. McNamara, "Argument Without End", (Public Affairs, 1999) pg. 332
30. Michael Lind, "Vietnam, The Necessary War" (The Free Press. 1999) pg. 16, 18
31. *Ibid* pg. 19
32. *Ibid* pg. 19
33. "Shadow Over the White House", pg. 95
34. William Ridings, "Rating the Presidents", (Carol Publishing Group, 1997) pg. 233
35. "Shadow Over the White House", pg. 97
36. "Vietnam Voices", pg. 187
37. "Argument Without End", pg. 334
38. *Ibid* pg. 336
39. *Ibid* pg. *337*
40. "Vietnam, the Necessary War", pg. 86
41. *Ibid* pg. 98
42. "Argument Without End", pgs. 359,360
43. *Ibid* pg. 360
44. *Ibid* pg. 362
45. *Ibid* pg. 362
46. "Shadow of the White House", pg. 103
47. "Vietnam Voices", pg. 361
48. "Vietnam, the Roots of Conflict", pg. 138

49. *Ibid* pg. 161

Chapter 3 Henry Kissinger

1.Walter Isaacson, "Kissinger", (Simon & Schuster, 1992) pgs. 79, 80
2. *Ibid* pg. 75
3. *Ibid* pg. 76
4. *Ibis* pg. 76
5. *Ibid* pg. 108
6. *Ibid* pg. 141
7. *Ibid* pg. 159
8."Vietnam Voices", pg. 394
9. Ibid pg. 399
10. Ibid pg. 407
11. "Kissinger", pgs. 160
12. *Ibid* pg. 161
13. *Ibid* pg. 166
14. *Ibid* pg. 167
15. *Ibid* pg. 169
16. "Years of Renewal", pg. 79
17. *Ibid* Pg. 77
18. Larry Berman, "No Peace, No Honor" (The Free Press, 2001), pg. 41.
19. *Ibid* pg. 42
20. "Kissinger", pg. 41
21. *Ibid...* pg. 42
22. *Ibid* pg. 245
23. *Ibid* pg. 248, 249
24. "No peace, No Honor" pg. 43, 44
25. "Kissinger", pg. 332
26. "No Peace, No Honor", pg. 50
27. "Years of Renewal", pg. 94
28. "No Peace, No Honor", pg. 56
29. *Ibid* .pg. 57
30. *Ibid* pg. 60
31. *Ibid* pg. 68
32. *Ibid* pg. 68
33. *Ibid* pg. *73*
34. *Ibid* pg. 74
35. *Ibid* pg. 78, 79

36. *Ibid* pg. 108
37. *Ibid* pg. 78
38. *Ibid* pg. 79
39. "Kissinger, a Biography", pg. 398
40. "No Peace, No Honor", pg. 123
41. "Kissinger, a Biography", pg. 415
42. "No Peace, No Honor", pg, 124
43. *Ibid* pg. 125
44. *Ibid* pg. 129
45 *Ibid* pg. 132
46. *Ibid* pg. 133
47. *Ibid* pg. 139
48. *Ibid* pg. 146
49 "Kissinger, a Biography", pg. 441
50 "No Peace, No Honor", pg. 148
51. "Kissinger, a Biography", pg. 447
52. *Ibid* pg.450
53. *Ibid* pg. 450
54. "No Peace, No Honor": pg.158
55. "Kissinger, a Biography", pg. 453
56. "No Peace, No Honor", pg. 169
57. *Ibid* pg. 170
58. *Ibid* pg. 172.
59. *Ibid* pg. 176
60. *Ibid* pg. 188
61. "Kissinger, a Biography" pg. 456
62. "No Peace, No Honor" pg. 189
63. *Ibid* pg. 316
64. "Kissinger, a Biography" pg.484
65. *Ibid* pg. 486
66. "No Peace, No Honor" pg.253
67. *Ibid* pg. 233
68. *Ibid* pg. 239
69. "Kissinger, a Biography" pg. 647
70. "No Peace, No Honor" Pg.110
71. *Ibid* pg.120\
72. *Ibid* pg. 143
73. *Ibid* pg. 154
74. *Ibid* pg. 158

75. *Ibid* pg. 158
76. *Ibid* pg. 176
77. *Ibid* pg. 208
78. *Ibid* pg. 225
79. *Ibid* pg. 254
80. *Ibid* pg. 253
81. "The Men We Left Behind" pg. 22
82. *Ibid* pg. 29
83. *Ibid* pg. 113
84. *Ibid* pg. 165
85. *Ibid* pg. 166
86. *Ibid* pg. 355
87. *Ibid* pg. 358
88. *Ibid* pg. 370
89. *Ibid* pg. 298
90. *Ibid* pg. 299
91.. *Ibid* pg. 299
92. *Ibid* pg. 299
93. *Ibid* pg. 310
94. *Ibid* pg. 320
95. *Ibid* pg. 321
96. *Ibid* pg. 322
97. *Ibid* pg. 325
98. *Ibid* pg. 325
99. *Ibid* pg. 328
100. *Ibid* pg. 346
101. *Ibid* pg. 347
102. *Ibid* pg. 348
103. *Ibid* pg. 349
104. *Ibid* pg. 355
105. *Ibid* pg. 356
106. *Ibid* pg. 359
107. *Ibid* pg. 370
108. *Ibid* pg. 371
109 "Kissinger, a Biography": pg. 733
110. *Ibid* pg. 732
111. *Ibid* pg. 748
112."Seeds of Fire" pg. 187
113. *Ibid* pg. 187

114. *Ibid* pg. 187
115. *Ibid* pg. 187
116. *Ibid* pg. 187
117. "Kissinger, a Biography" pg. 513
118. *Ibid* pg. 513
119. *Ibid* pg. 514
120. *Ibid* pg. 515
121. *Ibid* pg. 515
122. *Ibid* pg. 516
123. *Ibid* pg. 517
124. *Ibid* pg. 518
125. *Ibid* pg. 519
126. *Ibid* pg. 519
127. *Ibid* pg. 520
128. *Ibid* pg. 521
129. *Ibid* pg. 521
130. *Ibid* pg. 522
131. *Ibid* pg. 536
132. *Ibid* pg. 540
133. *Ibid* pg. 541
134. *Ibid* pg. 542
135. Richard Bernstein and Ross H. Munro, "The Coming Conflict With China" (Knopf, 1997) pg. 116
136. *Ibid* pg. 117
137. *Ibid* pg. 119
138. *Ibid* pg. 119
139. *Ibid* pg. 120
140. *Ibid* pg. 121
141. *Ibid* pg. 123
142. *Ibid* pg. 126

Chapter 4 "Jimmy Carter"

1."American Presidents", pg. 389
2. *Ibid* pg. 394
3. *Ibid* pg. 400, 401
4. *Ibid* pg. 401

5. *Ibid* pg. 402
6. *Washington Times, March 1 -7, 1999,* article by Robert Morton
7. *Ibid*
8. "The Men We Left Behind", pg. 172
9. *Ibid* pg. 174, 175
10. *Ibid* (State Dep't. Telegram, Sept. 6, 1976 (note 32)
11. "Campus Report", May 2002, article by Sara Russo.
12. "The American Presidents", pg. 263
13. Mona Charen, "Useful Idiots"(Regnery Publishing, 2003) pg.81,82
14. *Ibid* pg. 82
15. *Ibid* pg. 84
16. *Ibid* pg. 84,85
17. *Ibid* pg. 85
18. *Ibid* pg. 195
19. *Ibid* pg. 204, 205
20. *Ibid* pg. 205
21. *Ibid* pg. 206
22. "NewsMax.Com", February, 2003, article by Paul Crespo. Pgs.12-16
23. *Ibid*
24. *Ibid*
25. *Ibid*
26 Ann Coulter, "Treason" (Crown Forum, 2003) pg.15

Chapter 5 "William Jefferson (Blythe)Clinton and Hillary Rodham Clinton

1.David Brock, "The Seduction of Hillary Rodham" (Free Press, 1996) pg. 321
2.Ambrose Evans-Pritchard, "The Secret Life of Bill Clinton" (Regnery, 1997) pg. 276
3. *Ibid* pg. 276
4. *Ibid* pg. 275
5.R. Emmett Tyrell, Jr. (Boy Clinton, Regnery, 1996) pg.36
6. *Ibid* pg. 37
7. *Ibid* pg. 37
8. *Ibid* pg. 39
9. "Useful Idiots" pg. 205

10. *Ibid* pg. 206
11. *Ibid* pg. 207
12. *Ibid* pg. 209
13. *Ibid* pg. 216
14. *Ibid* pg. 217, 218
15. *Ibid* pg. 219
16. *Ibid* pg. 221
17. *Ibid* pg. 222
18. *Ibid* pg. 222, 223
19. *Ibid* pg. 223
20. *Ibid* pg. 224
21. *Ibid* pg. 225
22. *Ibid* pg. 211
23. Terry Reid and John Cummings, "Compromised" (Clandestine Publishing) pg. 68
24. *Ibid* pg. 45
25. "The Secret Life of Bill Clinton" pg.316
26. *Ibid* pg. 318
27. *Ibid* pg. 320
28. *Ibid* pg. 320
29. *Ibid* pg. 340, 341
30. Ibid pg. 342
31. Ibid pg. 309
32. Ibid pg. 310
33. Ibid pg. 311
34. Sen. Leahy, Senate floor speech
35. "Boy Clinton", pg. 125
36. *Ibid* pg. 125
37. *Ibid* pg. 128
38. *Ibid* pg. 129
39. *Ibid* pg. 121, 122
40. *Ibid* pg. 122
41. ":Secret Life of bill Clinton" pg. 141
42. *Ibid* pg. 142, 143
43. *Ibid* Pg. 262
44. *Ibid* pg. 263
45. *Ibid* pg. 264
46. *Ibid* pg. 265
47. *Ibid* pg. 267

48. Ibid pg. 268
49. Ibid pg. 268
50. Ibid pg. 269
51. Ibid pg. 269
52. Ibid pg. 270
53. Ibid pg. 272
54. Ibid pg. 273
55. Ibid pg. XV
56. Ibid pg. XVI
57. Richard Bernstein and Ross H. Munro, "The Coming Conflict With China" (Knopf Inc. 1997) pg. 207, 208
58. Edward Timperlake and William C. Triplett II, "Year of the Rat" (Regnery, 1998) pg. 217
59. *Ibid* pg. 218, 219
60. *Ibid* pg. 221
61. *Ibid* pg. 223
62. *Ibid* pg. 224
63. *Ibid* pg. 225
64. Edward Timperlake and William C. Triplett II, "Red Dragon Rising" (Regnery , 1999) pg.. 195
65. *Ibid* pg. 198
66. *Ibid* pg. 199
67. Bob Woodward, "Shadow" (Simon & Schuster, 1999) pg. 291
68. *Ibid* pg. 339
69. *Ibid* pg. 341
70. "Year of the Rat" pg. 226
71. *Ibid* pg. 227
72. Lt. Col. Robert Patterson USAF (Ret.) "Dereliction of Duty" pg. 113
73. *Ibid* pg. 114, 115
74. *Ibid* pg. 93
75. *Ibid* pg. 116, 117
76. Ibid pg. 118
77. Ann Coulter, "Treason" (Crown Forum, 2003) pg. 211
78. David Limbaugh, "Absolute Power" (Regnery, 2001) pg. 251, 252, 253
79. "The Seduction of Hillary Rodham" pg.17
80. *Ibid* pg. 17
81. *Ibid* pg. 92
82. *Ibid* pg. 96
83. *Ibid* pg. 98

84. *Ibid* pg. 100
85. *Ibid* pg. 101
86. *Ibid* pg. 112
87. *Ibid* pg. 113
88. *Ibid* pg. 115
89. *Ibid* pg. 116
90. *Ibid* pg. 313
91. *Ibid* pg. 314
92. *Ibid* pg. 315
93. *Ibid* pg. 316
94. *Ibid* pg. 317
95. *Ibid* pg. 332
96. *Ibid* pg. 359
97. Barbara Olson, "Hell To Pay" (Regnery, 1999) pg. 3
98. *Ibid* pg. 5
99. *Ibid* pg. 49
100. *Ibid* pg. 51
101. *Ibid* pg. 54
102. *Ibid* pg. 56
103. *Ibid* pg. 62
104. *Ibid* pg. 103
105. *Ibid* pg. 105
106. *Ibid* pg. 135
107. *Ibid* pg. 164, 165
108. *Ibid* pg. 169
109. *Ibid* pg. 245
110. *Ibid* pg. 316
111. ABC's "20/20" (Janet Napolitano, U.S. Attorney for Arizona, now Governor of Arizona)
112. "The Seduction of Hillary Rodham" pg. 316
113. *Ibid* pg. 317
114. *Ibid* pg. 332
115. *Ibid* pg. 359
116. "Hell To Pay" pg.3
117. *Ibid* pg. 5
118. *Ibid* pg. 50
119. *Ibid* pg. 51
120. *Ibid* pg. 54
121. *Ibid* pg. 56

122. *Ibid* pg. 62
123. *Ibid* pg. 102
124. *Ibid* pg. 105
125. *Ibid* pg. 135
126. *Ibid* pg. 165, 166
127. *Ibid* pg. 169
128. *Ibid* pg. 248
129. *Ibid* pg. 316
130. Carl Limbacher, "Hillary's Scheme" (Crown Forum, 2003) pg. 80
131. *Ibid* pg. 60
132 *Ibid* pg. 64
133. *Ibid* pg. 71
134. Gary Aldrich, "Unlimited Access" pg. 139
135. "Hillary's Scheme" pg. 78
136. "Hell To Pay" pg. 317

Chapter 6 "Congress"

1.Ronald Kessler, "Inside Congress" (Pocket Books, 1997) pg. 4
2.Ibid pg. 4
3.Ibid pg. 52
4.Ibid pg. 53
5.Ibid pg. 55
6.Ibid pg. 66
7.Ibid pg. 67
8.Charles Lewis, The Buying of Congress" (Avon Books, 1997) pg. 3,4,5
9.Ibid pg. 6
10.Ibid pg. 16
11."Inside Congress" pg. 128
12."The Buying of Congress" pg. 17
13. *Ibid* pg. 22
14. *Ibid* pg. 30
15. *Ibid* pg. 32
16. *Ibid* pg. 38
17."Inside Congress" pg. 79
18. *Ibid* pg. 80
19. *Ibid* pg. 81

20. *Ibid* pg. 83
21. *Ibid* pg. 84
22. *Ibid* pg. 86
23. *Ibid* pg/ 107
24. *Ibid* pg. 107
25. *Ibid* pg. 109
26. *Ibid* pg. 119
27. *Ibid* pg. 122
28. "The Buying of Congress" pg. 250
29. *Ibid* pg.250
30. *Ibid* pg. 241
31. *Ibid* pg. 304, 305
32. *Ibid* pg. 307
33. *Ibid* pg. 310
34. *Ibid* pg. 311
35. *Ibid* pg. 311
36. *Ibid* pg. 312
37. "Inside Congress: pg. 142
38. *Ibid* pg.143
39. *Ibid* pg. 182
40. *Ibid* pg. 183
41. "The Buying of Congress" pg.16
42. "Inside Congress" pg. 14
43. *Ibid* pg. 242
44. *Ibid* pg, 242
45. *Ibid* pg. 245
46. *Ibid* pg. 249
47. "The Breach" pg. 18
48. *Ibid* pg. 49
49. *Ibid* pg. 80, 81
50. *Ibid* pg. 89
51. *Ibid* pg. 91
52. *Ibid* pg. 144
53. *Ibid* pg. 162
54. *Ibid* pg. 253
55. *Ibid* pg. 276
56. *Ibid* pg. 302
57. *Ibid* pg. 302
58. *Ibid* pg. 318

59. *Ibid* pg. 327
60. *Ibid* pg. 327
61. *Ibid* pg. 335
62. *Ibid* pg. 336
63. *Ibid* pg. 341
64. *Ibid* pg. 351
65. *Ibid* pg. 377
66. *Ibid* pg. 410
67. *Ibid* pg. 415
68. *Ibid* pg. 416

Chapter 7 "Organizations"

1.Sydney Hook, Vladimir Bukovsky, Paul Hollander, "Soviet Hypocrisy and Western Gullibility" University Press, 1981 Pg 10
2.Ibid pg. 6
3.Robert H. Bork, "The Tempting of America" (The Free Press., 1990) pg. 243
4.Ibid pg. 244
5.Ibid pg. 245, 246
6.Ibid pg. 246
7.Michael Savage. "The Savage Nation" (WND Books, 2002) pg. 161, 162
8.Sean Hannity, "Let Freedom Ring" (Regan Books, 2002) pg. 139
9.Ibid pg. 141
10.The Savage Nation pg. 161
11.Sally D. Reed, "NEA: Propaganda Front of the Radical Left", (National Council for Better Education, 1984) pg.3
12. Ibid pg. 10
13. Ibid pg. 13
14. Ibid pg. 15
15. Ibid pg. 19
16. Ibid pg. 32, 33
17. Ibid pg. 42
18. Ibid pg. 53
19. Ibid pg. 50,51
20.Rip Caudill and Don Swarthout, "Wake Up America" (booklet) pg. 58
21.Ibid pg. 59
22.Ibid pg. 40

23.John A. Stormer, "None Dare Call It Education" (Liberty Bell Press, 1998) pg.170
24.Ibid pg. 170
25.Ibid pg. 171
26.Ibid pg. 169
27.Ibid pg. 174
28.Ibid pg. 178
29.Ibid pg. 180
30.Bill O'Reilly, "Who's Looking Out for Your?" (Random House, 2003) pg. 120
31.Ibid pg. 121
32."None Dare Call It Education" pg. 8
33.Ibid pg. 10
34.Ibid pg. 11
35.Ibid pg. 15
36.Laura Ingram, "Shut Up & Sing" (Regnery, 2003) pg.235
37.Ibid pg. 237
38.Ibid pg. 238
39.Ibid pg. 239
40.Ibid pg. 245
41."The Savage Nation" pg. 156, 157

Chapter 8 "Media"

1.L.Brent Brozell and Brent Baker, "And That's the Way It Isn't" (Media Research Center, 1990) pg. 5
2.Ibid pg. 20
3.Ibid pg. 32
4.Ibid pg. 33
5.Ibid pg. 42
6.Ibid pg. 44
7.Ibid pg. 54
8.Barnard Goldberg, "Bias" (Regnery, 2002) pg. 19
9.Ibid pg. 25
10.Ibid pg. 102
11.Ibid pg. 118
12.Ibid pg. 123
13.Ibid pg. 222

14.Tim Graham, "Patterns of Deception" (Media Research Center, 1998) pg. 82
15.Ibid pg. 115
16.Ibid pg. 135
17.Ibid pg. 138
18.Ibid pg. 156
19.Ibid pg. 157
20.Ibid pg. 233
21.William A. Rusher, "The Coming Battle for the Media" (Morrow, 1988) pg. 83
22.Ibid pg.88
23.Reed Irvine, Joseph C. Goulden, Cliff Kincaid "The News Manipulators" (Book Distributors, Inc. 1993) pg. 27
24.Ibid pg. 219
25.Ibid pg. 123
26.Ibid pg. 141
27.Ibid pg. 143
28.Ibid pg. 160
29.Ibid pg. 189
30.Ibid pg. 277
31.Ibid pg. 290
32.Ibid pg. 292
33."Mind Siege" pg. 54
34. *Ibid* pg. 192
35.Ibid pg. 192
36. *Ibid* pg. 194
37. S. Robert Lichter, Stanley Rothman, Linda S. Lichter, "The Media Elite", (Adler & Adler, 1986) pg. 294
38.William McGowan, "Coloring the News" (Encounter Books, 2001) pg. 9
39. *Ibid* pg. 35
40.
I*bid* pg. 137
41. *Ibid* pg. 137
42. *Ibid* pg. 161
43. *Ibid* pg. 163
44 *Ibid* pg. 226
45. *Ibid* pg. 227
46. *Ibid* pg. 246

Glossary

AAPSO: Afro-Asian People's Solidarity Org.
ACLU: American Civil Liberties Union
AFL: American Federation of Labor
AFT: American Federation of Teachers
AID: Agency for International Development
ANC: African National Congress
An-Namese: Historical term for Vietnamese people
ARVN: Army of Vietnam (South)
ATF: (Treasury) Alcohol, Tobacco, and Firearms
ATL: Association of Trial Lawyers
Black Caucus: U.S. Congressional Group
Bonz: Buddhist Monk
Bull Moose Party: Progressive party (1912-1924)
CAS: Term for C.I.A. sourced
C/NPAC: Commander in Chief, Pacific
CIO: A labor union
Contras: Nicaraguans resisting Sandinistas
COS: Chief of Station
CPC: Christian Peace Conference
CPUSA: Communist Party, U.S.A.
CRDS: An A.I.D. program
CWG: Campaign for World Development
DCI: Director of Central Intelligence
DEA: Drug Enforcement Agency
DIA: Defense Intelligence Agency
DNC: Democratic National Committee
DOD: Department of Defense
DOE: Department of Education
DOR: Declaration of rights of man
DRV: (N.Vietnam) Democratic Republic of Vietnam
DSA: Democratic Socialists of America
Emily's List: A radical feminist cause
FAA: Federal Aviation Authority
FAN: Army neutral forces, Laos
FAR: Royal Laos Forces
FEMA: Federal Emergency Management Assoc.

GAO: Government Accounting Office
GATT: General Agreement on Trade Tariffs
Hermes: A God who served as messenger & herald for other Gods
Hmong: Hill tribesmen in Laos & South Vietnam
Ho Chi Minh Trail: Supply route from N. Vietnam through Laos
Huckster: One who uses questionable methods to sell
Humanism: Emphasis on secular concerns
I&N: Immigration and Naturalization
IAD: International Association of Democratic Lawyers
IIP: International Institute for Peace
IPA: International Peace Academy
IUS: International Union of Students
Liberal Left: Strong support of Socialist style of government programs: calling for heavy taxes to support social programs
MACV: Military Activities Command in Vietnam
McCarthyism: Pejorative tag against Sen. Joseph McCarthy, first used by the Communist party.
MEO: Montagnard Hill Group
NAACP: National Association for Advancement of Colored People
NAFTA: North American Free Trade Assoc. (1994)
NAMBLA: North American Men's Boy's Assoc.
NATO: North American Treaty Organization
NCLB: No Child Left Behind Act. (1994)
NEA: Near East/Asia
NEA: National Education Assoc.
NLF: National Liberation Front (N. Vietnam political organization)
NOW: National Organization of Women
NRDC: National Resources Defense Council
NSC: National Security Council
NVN: North Vietnam
PAC: Political Action Committee (fund raising vehicle)
Pathet Lao: Communist Laotian Political & Military
PAVN: People's Army of Vietnam (N. Vietnam)
PLA: People's Liberation Army
PLO: Palestinian Liberation Organization
Pluralism: A state of society in which members of diverse ethnic, racial, religious, or social groups mainstream an autonomous participation in development of their traditional culture within the confines of a common civilization.
Proletariat: Working class
PRU: Provincial Reconnaissance Units
PTA: Parent Teacher Assoc.
PUSH/EXCEL: Jesse Jackson's organization (e.g. Rainbow Coalition) to obtain funds
ROTC: Reserve Officers Training Corp.
SACP: South African Communist Party
Sandinistas: Communist front group, Nicaragua

SAT/GPA SCORES: Standard tests given to students intending to attend college/ universities
SEA SWALLOWS: Father Hoa's bodyguards
SEATO: South East Asian Treaty Organization
Secularists: Indifferent to religion. Antagonistic to traditional religion.
S.D.I:. National Self-Defense Initiative
SVN: South Vietnam
TET: Traditional national holiday in Vietnam
ULLICO: Union Labor Life Insurance Co.
UN/UNGA: United Nations/National General Assembly
UNFPA: United Nations Population Fund
USOM: United States Over Seas Mission
USPC: United States Peace Council (Affiliate of World Peace Council)
Viet Cong: Pro N. Vietnamese paramilitary forces operating in S. Vietnam
Viet Minh: N. Vietnam's paramilitary forces operating in S. Vietnam
WCC: World Council of Churches
WFTU: World Federation of Trade Unions
WFDY: World Federation of Democratic Youth
WMD: Weapons of mass destruction
WPC: World Peace Council

Index

Printed in the United States
21023LVS00003B/208-225

9 781932 762136